W0232228

STRUCTURE AND FUNCTION OF PLASMA PROTEINS
VOLUME 1

STRUCTURE AND FUNCTION OF PLASMA PROTEINS

VOLUME 1

Edited by

A. C. Allison

Clinical Research Centre
Watford Road
Harrow
Middlesex

PLENUM PRESS · LONDON AND NEW YORK

Library of Congress Catalog Card Number: 72-95071

ISBN-13: 978-1-4684-2678-6   e-ISBN-13: 978-1-4684-2676-2
DOI: 10.1007/978-1-4684-2676-2

Copyright © 1974 by Plenum Publishing Company Ltd
Plenum Publishing Company Ltd
Softcover reprint of the hardcover 1st edition 1974

4a Lower John Street
London WIR 3PD
Telephone 01-439-2364

U.S. Edition published by
Plenum Publishing Corporation
227 West 17th Street
New York, New York 10011

All Rights Reserved
No part of this book may be reproduced, stored in a retrieval system,
or transmitted, in any form or by any means, electronic, mechanical,
photocopying, microfilming, recording or otherwise, without written
permission from the Publisher

Contributors to Volume 1

Chester A. Alper   Centre for Blood Research,
                          800 Huntington Avenue,
                          Boston, Massachusetts 02115, U.S.A.

D. E. G. Austen   Oxford Haemophilia Centre,
                          Churchill Hospital, Headington,
                          Oxford OX3 7LJ, England.

Audrey S. Fosbrooke   Institute of Child Health, University of London,
                          30 Guilford Street, London, W.C.1., England.

Geoffrey Franglen   Department of Chemical Pathology
                          St. George's Hospital Medical School,
                          Hyde Park Corner, London SW1, England.

J. S. Garrow   Clinical Research Centre, Watford Road,
                          Harrow, Middlesex HA1 3UJ, England.

E. Giblett   King County Central Blood Bank, Seattle,
                          Washington, U.S.A.

A. Koj   Department of Animal Biochemistry,
                          Institute of Molecular Biology,
                          Jagiellonian University, Cracow, Poland.

Ham Heng Liem   Scripps Clinic & Research Foundation,
                          La Jolla, California 92037, U.S.A.

June K. Lloyd   Institute of Child Health, University of London,
                          30 Guilford Street, London, W.C.1., England.

Ursula Muller-   Scripps Clinic & Research Foundation,
Eberhard   La Jolla, California 92037, U.S.A.

E. Regoeczi   Department of Pathology, McMaster
                          University, Hamilton, Ontario, Canada.

C. R. Rizza   Oxford Haemophilia Centre, Churchill Hospi-
                          tal, Headington, Oxford OX3 7LJ, England.

G. T. Stevenson   Tenovus Research Laboratory,
                          General Hospital, Southampton SO9 4XY,
                          England.

Preface

Plasma proteins are of interest from many points of view. Biochemists have separated and purified numerous plasma proteins and studied their physical properties, aminoacid composition and sequence, the carbohydrate components of some, and binding of metals, hormones and other materials. Much work has also been carried out on the synthesis, rates of turnover and degradation of plasma proteins.

Many plasma proteins show inherited variations, some of which (e.g. those of heptoglobins and transferrins) are common in various human populations while others (e.g. absence of lipoproteins or immunoglobins) are rare but important because of their association with clinical syndromes. Since blood is the most accessible bodily constituent, geneticists have made good use of serum protein differences as genetic markers in family and population studies.

Physiologists have long been interested in plasma proteins in relation to colloid osmotic pressure, transport of lipids, iron, hormones and other materials, the activities of renal glomeruli and tubules, the function of the liver, and many other bodily activities. Plasma proteins are also widely studied in relation to malnutrition and undernutrition, particularly that associated with defective intake of protein.

One of the routine activities of clinical chemistry laboratories is the analysis of plasma proteins. Although used primarily in relation to the diagnosis and assessment of prognosis of liver and kidney diseases and lymphoreticular malignancies, plasma proteins can provide information of value in many disorders. Detailed studies of particular groups of plasma proteins, such as the blood clotting factors, immunoglobulins and complement components, are of basic importance to haematologists and immunologists.

Developmental biologists are interested in the ontogeny of the plasma proteins and the reappearance in malignant disease of proteins which are normally confined to the foetus. Regulation of the synthesis of plasma proteins, for example acute-phase reactants, is a problem of general interest as well as importance in limiting the progress of inflammation. Several protease inhibitors circulate in the plasma and their role *in vivo* is still imperfectly understood.

Plasma proteins have been used, for the most part empirically, in media for culturing cells and tissues, and it is becoming clear that their interactions with various cell types are complex. Many stimulatory and inhibitory effects of particular plasma protein components on cell motility, division, enzyme synthesis and other aspects of cell function have been reported. Indeed, study of the interactions of various plasma proteins with different cell types are now a major branch of cell biology, as well as having an important bearing on the pathogenesis of such common diseases as atherosclerosis and cancer. Specific interactions of plasma proteins with cells can be illustrated by the attachment of immunoglobulins and complement components to leucocytes and the elimination by the liver of plasma proteins with certain sugar groups exposed by removal of terminal sialic acid groups.

As a result of all these studies an enormous body of knowledge about plasma proteins has accumulated. Only a small part of this knowledge has penetrated to textbooks of biochemistry, physiology and medicine. Much of the information is either in large, expensive and inaccessible handbooks or in volumes on specialized subjects. The need was recognized for a convenient handbook on plasma proteins in which the main observations would be clearly set out with enough detail for most purposes, together with references for interested readers to obtain access to more detailed studies on any topic.

For the convenience of authors and readers the subject has been divided into two volumes, each with ten chapters. Each chapter will provide some guide to techniques appropriate for the isolation or characterization of the proteins, and one chapter in the second volume is devoted to methods. This handbook is intended for senior and postgraduate students in biochemistry, medicine and other subjects as well as hospital biochemists, physicians and research workers. I have confidence that the knowledge and skill of the authors in presenting so complex a subject in so lucid a fashion will appeal to a wide range of readers.

A. C. Allison
March 1974

Contents for Volume 1

Chapter 1

Plasma Lipoproteins

June K. Lloyd and Audrey S. Fosbrooke

*Department of Child Health,
Institute of Child Health,
University of London*

Lipoproteins are macromolecules in which the water insoluble lipids are transported in plasma. Four main lipoprotein classes have been defined in human plasma. Each of these contains esterified and non-esterified cholesterol, phospholipid and triglyceride, though in differing and characteristic proportions. Non-esterified fatty acid (NEFA) is transported as a complex with albumin; the albumin-fatty acid complex is not, however, conventionally regarded as one of the lipoproteins. The terminology used for the classification of the lipoprotein classes depends upon the methods used for their separation; a description of the methodology is thus a prerequisite to any discussion of terminology or physiology.

1.1. METHODOLOGY

1.1.1. SEPARATION OF LIPOPROTEINS

The main methods used for the separation of lipoproteins depend upon either the electrical charge (electrophoresis), the density (ultra-centrifugation), the immunochemical characteristics of the protein moiety, the selective precipitation of individual lipoproteins by chemical reagents, or the size of the particle.

(a) *Electrophoresis*

A variety of techniques have been described; paper, cellulose acetate and agarose are the most commonly used support media, with subsequent staining of the lipid moiety of the lipoprotein with Oil Red O or Sudan Black.

Electrophoretic separation on paper in barbitone buffer at pH 8.6[1,2] permits identification of four lipoprotein bands (Fig. 1.1). The fraction which remains at the origin consists of chylomicrons, betalipo-protein has the mobility of beta-globulin, alphalipoprotein has the mobility of alpha$_1$-globulin and pre-betalipoprotein moves just ahead of betalipoprotein, from which it is not completely separated. The simultaneous running of a duplicate strip which can be stained for protein may aid identification of the lipoprotein bands.[1] The addition of 1% albumin to the buffer is claimed to improve resolution of the beta and pre-beta fractions.[2]

Cellulose acetate[3] and agarose[4] both have the advantage over paper in that smaller samples of serum can be loaded, the time taken for separation reduced from 16 hours (overnight) to a few hours, and that improved separation is obtained between beta and pre-betalipoproteins.

Attempts have been made to make electrophoretic techniques quantitative by measurement of the uptake of the fat stain by dye elution[1] or by scanning[5] but because of the inherent problems arising from differential dye uptake by the various lipids making up the lipoproteins, electrophoresis should probably be used to give qualitative information only.

(b) *Ultracentrifugation*

The various classes of lipoproteins differ in the proportion of lipid compared with protein, and the greater the proportion of lipid, the lower the density of the lipoprotein complex. The differing densities allow separation of lipoproteins by the application of accelerated gravitational force, and in the ultracentrifuge forces of the order of 100 000 × gravity (g) can be obtained. Under such conditions lipoproteins will either float or sediment according to the differences between the density of the lipoprotein and that of the fluid environment.

There are two types of ultracentrifuge: the analytical and the preparative. The analytical ultracentrifuge is used for the determination of the rate of flotation of lipoproteins at a fixed density of 1.063 g/ml.[6] The flotation of lipoprotein fractions through the medium is recorded photographically in the form of Schlieren patterns, and the rate of flotation is measured in Svedberg units (10^{-13} cm/sec/dyne/g); a subscript to denote flotation is added to give S_f units. The lipoproteins of lowest density, that is those carrying the most lipid relative to protein, have the highest flotation rates. The analytical ultracentrifuge is also used for determining total lipoprotein concentrations by reference to the area under the curve in the Schlieren pattern.[7] The instrument does not, however, give information about lipoprotein composition.

The preparative ultracentrifuge is used to prepare fractions which contain lipoproteins of a particular density; these fractions are obtained in sufficient quantity for detailed analyses of composition to be made. The main advantage of this method is that it allows definition of the composition of plasma lipoproteins in addition to giving information about concentrations. Lipoproteins are inherently somewhat unstable and, therefore, the use of lengthy procedures to achieve complete fractionation is accompanied by the risk of physico-chemical changes. Such changes may be potentiated by exposure to dilution with the various salt solutions which are used to achieve the required density.

SERUM LIPOPROTEIN ELECTROPHORESIS

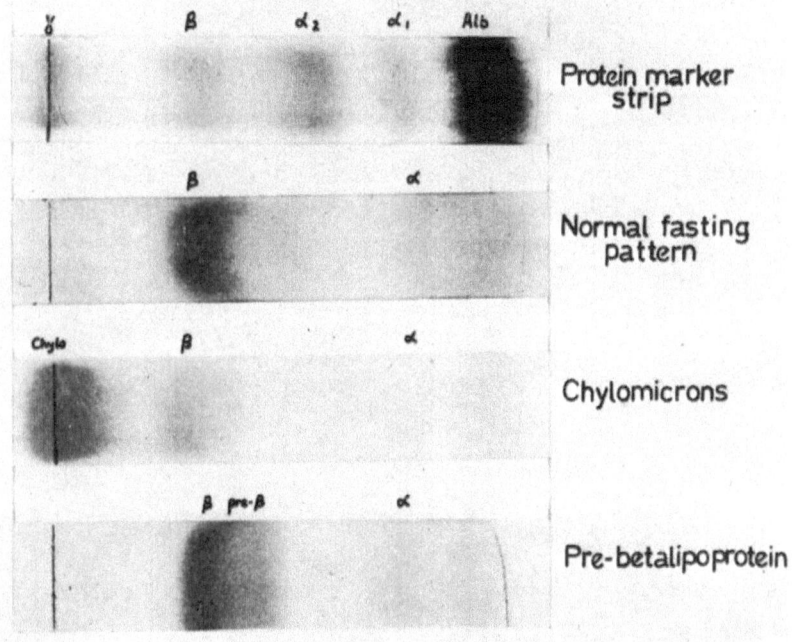

FIG. 1.1. Paper electrophoretic separation of serum lipoproteins. (Reproduced from Archives of Disease in Childhood, 1968, **43,** 393 by permission of the editors.)

However by careful selection of the conditions of time and temperature for ultra-centrifugation, and in some instances by the addition of certain chemical reagents, it is possible to minimise the risks of degradation. Various schemes of preparative ultracentrifugation have been described;[8] sequential spins at progressively increasing density, with collection of the required lipoprotein fraction at the meniscus after each spin, may require several days of ultracentrifugation for complete fractionation. The use of density gradient techniques[9] reduces the time required for ultracentrifugation. Although the preparative ultracentrifuge is the only method available for complete analysis of plasma lipoproteins, its use is limited by its cost together with the size of blood sample required (about 10 ml), the sophisticated technical support needed, and the fact that the method is not amenable to automation.

(c) *Immunochemical Methods*

Human serum lipoproteins behave as protein antigens and can be used for the production of antibodies in various animal species. When polyvalent antihuman antiserum is prepared, specific antibodies are produced which react with betalipoprotein and with alphalipoprotein. The use of a purified lipoprotein is necessary for the production of a single specific aniserum.

Several techniques have been described for the immunochemical estimation of betalipoprotein, but because the specific antiserum also reacts with pre-betalipoprotein the latter fraction must be removed by prior ultracentrifugation, or misleading results will be obtained, especially with abnormal plasma. A satisfactory immunoassay for betalipoprotein, based on radial immunodiffusion in agarose of serum from which pre-betalipoprotein was removed by ultracentrifugation, has been described.[10]

Immunochemical methods are also useful for the characterisation of different lipoprotein species. In combination with electrophoresis both qualitative[11] and quantitative[12] information can be obtained.

(d) *Chemical Precipitation Methods*

Lipoproteins react with certain polyfunctional anions to form insoluble complexes. Heparin in the presence of manganese ions,[13] or dextran sulphate in the presence of magnesium ions,[14] or sodium phosphotungstate in the presence of magnesium ions[15] have all been described for the selective precipitation of beta together with pre-betalipoproteins. The amount of precipitate formed has been used for the estimation of the total beta and pre-betalipoprotein concentration by a turbidimetric method;[14] the main problems of this type of method are the failure to differentiate between beta and pre-betalipoproteins, together with the difficulties of standardisation. Alphalipoproteins are not precipitated under the conditions described and can be estimated directly in the supernatant; this forms the basis of the method described by Fredrickson *et al.*[13]

(e) *Gel Filtration*

It is possible to make use of materials which behave as molecular sieves (for example various cross-linked dextrans) to resolve mixtures of lipoproteins, as well as the other plasma proteins, according to their molecular size. Columns of Sephadex G200 have been used to produce fractions containing betalipoproteins and alphalipoproteins.[16]

(f) *Nephelometry*

Some lipoproteins are sufficiently large to scatter light and nephelometry has been used to measure the concentration of these particles which include pre-betalipoproteins and chylomicrons. Stone and Thorp[17] have described a method in which ultrafiltration through cellulose ester membranes is used to separate chylomicrons from pre-betalipoprotein prior to their estimation by nephelometry.

1.1.2. ANALYSIS OF THE LIPID MOIETY OF LIPOPROTEINS

The chemical nature of the main plasma lipids is indicated in Fig. 1.2. Analytical methods for estimating plasma lipids were developed before techniques became available for the separation and estimation of lipoproteins. Estimation of the concentration of individual lipids in plasma gives only limited information regarding the concentration of individual lipoproteins, but when taken in conjunction with qualitative information on the lipoproteins separated by electrophoresis it is possible to make a meaningful interpretation of the lipoprotein status. Analysis of the lipids in lipoprotein fractions isolated by preparative ultracentrifugation is necessary for the determination of the composition of the fractions; also the lipid content is used as a measure of the lipoprotein concentration for each density class. The methods used for the estimation of lipids in whole plasma and in isolated lipoprotein fractions are essentially the same.

(a) *Lipid Extraction*

It is doubtful whether methods of chemical analysis are sufficiently specific to allow omission of the preparation of a lipid extract. The treatment of plasma with either ethanol/diethyl ether, chloroform/ methanol, or isopropanol, results in denaturation of the lipoproteins and quantitative release of the bound lipids into solution. Aliquots of this solution are used for the analysis of individual lipids.

(b) *Cholesterol*

Satisfactory colorimetric analyses based on the reaction with ferric chloride/sulphuric acid and on acetic anhydride/sulphuric acid are available for total (esterified plus unesterified) cholesterol. Versions

FATTY ACIDS

   eg. Palmitic Acid (16 carbon atoms, no double bonds ; C 16:0)

CH_3 CH_2 CH_2 CH_2 CH_2 CH_2 CH_2 CH_2 CH_2 CH_2 CH_2 CH_2 CH_2 CH_2 CH_2 COOH

TRIGLYCERIDES   : Glycerol + 3 Fatty Acids (FA)

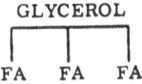

GLYCEROL

FA   FA   FA

PHOSPHOLIPIDS   : Two Main Classes

Glycerophosphatides : Glycerol +
                    2 Fatty Acids

Sphingolipids : Sphingosine +
             1 Fatty Acid

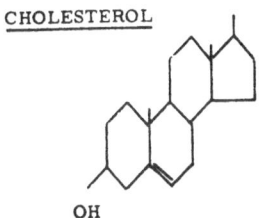

eg. GLYCEROL

FA   FA   Phosphate
           + Choline

Phosphatidyl Choline

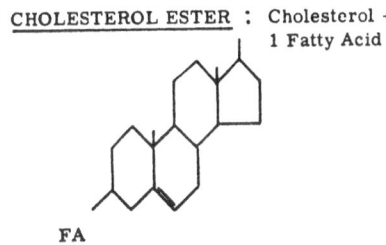

eg. SPHINGOSINE

FA   Phosphate
     + Choline

Sphingomyelin

CHOLESTEROL

OH

CHOLESTEROL ESTER : Cholesterol +
                              1 Fatty Acid

FA

FIG. 1.2. Chemical nature of main plasma lipids. (Reproduced from Archives of Disease in Childhood, 1968, **43**, 393 by permission of the editors.)

of these methods suitable for the autoanalyser are available, and precision has been improved by the use of automated methods. Estimation of the relative amounts of esterified and unesterified cholesterol can be made using digitonin precipitation of unesterified cholesterol, or perhaps most conveniently by the use of thin-layer or column chromatography to separate the two fractions prior to their colorimetric determination.

(c) *Triglyceride*

The analysis of triglyceride has in the past presented difficulties because of the lack of a direct colorimetric reaction. The development of alternative techniques such as gas-liquid chromatography, fluorimetry, infra-red absorptiometry and enzymology now enables triglyceride to be reliably measured. An automated version using fluorimetry is available.

(d) *Phospholipids*

Lipid phosphorus can be determined after oxidation to inorganic phosphate by a colorimetric reaction using a "molydenum-blue" procedure, and the application of an average molecular weight factor allows calculation of phospholipid. The same method can be used for estimation of the different phospholipids after their separation by chromatography. The major plasma phospholipids are phosphatidyl choline, sphingomyelin, lysophosphatidyl choline, and phosphatidyl ethanolamine.

(e) *Fatty Acids*

Non-esterified fatty acids (present as an albumin complex) comprise only a small fraction of total plasma fatty acids and can be estimated by a number of procedures including titrimetric methods, colorimetry and gas-liquid chromatography. Nearly all the plasma fatty acids are present in esterified cholesterol, in the various glycerides of which triglyceride constitutes about 90–95 % of the total, and in the phospholipids. The estimation of total esterified fatty acid concentration is of limited value. The introduction of gas-liquid chromatography, which allows separation and estimation of the individual fatty acids has enabled studies of fatty acid composition to be made. Because the lipid classes have different fatty acid compositions (Table 1.1) separation of the

TABLE 1.1
Composition of the plasma esterified fatty acids†

Major fatty acids (>5 % total)			Triglyceride	Cholesterol Ester	*PC	S	PE
Trivial name	C Atoms	Double bonds	(Fatty acids, g/100g total fatty acids)				
palmitic	16	0	31	14	35	40	30
palmitoleic	16	1	5				
stearic	18	0	5		15	10	15
oleic	18	1	44	25	14		9
linoleic	18	2	10	45	21		9
arachidonic	20	4		6	9		20
behenic	22	0				16	
	22	6					13
lignoceric	24	0				9	
nervonic	24	1				15	

*PC = phosphatidyl choline
  S = sphingomyelin
  PE = phosphatidyl ethanolamine
 † Mean values obtained in our laboratory from a group of healthy children and young adults.

individual lipids should be made before their analysis by gas-liquid chromatography. The method can be made quantitative by the addition of an unphysiological fatty acid as an internal standard.

Table 1.1 shows the percentage distribution of esterified fatty acids found in the various lipid classes in plasma of subjects eating a normal "Western-type" diet. Cholesterol ester is especially rich in linoleic acid, triglyceride contains mainly palmitic and oleic acids, and phosphatidyl choline contains mainly palmitic and linoleic acids. Sphingomyelin contains a large proportion of long-chain saturated acids, whereas phosphatidyl ethanolamine has the highest content of long chain polyunsaturated acids.

1.1.3. ANALYSIS OF THE PROTEIN MOIETY OF LIPOPROTEINS

The main difficulty in determining the protein concentration of lipoproteins lies in the preparation of the fraction free from the other plasma proteins. Prolonged ultracentrifugal procedures are required to prepare the lipoprotein fractions and their purity should be checked by immunochemical methods. Protein determination can then be made by the sensitive Lowry colorimetric method or the Kjeldahl method. Estimation by measurement of ultraviolet absorption at 280 mμ is only semi-quantitative, probably because of interference in the absorption spectrum by the lipid moiety. The amino acid composition of the protein can be determined by standard techniques of hydrolysis and chromatography, and endgroup analyses carried out by the appropriate techniques.

1.2. TERMINOLOGY, PHYSICAL CHARACTERISTICS AND CHEMICAL COMPOSITION

The nomenclature used to define the four main plasma lipoprotein classes is derived either from their electrophoretic mobility or their behaviour in the ultracentrifuge. The relationships between the various terminologies are shown in Fig. 1.3. Because of variations between the methods used for the analysis of lipoproteins it is important that results are accompanied by full details of the methodology, and caution should be exercised in comparing results obtained by different workers.

The chemical composition of plasma lipoproteins has been determined after isolation by preparative ultracentrifugation. The main characteristics of the four classes are summarised in Table 1.2. The values obtained by different workers are in broad agreement and the differences between various studies may arise from variations in the conditions used for isolation and analysis.

The identification of peptides with different carboxy-terminal amino acids has led to the use of nomenclature in which the apoprotein is

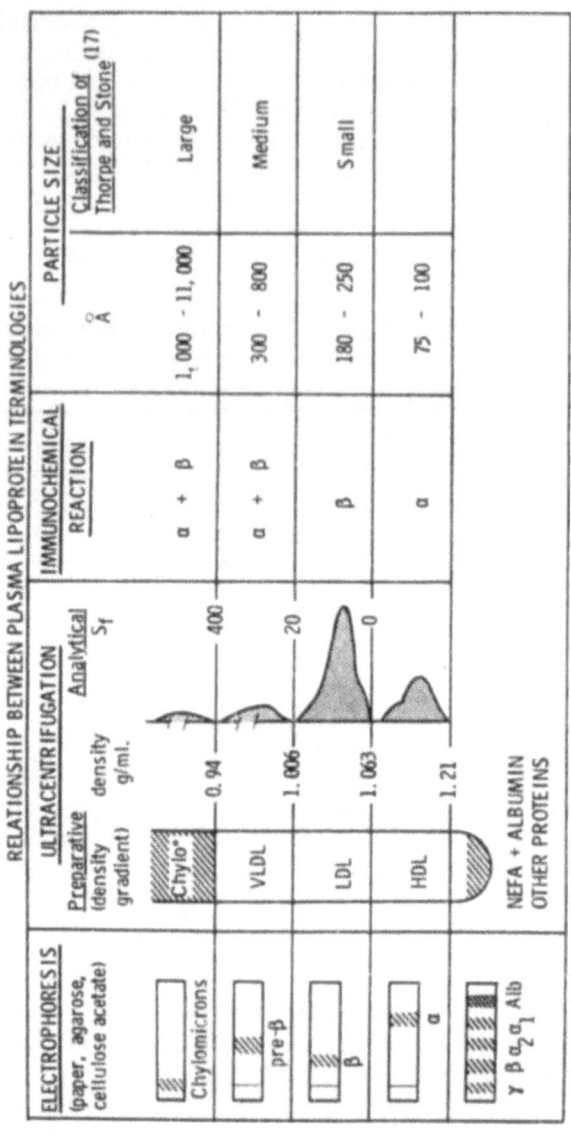

pre-β = pre-betalipoprotein   VLDL = very low density lipoprotein

β   = betalipoprotein   LDL = low density lipoprotein

α   = alphalipoprotein   HDL = high density lipoprotein

Chylomicrons are usually removed before ultracentrifugation.

Fig. 1.3. Comparison of lipoprotein terminologies.

designated according to the nature of its C-terminal group. Difficulties with this system, however, arise when different peptides have the same C-terminals.[17a] Furthermore, characterisation of many of the peptides is still incomplete, reinvestigation has already resulted in correction of previous observations,[17b,31] and other peptides may yet be discovered. We agree that the suggested nomenclature is provisional and temporary, and that a more acceptable and stable convention is required.[69]

TABLE 1.2
Composition of plasma lipoproteins (data of Oncley[24])

	Average percentage composition by weight				
Lipoprotein	Protein	Choles-terol	Choles-terol ester	Phos-pholipid	Trigly-ceride
Chylomicrons	2	2	5	7	84
Pre-betalipoprotein (VLDL)	8	7	13	19	51
Betalipoprotein (LDL)	21	8	37	22	11
Alphalipoprotein (HDL)	50	3	14	22	8

VLDL = very low density lipoprotein
LDL = low density lipoprotein
HDL = high density lipoprotein

1.2.1. CHYLOMICRONS

The term chylomicrons is used to refer to the large lipid particles in lymph or plasma which have an S_f greater than 400 and contain triglycerides primarily derived from dietary fat.[18] The density of chylomicron fractions as determined both by direct measurement and by calculation from the chemical composition is 0.94 g/ml. The particles are spherical and vary widely in diameter from approximately 1000–11 000 Å, although most of the chylomicrons during fat absorption are probably of the order of 1500–4000 Å.

Triglyceride is the predominant lipid in chylomicrons accounting for about 85% of the total lipid; its fatty acid composition shows a close similarity to that of the dietary fat.

Chylomicrons undergo certain chemical changes as they pass from their site of formation in the intestinal epithelial cell through the lymphatics to the blood stream. Interaction with plasma lipoproteins results in differences in composition between lymph chylomicrons and plasma chylomicrons. The triglyceride fatty acid composition of plasma chylomicrons does not parallel that of the dietary fat quite as closely as does that of lymph chylomicrons. This difference reflects the contribution of endogenous fatty acids to plasma chylomicrons.

The cholesterol content of plasma chylomicrons is greater than that of lymph chylomicrons. Amounts up to 8.0% have been found in human plasma chylomicrons[19] whereas the values quoted for lymph chylomicrons range from 0.5–5.1%.[20] Approximately 50% of the cholesterol in plasma chylomicrons is esterified. The amount of cholesterol and the proportion esterified is influenced by the quantity of dietary cholesterol and by the amount and type of dietary triglyceride.[18] The fatty acid composition of the cholesterol ester reflects that of the diet although not as closely as does that of the triglyceride.

Values for the phospholipid content of chylomicrons isolated from either lymph or plasma vary rather widely within the range of 2–10%.[18] The principal phospholipid is phosphatidyl choline, as in the other plasma lipoproteins. Lymph chylomicrons, however, contain more phosphatidyl ethanolamine than the phospholipids of whole plasma. The fatty acid composition of the individual phospholipids in chylomicrons is less closely related to that of the dietary fat than is the case for either triglyceride or cholesterol ester. The fatty acid composition of phosphatidyl choline changes only slowly in response to dietary changes and probably reflects dietary fat composition over a period of days or weeks.

The small amount of protein present in chylomicrons has not yet been completely characterised. Plasma chylomicrons contain about 2% of protein. Studies of N-terminal amino acids of the protein residues indicated that at least 3 proteins are present in chylomicrons, and one of these appeared identical with that of alphalipoprotein.[21] Immunochemical studies have demonstrated reactions with both beta and alphalipoprotein antisera. The interpretation of this data is, however, difficult as it has been demonstrated that lipid particles can readily adsorb both alpha and betalipoproteins[22] and, therefore, the demonstration of these lipoproteins in plasma chylomicrons does not necessarily indicate that either is an integral part of the apoprotein of chylomicrons.

1.2.2. BETALIPOPROTEIN (LOW-DENSITY LIPOPROTEIN)

Betalipoprotein corresponds in the ultracentrifuge to the fraction with S_f 0–20 (some workers prefer to limit the range to S_f 0–12 or 3–9) and has a density between 1.006–1.063 g/ml. The particles are spherical and their size has been estimated by a variety of techniques including ultracentrifugation, gel-filtration and electron-microscopy; a review of the data indicates an average diameter between 185–258 Å.[23] The molecular weight is about $3 \times 10.^6$

Betalipoprotein contains about 78% of lipid, 21% of protein and 1% of carbohydrate.[24] The major lipid is cholesterol, comprising about 40% of the total lipid; 65–70% of the cholesterol is esterified and the major fatty acid is linoleic acid with proportions varying between

40–60% depending upon the nature of the dietary fat. Phospholipids account for about 28% of betalipoprotein lipid. Phosphatidyl choline comprises about 64% of the phospholipid and sphingomyelin most of the remainder (26%); only small amounts of lysophosphatidyl choline, phosphatidyl ethanolamine, serine, and inositol are present.[25] The fatty acid composition of the individual phospholipids differs markedly one from another, phosphatidyl choline containing the highest proportion of linoleic acid (around 20%) and sphingomyelin containing a considerable proportion of the longer chain fatty acids.

Triglyceride is a minor component of betalipoprotein, comprising about 10% of the total lipid. The fatty acid composition reflects its endogenous origin.

The protein moiety isolated from betalipoprotein (beta apoprotein) has been analysed by a variety of techniques. Results of analysis of the amino acid composition obtained by different workers have been compared by Margolis[23] and are in good agreement considering the different preparative techniques employed. Characterisation of the peptide subunits has not yet been completed although it appears that at least 1 peptide has N-terminal glutamate and that serine represents the C-terminal of 1 peptide. The carbohydrate which constitutes 5–9% of the apoprotein includes galactose, fucose, mannose, glucosamine and sialic acid.[23]

1.2.3. PRE-BETALIPOPROTEIN (VERY LOW-DENSITY LIPOPROTEIN)

Pre-betalipoprotein corresponds in the ultracentrifuge to the fraction with S_f 20–400 and has a density between 0.94–1.006 g/ml. The particles are spherical and there is a wide variation in size; an average diameter of 414 Å was found using gel-filtration techniques,[23] and a range of 320–800 Å has been reported by electronmicroscopy.[26] Because of the wide size distribution no single molecular weight can be given; the average molecular weight of 15×10^6 was calculated by Margolis[27] with a range of 5×10^6–100×10^6.[26]

Although there is considerable heterogeneity, the chemical composition of pre-betalipoprotein is distinct from that of betalipoprotein; about 8% is protein and 90% lipid.[24] A small amount of carbohydrate, similar in composition to that in beta-lipoprotein, is present in the protein moiety. The predominant lipid is triglyceride, accounting for about 55–60% of the total. Palmitic and oleic acids are the major fatty acids. The fatty acid composition is influenced only indirectly by the diet and is similar to that of the triglyceride in betalipoprotein and in alphalipoprotein. Cholesterol and phospholipid comprise about 17% and 20% respectively of the total lipid. About 60% of the cholesterol is esterified; the major fatty acid is linoleic acid but there are relatively greater concentrations of palmitic and oleic acids than in betalipoprotein. The percentage distribution of the individual phospholipids is similar to that found in betalipoprotein although the proportion of sphingomyelin is somewhat less, and that of the noncholine phospholipids is slightly greater, in pre-beta than in betalipoprotein.[23] Their fatty acid composition does not differ from that found in betalipoprotein.

The amino acid composition of the apoprotein of pre-betalipoprotein has been reported to be identical with that of betalipoprotein,[28] but Levy et al.[29] detected small but significant differences in certain components and in particular pre-beta apoprotein contained less cystine than beta apoprotein. The major N-terminal amino acids are serine and threonine,[30] and serine and alanine have been confirmed as C-terminals.[31] There is, therefore, evidence for differences between the peptides of beta and pre-beta apoproteins. Many immunochemical studies have established antigenic similarity between pre-betalipoprotein and betalipoprotein,[23] and in addition pre-betalipoprotein has been found to contain proteins which react with antisera to alphalipoprotein.[13] It has also been established that a third antigenically distinct apoprotein, designated apoprotein C, is present in appreciable amounts in pre-betalipoprotein.[32] Apoprotein C has a different peptide pattern from beta or alpha-apoproteins, and has threonine and serine as N-terminal amino acids. Pre-betalipoprotein, therefore, contains several different proteins or peptides, and further studies are needed to define the interrelationships with the apoproteins of the other lipoproteins.

1.2.4. ALPHALIPOPROTEIN (HIGH-DENSITY LIPOPROTEIN)

Alphalipoprotein corresponds to high-density lipoprotein (HDL) and has a density between 1.063–1.21 g/ml. Within this density range two major subclasses have been described, a fraction with density 1.063–1.12 g/ml known as HDL_2, and a fraction with density 1.12–1.21 g/ml known as HDL_3. Electronmicrographs indicate that alphalipoprotein is made up of subunits; the number of subunits varies and it has been suggested that HDL_2 contains particles with 4–6 subunits whereas HDL_3 has particles with 3–4 subunits.[32] Diameters of the particles are about 100 Å for HDL_2 and 75 Å for HDL_3, and the corresponding molecular weights have been estimated as 3.6×10^5 and 1.7×10^5. The lability of the subunit structure gives rise to considerable analytical problems; the preparation of the high-density sub-classes requires prolonged ultracentrifugation and it has already been shown that some of the differences between HDL_2 and HDL_3 are artefacts.[33] It is not known to what degree different classes of alphalipoprotein are actually present in circulating plasma.

Alphalipoprotein contains about 50% protein, and of the lipid components the major fraction is phospholipid which comprises 44% of the total lipid. Phosphatidyl choline accounts for about 75% and sphingomyelin for about 12% of the phospholipid.[25] The fatty acid composition of the individual phospholipids is similar to that in beta-lipoprotein. Cholesterol (of which about 75% is esterified) comprises 22% and triglyceride 16% of the total lipid. The fatty acid composition of cholesterol ester and triglyceride is the same as in betalipoprotein.

The amino acid composition of alphalipoprotein has been determined by a number of workers.[34] Aspartic acid appears to be the only N-terminal amino acid. The work of Shore and Shore[35] demonstrated the presence of equal amounts of C-terminal threonine and glutamine, and these workers isolated two peptides of similar molecular weight, but differing in their amino acid composition and C-terminal sequence. More recently, however, Kostner and Alaupovic[17b] have reported that apolipoprotein-threonine actually contains glutamine as the C-terminal amino acid, and thus the C-terminal residues of the two major peptides may be the same.

Alpha apoprotein contains about 3% of carbohydrate as glucosamine, methylpentose, galactose, mannose and sialic acid.[34]

1.3. PHYSIOLOGY

1.3.1. SYNTHESIS

The liver is the main site of synthesis of all plasma lipoproteins except chylomicrons. It is probable, however, that the cells of the intestinal mucosa also synthesise lipoproteins other than chylomicrons.[34a]

Whilst there is general agreement that a fundamental step in the synthesis of lipoproteins is the formation of the apoprotein, many of the factors involved in the subsequent binding of lipid and in the release of the macromolecule into the plasma are not fully understood. For example, it is not yet established whether a complete lipoprotein macro-molecule is secreted by the cell, or whether lipid is attached after the apoprotein is released. The latter proposition was supported by the experiments of Roheim et al.[37] which demonstrated the presence of an apoprotein in plasma (of rats) which combined with lipid in the liver and was then released into the circulation as lipoprotein. Marsh now suggests that in liver cells synthesis of peptides on the ribosomes is followed by the addition of lipid at an intracellular site.[37a] Attachment of the carbohydrate moiety takes place in the region of the Golgi apparatus. The nature of the lipoprotein product released into the plasma by the liver cell is not established with certainty. Marsh[37a] favours the hypothesis that pre-betalipoprotein may be the initial product. Bilheimer, Eisenberg and Levy[37a] have shown that the apo-proteins of pre-betalipoprotein constitute a precursor in vivo as well as in vitro of the apoproteins of both betalipoprotein and alphalipo-protein. The half-life of betalipoprotein was found by Gitlin[38] to be 3.3 days and that of alphalipoprotein to be 4.6 days. From these ob-servations and other studies reviewed by Marsh,[36] the rate of synthesis of beta-lipoprotein appears to be greater than that of alphalipoprotein. There is no evidence for the storage of lipoproteins or their apoproteins in the hepatic cell, and the amount of lipoprotein in the liver at any one time is small. The administration of inhibitors of protein synthesis to experimental animals results in a fall in plasma lipoprotein concentrations and accumulation of lipid, mainly as triglyceride, in the liver[39] and synthesis of the protein moiety is undoubtedly essential for the release of lipid into the circulation.

Further areas of uncertainty relate to the source of the lipid moiety and the extent to which one class of lipid can substitute for another.[36] Changes in the lipid composition of lipoproteins take place within the plasma. Esterification of cholesterol proceeds intravascularly under the influence of plasma lecithin-cholesterol acyl transferase, and removal of triglyceride from the plasma is effected by lipoprotein lipase, a membrane-bound tissue enzyme. The role of lipid in regulating the rate of lipoprotein formation and release is not well defined. Raised concentrations of plasma non-esterified fatty acids result in their increased hepatic uptake, followed by triglyceride synthesis and an increase in the plasma concentration of triglyceride-rich lipoproteins. This does not necessarily imply increased hepatic synthesis of apoprotein.[40] It could be explained by the addition of lipid to the already circulating apoprotein described by Roheim et al.[37]

Investigations of the effect of cholesterol on lipoprotein synthesis suggest that increased concentrations of hepatic cholesterol do not result in increased output of betalipoprotein although some conflicting results have been obtained.[41]

The absorptive cells of the intestinal epithelium are the source of chylomicrons and the available evidence suggests that the entire particle, including the protein moiety, is synthesised within the intestinal cell.[18] After resynthesis of triglyceride from absorbed fatty acids and monoglyceride, the lipid droplet is coated with protein and phospholipid; some of the unesterified cholesterol also stays at the surface of the particle but the remainder, together with the esterified cholesterol, dissolves in the interior. Extrusion of the chylomicron particle into the interstitial space probably takes place by reverse pinocytosis, and the chylomicrons are then transported by intestinal lymphatics to the thoracic duct and thence to the systemic blood stream. The most important factor governing the rate of synthesis of chylomicrons is the amount of fat being absorbed. The nature of the dietary fat may also influence chylomicron formation; during the absorption of polyunsaturated fatty acids nearly all the absorbed lipid appears in chylomicrons, whereas during the absorption of saturated fatty acids a significant amount of these lipids appear in pre-betalipoprotein.[42] The rate of synthesis of phosphatidyl choline, the main phospholipid in chylomicrons, has been shown to be increased during fat absorption[43] and this factor may also govern both the rate of formation and size of the particles. Protein synthesis is also important and, in animals, the administration of protein inhibitors prevents chylomicron formation and triglyceride accumulates within the cell.[44]

Although it is firmly established that pre-betalipoprotein is synthesised in the liver it is now evident that the small intestine is important as a site of pre-betalipoprotein formation.[44] Pre-betalipoprotein in intestinal lymph in fasting animals is derived from endogenous lipid but during fat absorption exogenous fatty acids, especially saturated fatty acids, can be transported in this fraction.

1.3.2. CATABOLISM

The main function of chylomicrons and pre-betalipoprotein is the transport of triglyceride. Chylomicrons, carrying the bulk of absorbed dietary glyceride, undergo a number of changes during their passage through the lymphatics and blood stream. The interaction of chylomicrons with the plasma lipoproteins present in lymph and blood results in a gain of cholesterol and protein and a loss of phospholipid by the chylomicrons. This change in composition is responsible for differences in behaviour on polyvinylpyrrolidone (PVP) columns; lymph chylomicrons float to the top in a 0–5% PVP gradient and have been named primary particles, whereas chylomicrons that have been circulating in the plasma sink to the bottom of the column and are named secondary particles.[45] The formation of secondary particles may be an essential step in the further breakdown of chylomicrons.

The half-life of chylomicrons in the plasma is short, about 5–15 minutes.[46] The triglyceride is hydrolysed by lipoprotein lipase and the fatty acids are taken up by the tissues where they undergo oxidation or are resynthesised into triglyceride. Lipoprotein lipase is located on the capillary walls of most, if not all, tissues and the lipolytic reaction probably occurs at the cell surface. Activity of the enzyme in the plasma is normally very low; however, the administration of intravenous heparin results in the liberation of the enzyme into the plasma, and this forms the basis for the estimation of plasma activity in man.[47] The half-life of prebetalipoprotein is comparable to that of chylomicrons and the removal of triglyceride is accomplished by the same mechanism. The nature and fate of the fragments remaining after the removal of triglyceride from chylomicrons and pre-betalipoprotein is not known. It has been suggested that the lipoprotein products may constitute a source of beta and possibly alphalipoproteins,[32] and there is evidence that the apoproteins of beta and alphalipoprotein can derive from prebetalipoprotein.[37b]

The functions of beta and alphalipoproteins are poorly defined. The absence of either lipoprotein in man is compatible with life, but both conditions are associated with disease states; these are described in a later section and the contribution they make to the further understanding of lipoprotein function is discussed. Both beta and alphalipoprotein play an important part in the esterification of cholesterol in plasma. The enzyme responsible for esterification, lecithin-cholesterol acyltransferase,[48] is probably closely associated with alphalipoprotein which also provides the substrates for the transesterification process. The unesterified cholesterol and phospholipid components of beta and alphalipoproteins are in dynamic equilibrium with the lipids in the membrane of red blood cells and probably also with the lipoproteins of other cell membranes. The half-lives of beta and alphalipoproteins are considerably longer (about 3 and 4 days respectively) than that of chylomicrons and pre-betalipoprotein, but little is known about the site or mechanism of their breakdown. Thyroxine has been shown to influence the rate of catabolism of betalipoprotein, the half-life being prolonged in hypothyroidism.[49]

1.3.3. FACTORS INFLUENCING CONCENTRATION AND COMPOSITION

Considerable variation occurs in the concentration and composition of the plasma lipoproteins between different healthy individuals, and also in the same individual studied under different conditions. A knowledge of the factors involved is important to ensure appropriate sampling conditions and correct interpretation of the results.

(a) *Age*

At birth the concentration of betalipoprotein is only about one third, and alphalipoprotein about one half, of the adult concentrations. There is no relationship between the concentrations in the infant at birth and those in the maternal blood. After birth lipoprotein concentrations rise rapidly and by the 6th day of life plasma cholesterol and phospholipid have approximately doubled. Thereafter concentrations rise slowly and normal childhood values are probably achieved by the latter half of the first year. Little change occurs throughout childhood although some authors report a slow upward trend in plasma cholesterol of about 2–3 mg/100 ml per year. During adolescence cholesterol concentrations tend to fall in both sexes. Throughout adult life beta and pre-betalipoprotein concentrations slowly rise; the average increase in plasma cholesterol is about 2–3 mg/100 ml per year.

(b) *Sex*

No significant differences have been found in lipoprotein concentrations between the sexes during infancy and childhood. In adult life, women have higher concentrations of alphalipoprotein and lower concentrations of pre-betalipoprotein than men. Plasma triglyceride levels are lower in women, but cholesterol levels are similar between the sexes until the age of 50–60 years after which women tend to have higher values. Fluctuations in lipoprotein concentrations, and hence in plasma lipids, occur during the menstrual cycle; hyperlipoproteinaemia due to increases in prebeta and alphalipoprotein occurs during the later months of pregnancy (Table 1.3) and levels do not return to normal for 3–6 months after delivery. The difference in alphalipoprotein concentration between the sexes can be accounted for by the effects of the sex hormones on lipoprotein metabolism.[50] Oestrogens increase the concentration of alphalipoprotein, and diminish the amount of cholesterol in both alpha and betalipoproteins; androgens have the opposite effect, decreasing alphalipoprotein and increasing betalipoprotein.

(c) *Diet*

Both the timing and composition of the preceding meal and the usual diet of the individual influence the concentration and composition of the plasma lipoproteins. Chylomicrons are only found after a meal containing fat; the course and duration of alimentary hyperlipidaemia is variable but peak levels are usually reached after 2–4 hours and clearing is not complete for 6–8 hours. Some increase in pre-betalipoprotein also occurs during the later stages of fat absorption but concentrations of beta and alphalipoprotein are not usually significantly altered.

The diet of the individual affects the concentrations of beta and pre-betalipoprotein and the fatty acid composition of lipids in all lipoprotein fractions. The most important constituents influencing lipoproteins are the amount and type of the fat, and the amount (and possibly the type) of the carbohydrate. Betalipoprotein concentrations and hence plasma cholesterol levels are lowered by a reduction in the intake of fat, and by diets in which there is an increase in the proportion of polyunsaturated fatty acids relative to saturated fatty acids. The most common polyunsaturated fatty acid in the diet is linoleic acid and the percentage of this acid in plasma cholesterol ester, triglyceride, and phosphatidyl choline increases when the intake of polyunsaturated fat is raised.

An increase in dietary carbohydrate results in increased hepatic lipogenesis; this may result in increased triglyceride synthesis with a consequent rise in plasma pre-betalipoprotein concentrations. There is some evidence that sucrose may promote more lipogenesis than glucose or starch. Increased lipogenesis is accompanied by changes in the fatty acid composition of plasma triglyceride; an increased proportion of palmitoleic acid is the most constant feature.

Change in body weight may also be associated with changes in the lipoprotein pattern. Weight gain due to excessive calorie intake is likely to be associated with increased lipogenesis. Weight loss may be associated with a fall in betalipoprotein concentration if dietary fat intake has been curtailed.

(d) *Other Factors*

Genetic factors are obviously important in the regulation of plasma lipoproteins; the genetically determined disorders are described in a later section.

Other factors which have been reported to affect lipoproteins include emotional stress, physical activity, trauma and even mild intercurrent illness. In some areas there is seasonal variation which may be related to dietary changes. The influence of alcohol, nicotine and the contraceptive pill also have to be considered in the interpretation of the lipoprotein pattern in an individual subject.

1.4. CONDITIONS OF SAMPLING AND NORMAL VALUES

Conditions of sampling must be standardised if the results are to be properly interpreted and comparisons made with the data of other workers. Unless a study is being made of the production or removal of chylomicrons, blood for analysis of lipoproteins must be obtained in the fasting state, that is at least 10 hours and preferably 12–16 hours after the previous meal. The subject should have been eating his usual diet during the preceding 1–2 weeks and have maintained a steady body weight. When the fatty acid composition of the lipids is being determined, a knowledge of the nature of the dietary fat is desirable. For women a record should be made of the date of the last menstrual period, and the taking of any contraceptive medication.

The blood should be drawn with the minimum of venous stasis. For analysis of serum, blood is allowed to clot at room temperature; the most satisfactory anticoagulant for plasma is disodium ethylenediamine-tetraacetate (EDTA), 1 mg/ml of blood. The serum or plasma sample should be kept at 0–4°C until analysed. Separation of the lipoproteins should be carried out the same day. Storage or freezing seriously affects chylomicrons and pre-betalipoproteins, and although valid analyses may be obtained on stored samples of normal plasma, pathological samples must be analysed without delay. Plasma or serum for lipid analysis may be stored at −20°C without deterioration.

Normal values should ideally be established in each laboratory undertaking lipoprotein analyses. Differences in both genetic and environmental factors invalidate the use of control data derived from populations differing in these respects. Even for the same population a change in dietary habit may alter the normal values obtained over a period of time. For example, the increasing consumption of polyunsaturated oils in the U.K. is affecting the fatty acid composition of plasma lipids and may also influence their concentration.

1.5. CLINICAL DISORDERS

1.5.1. CLASSIFICATION

Abnormalities of the concentration or composition of lipoproteins may result from a primary genetic defect, or occur as a secondary manifestation in association with a variety of disease states (Table 1.3). Because the mechanisms responsible for most of the disorders are poorly understood, the terminology is descriptive of the plasma lipoprotein and/or lipid abnormalities. The classification given in Table 1.3 is based on abnormalities in the lipoprotein pattern as defined by paper electrophoresis; the alternative terminology in common usage for the primary disorders is also given. In many disorders abnormalities are found in more than one lipoprotein class.

1.5.2. HYPERLIPOPROTEINAEMIA

(a) *Primary Hyperchylomicroaemia* (Type I[13,13a] Fat-induced hypertriglyceridaemia; exogenous hypertriglyceridaemia).

Primary hyperchylomicronaemia is probably the least common, but best defined (in that the underlying enzyme defect is known) of the hyperlipoproteinaemic states. It is due to deficiency of the enzyme lipoprotein lipase and is inherited as an autosomal recessive. Clinically the presenting features may include eruptive xanthomata, hepatosplenomegaly, lipaemia retinalis, and attacks of abdominal pain and vomiting. The disorder is usually diagnosed during childhood, but may occasionally be symptomless and only diagnosed at a later age by the accidental finding of turbid plasma. At present there is no firm evidence that the incidence of atherosclerosis is increased in adult life.

The plasma is characteristically turbid even in the fasting state due to the increase in chylomicrons. Paper electrophoresis and ultracentrifugation demonstrate that there is also some increase in pre-beta-lipoprotein and that concentrations of beta- and alphalipoprotein are both reduced. Analysis of the plasma lipids reflects the composition of chylomicrons; there is a great increase in triglyceride with values of the order 5000–10 000 mg/100 ml, and less marked increases in total cholesterol and phospholipid. The ratio of triglyceride to cholesterol is of the order of 8:1 to 10:1. About 50% of the cholesterol is unesterified due to the greater proportion of unesterified cholesterol in chylomicrons; this decrease in esterified cholesterol does not indicate impaired hepatic function. The enzyme defect is demonstrated by the estimation of plasma post-heparin lipolytic activity (PHLA).[47] Delayed clearing of dietary fat can be shown by determination of plasma triglyceride or turbidity after a meal containing fat. Both these investigations must be done within a few days of the commencement of a low-fat diet because normal subjects on low-fat diets for longer than 1 week may show abnormal responses.[47] During the episodes of abdominal pain there may be evidence of pancreatitis with raised plasma amylase levels.

The diagnosis is established by the identification of the lipoprotein pattern, by the demonstration of low levels of plasma PHLA, and by the response to reduction in dietary fat. A virtually fat-free diet (<2.0 g/day) renders the plasma optically clear and greatly reduces the triglyceride level within about a week. Although dietary treatment abolishes hyperchylomicronaemia in fasting plasma, the increase in pre-betalipoprotein may become more marked; this may be the result of increased lipogenesis due to the relatively high carbohydrate content of low-fat diets, together with impaired clearing of the endogenous triglyceride. Reduction in beta and alphalipoprotein persist in the treated state so that the cholesterol concentrations may be as low as 100 mg/100 ml. The degree of dietary fat restriction required for adequate treatment is often difficult to maintain on a long-term basis. The diet can be made more palatable by the use of medium-chain triglyceride; this fat, containing fatty acids with carbon chain lengths of 8 and 10 is absorbed directly into the portal vein, thus bypassing the chylomicron route.

The findings in heterozygotes are not consistent. A proportion of first degree relatives have low levels of plasma PHLA,[47] and in some individuals there may be abnormalities of the plasma lipoproteins with moderate elevation of serum triglyceride, an increase in pre-betalipoprotein, and delay in clearing dietary fat.

(b) *Primary Hyperbetalipoproteinaemia* (Type II;[13,13a] familial hypercholesterolaemia)

Primary hyperbetalipoproteinaemia is possibly the most common of the genetically determined disorders of lipoproteins, although the gene frequency has not yet been established. It is inherited as an autosomal dominant with variable clinical expression in heterozygotes, who usually have no clinical manifestations until early adult life when xanthomata and corneal arcus may occur. The risk of ischaemic heart disease, which may be the presenting clinical feature, is increased in both sexes, though it occurs more frequently and at an earlier age in men.[51] In the homozygous state xanthomata and ischaemic heart disease occur in childhood and death in adolescence is common.

The mechanism responsible for the increased concentration of betalipoprotein has not been defined, but there is some evidence that it concerns the catabolism of betalipoprotein.[52] Fasting plasma is clear; electrophoresis and ultracentrifugation show an increase in the betalipoprotein fraction. Cholesterol concentrations are of the order of 300–500 mg/100 ml in heterozygotes, and 700–1000 mg/100 ml in homozygotes. (These are values we have observed in the U.K. and may not be applicable to other communities.) Phospholipid concentrations are also raised but triglyceride concentrations are usually normal; however, moderate increases in triglyceride and raised levels of pre-betalipoprotein have been reported in some patients.[13a] The composition of the betalipoprotein is abnormal with increase in the proportion of cholesterol ester and decrease in the proportion of triglyceride.[52a,52b]

The diagnosis is established by the finding of hypercholesterolaemia, increased betalipoprotein, and the demonstration of the same lipoprotein abnormality in at least one first degree relative. Diagnosis in subjects with "borderline" serum cholesterol concentrations may be difficult; however precisely normal values are defined, no definite

TABLE 1.3. Classification of lipoprotein disorders

Lipoprotein	Type of Pattern[13]	Hyperlipoproteinaemia Primary	Hyperlipoproteinaemia Secondary	Hypolipoproteinaemia Primary	Hypolipoproteinaemia Secondary
Chylomicron	I	Hyperchylomicronaemia (Fat-induced hypertriglyceridaemia; Exogenous hypertriglyceridaemia)	Alcoholism Diabetes Mellitus Dysglobulinaemia Pancreatitis		Fat malabsorption
Betalipoprotein (LDL)	II	Hyperbetalipoproteinaemia (Familial hypercholesterolaemia)	Diabetes Mellitus Glycogenosis (type III) Hypothyroidism Nephrotic syndrome *Obstructive liver disease	Abetalipoproteinaemia Hypobetalipoproteinaemia	Chronic anaemia Fat malabsorption Hepatocellular failure Hyperthyroidism
Abnormal betalipoprotein (Broad-beta)	III	Broad-beta disease*	Dysglobulinaemia Hypothyroidism		
Pre-betalipoprotein (VLDL)	IV	Hyperpre-betalipoproteinaemia (Carbohydrate-induced hypertriglyceridaemia; endogenous hypertriglyceridaemia)	Alcoholism Diabetes Mellitus Dysglobulinaemia Glycogenosis (type I & VI) Hypothyroidism Nephrotic syndrome *Obstructive liver disease Oral contraceptives Pancreatitis Pregnancy		
Chylomicrons + Pre-betalipoprotein	V	? separate entity or stages of type I or IV	Alcoholism Diabetes Mellitus Dysglobulinaemia Pancreatitis		
Alphalipoprotein			Alcoholism Dysglobulinaemia Obstructive liver disease Pregnancy	Familial alphalipoprotein deficiency	Conditions in which pre-beta-lipoprotein is increased. Familial lecithin-cholesterol acyl transferase deficiency. Hepatocellular failure.

* Lipoprotein of abnormal composition.

cut-off point can be set to divide familial hypercholesterolaemia from normal, although it has been suggested that the estimation of betalipoprotein cholesterol is a somewhat better discriminant than serum cholesterol alone.[52c] Other diseases known to be associated with elevation of betalipoprotein must be excluded (Table 1.3). The condition can be diagnosed during childhood after the age of 1 year.[54] Suggestions that cholesterol concentrations at birth could be diagnostic[53] have not been confirmed.[54]

In the heterozygote, plasma betalipoprotein and cholesterol can be reduced by diets low in saturated fat, which may be made more palatable by the use of oils and fats rich in polyunsaturated fat.[55] For most patients additional therapy[56] is required in order to achieve normal plasma levels. Cholestyramine, an unabsorbable ion-exchange resin which binds bile acids, is probably the most effective hypocholesterolaemic agent; nicotinic acid and clofibrate (chlorophenoxyisobutyrate) are also used. In the homozygote treatment with diet and a combination of two or more drugs results in appreciable lowering of cholesterol levels but normal values are not achieved. Xanthomata, however, have resolved with such treatment.[55]

It is not yet known with certainty whether lowering the plasma betalipoprotein concentration can prevent or delay the onset of ischaemic heart disease. To achieve the maximum preventive effect, treatment should be instituted before heart disease is clinically evident, and probably even in childhood.

(c) "Broad-beta Disease" (Type III[13,13a])

Fredrickson et al.[13] have defined a lipoprotein disorder characterised by a broadened betalipoprotein band on electrophoresis and by an excess of very low-density lipoprotein on ultracentrifugation. Neither the genetic defect, nor the mode of inheritance are yet known. Clinical manifestations appear to be delayed until adult life. Xanthomata may be tuberous, tendinous or eruptive, and xanthomata in the palmar skin creases are said to be characteristic. Premature atherosclerosis is likely to develop; both peripheral occlusive vascular disease and coronary heart disease occur.

Fasting plasma is opalescent or turbid due to an increase in triglyceride. Electrophoresis shows the characteristic "broad-beta band". On ultracentrifugation the concentration of very low-density lipoprotein is increased, whilst low-density (S_f 3–9, betalipoprotein) is decreased. The composition of the very low-density lipoprotein is abnormal with an increase in the cholesterol content and a decrease in triglyceride.[57] The plasma lipids reflect the lipoprotein abnormality; concentrations of cholesterol and triglyceride are variable but both are increased to a comparable degree. Glucose tolerance may be abnormal.

This lipoprotein disorder responds well to treatment. When the patient is obese, weight reduction may be sufficient to reduce plasma cholesterol and triglyceride concentrations to normal. Clofibrate is the drug of choice and its use can be expected to result in normal plasma lipid levels. The abnormal lipoprotein pattern, however, persists. Xanthomata will resolve and peripheral blood flow may increase.[58]

(d) *Primary Hyperpre-betalipoproteinaemia* (Type IV;[13,13a] endogenous hyper-triglyceridaemia; "carbohydrate-induced" hypertriglyceridaemia)

The incidence of primary hyperpre-betalipoproteinaemia is not known. This lipoprotein pattern is common. Many different mechanisms may be responsible for an increase in pre-betalipoprotein,[59] and it is often difficult to identify with certainty the presence of a primary disturbance. The majority of patients so far described have been adults. Eruptive xanthomata may occur, many patients are obese, and glucose tolerance is frequently abnormal. The incidence of premature atherosclerosis appears to be increased.

Fasting plasma is opalescent or turbid. Electrophoresis demonstrates increased pre-betalipoprotein, and some chylomicrons may be present because both endogenous and exogenous triglyceride share the same clearing mechanism. Ultracentrifugation shows an excessive amount of very low-density lipoprotein. Plasma triglyceride concentrations are always raised and although cholesterol is often increased it is not invariably above the normal range.

In some patients the increased synthesis of pre-betalipoprotein may be due to excessive hepatic lipogenesis. The fall in plasma triglyceride in response to low-carbohydrate diets has formed the basis for the use of the term "carbohydrate-induced" hypertriglyceridaemia. An increase in fasting plasma triglyceride has, however, been shown to be a normal response to a high-carbohydrate diet in both children and adults, and carbohydrate "inducibility" is not a reliable diagnostic test. The diagnosis is established by the demonstration of the lipoprotein pattern and the exclusion of secondary causes of hyperpre-betalipoproteinaemia. Family studies should be made and may demonstrate other affected members.

Treatment of the patient who is obese with a low calorie diet may restore the lipoprotein pattern to normal. For patients of normal weight the proportion of carbohydrate in the diet should be reduced, and it may be advantageous to replace some of the ordinary dietary fat with polyunsaturated oils. If dietary treatment is unsuccessful, clofibrate is the drug of choice; nicotinic acid may also be helpful.

(e) *Type V hyperlipoproteinaemia*[13,13a]

Fredrickson *et al.*[13] have described a lipoprotein pattern characterised by elevation of both chylomicrons and pre-betalipoprotein which they have designated as the Type V disorder. This pattern undoubtedly occurs as a secondary manifestation in a number of conditions (Table 1.3);

it is still uncertain whether it represents a primary disorder. The pattern may also be seen in some patients with primary hyperchylo-micronaemia (Type I) or primary hyperpre-betalipoproteinaemia (Type IV).

Clinical manifestations include eruptive xanthomata, attacks of abdominal pain, and hepatosplenomegaly. The patients are often obese. Post-heparin lipolytic activity is usually normal, and glucose tolerance is often abnormal.

Fasting plasma is turbid; electrophoresis and ultracentrifugation demonstrate excess chylomicrons and pre-betalipoprotein. Concentrations of both triglyceride and cholesterol are raised, through the increase in triglyceride is more marked and the ratio of triglyceride to cholesterol is of the order of 4:1.

Treatment consists of weight reduction if the patient is obese. Reduction in dietary fat may be required to control symptoms. Clofibrate has proved effective in some patients.

(f) Secondary Disorders

The main conditions associated with secondary increases of the plasma lipoproteins are given in Table 1.3. The type of lipoprotein pattern may differ between individuals with the same disease, and also in the same individual at different stages of the disease.

Poorly controlled *diabetes mellitus* is usually, although not invariably, associated with hyperlipoproteinaemia. The most common abnormality is excess of pre-betalipoprotein. The increased hepatic synthesis of triglyceride results from the high levels of plasma non-esterified fatty acids, caused by increased mobilisation of adipose tissue lipid. Hyper-chylomicroaemia may occur due to impaired clearing because of endogenous hypertriglyceridaemia, or due to diminished lipoprotein lipase activity in insulin deficiency.[60] Increase in betalipoprotein is found less commonly. In insulin-dependent patients plasma lipoproteins return to normal with adequate treatment. In maturity-onset diabetes hyperlipoproteinaemia often responds to weight control and may be reduced by oral hypoglycaemic agents.

The type of hyperlipoproteinaemia in the *hepatic glycogenoses* varies;[61] Fernandes and Pikaar[61] have described increased prebeta-lipoprotein in type I (glucose-6-phosphatase deficiency) and increase in betalipoprotein in types III (de-branching enzyme deficiency) and VI (phosphorylase deficiency). The differentiation is not, however, complete and the lipoprotein pattern cannot be used to distinguish between the different types of disorder. The use of different dietary regimes to control the hyperlipoproteinaemia has also been described by Fernandes and Pikaar.[61]

Hypothyroidism causes hyperbetalipoproteinaemia due to decreased catabolism of betalipoprotein.[49] Increased prebetalipoprotein can also occur[62] and in some patients a broad-beta pattern may be found.

The *nephrotic syndrome* is usually accompanied by hypercholesterolaemia. The lipoprotein pattern is variable and probably related to the degree of proteinuria.[63,64] An inverse relationship has been reported between the concentrations of plasma albumin and triglyceride.[64] Increases of betalipoprotein, pre-betalipoprotein and chylomicrons may be found. It is postulated that increased hepatic synthesis of beta (and possibly pre-beta) lipoprotein occurs as part of a general hepatic response to albumin loss. Increase in the triglyceride-rich lipoproteins (pre-betalipoprotein and chylomicrons) may also result from impaired clearing due to lipoprotein lipase deficiency.[65] Plasma lipoproteins return to normal when the proteinuria is controlled.

Obstructive liver disease (biliary cirrhosis) is associated with an unusual type of hyperlipoproteinaemia characterised by a marked increase in cholesterol, but there is an even greater increase in phospholipid so that the ratio of cholesterol to phospholipid may be as low as 0.5. The proportion of cholesterol that is unesterified is increased and phosphatidyl choline accounts for the increase in total phospholipid. Triglyceride concentrations may be moderately elevated. The electrophoretic pattern may be misleading because of the poor dye uptake of unesterified cholesterol and phospholipid. Ultracentrifugal studies indicate an increase in low density lipoproteins (density range 1.006–1.063 g/ml). Within this fraction three immunochemically distinct lipoproteins have been identified; betalipoprotein, alphalipoprotein, and a lipoprotein designated lipoprotein X.[66,66a] Lipoprotein X has a unique composition consisting of 6% protein, 65% phospholipid, 25% cholesterol and 3% triglyceride; 93% of the cholesterol is unesterified. Chemical and immunochemical studies suggest that apoprotein X may be identical with apoprotein C, the apoprotein normally found in pre-betalipoprotein. The unusual distribution of lipids in plasma in obstructive liver disease is due to accumulation of lipoprotein X. Treatment of the hyperlipoproteinaemia with cholestyramine may be successful in lowering cholesterol and phospholipid concentrations, but does not affect the underlying hepatic disorder and the abnormal lipoprotein pattern persists. Increased concentrations of alphalipoprotein are occasionally found in patients with liver disease.

In *dysglobulinaemia* hyperlipoproteinaemia may occur due to the accumulation of soluble complexes of antibodies with lipoproteins.[67] The main disorder associated with this type of hyperlipoproteinaemia is myeloma due to either IgA or IgM myeloma protein. The lipoprotein pattern is variable; increases in chylomicrons, pre-betalipoprotein and alphalipoprotein have been reported, and in some cases there has been a broad-beta band.

Pancreatitis is usually associated with increased chylomicrons and pre-betalipoprotein. *Alcoholism*, often associated with pancreatitis or liver disease, causes similar abnormalities; alphalipoprotein may be increased.

In *pregnancy* concentrations of pre-betalipoprotein and alphalipoprotein are increased, especially in the later months, and levels do not return to normal until 3–6 months after delivery.

Oral contraceptives containing oestrogens and progesterone may result in increased concentrations of triglyceride and prebetalipoprotein.[68] Increase in fasting plasma insulin and decrease in plasma post heparin lipolytic activity have been reported, and it has been suggested that the hypertriglyceridaemia results both from increased hepatic synthesis under the influence of insulin, and decreased removal due to the reduced lipoprotein lipase activity.

1.5.3. HYPOLIPOPROTEINAEMIA

(a) *Abetalipoproteinaemia*[69,70]

This disorder is inherited as an autosomal recessive, and about 40 cases have been reported in the literature. The gene defect probably concerns the synthesis of beta apoprotein. Steatorrhoea and acanthocytosis of the red cells are present from birth, and in later childhood a pigmentary retinopathy and ataxic neuropathy develop which result in progressive crippling in adolescence and adult life. Heterozygotes show no clinical or haematological abnormalities.

The plasma is clear, even after a meal containing fat, as neither chylomicrons nor pre-betalipoprotein can be formed. Absence of betalipoprotein can be demonstrated by electrophoretic, ultracentrifugal and immunochemical techniques. Alphalipoprotein is reduced to about one half of its normal concentration. The plasma lipids reflect the lipoprotein abnormalities; cholesterol is reduced to about 20–50 mg/100 ml, phospholipid to about 50–100 mg/100 ml and triglyceride to below 10 mg/100 ml. The proportion of cholesterol that is esterified is normal. The distribution of the individual phospholipids is abnormal with a relative decrease in phosphatidyl choline and increase in sphingomyelin. The proportion of linoleic acid present in all plasma lipids is extremely low. Carotenoids and vitamin E (both carried by betalipoprotein) cannot be detected in plasma; vitamin A concentrations are also low.

The intraluminal phase of fat digestion is normal and intestinal biopsy shows normal villous architecture, but the mucosal cells are distended with triglyceride due to the failure of chylomicron formation. Nevertheless a considerable proportion of dietary fat is absorbed (about 50–70%), presumably via the portal venous route as non-esterified fatty acids bound to albumin. Accumulation of triglyceride also occurs in the liver cells due to failure of pre-betalipoprotein formation but the amount of lipid is not sufficient to cause clinical hepatic enlargement or result in abnormal liver function tests. The inability to synthesise both chylomicrons and pre-betalipoprotein in abetalipoproteinaemia suggests that beta-apoprotein is necessary for the formation of all low-density lipoproteins.

Red cell survival is somewhat shortened but haemolytic anaemia is not a feature; in vitro haemolysis (auto-haemolysis and peroxide haemolysis) is increased and this abnormality can be corrected by vitamin E treatment. The lipid composition of the red cell membrane is abnormal with a relative decrease in phosphatidyl choline and increase in sphingomyelin; these changes are secondary to the plasma lipoprotein abnormality. The pathology of retinal and nervous tissue is poorly documented. As all cell membranes are essentially lipoproteins it is possible that the composition of retinal and nerve cell membranes is abnormal and that this may be the cause of the impaired function.

The diagnosis should be considered when the plasma cholesterol is below 50 mg/100 ml or when acanthocytes are seen in a fresh, wet, undiluted blood film. Confirmation by electrophoretic or immunochemical examination of the plasma lipoproteins is essential; acanthocytes may be found in other conditions. Examination of heterozygotes shows that the majority have normal lipoproteins, though in a few families reduced levels of betalipoprotein have been found.

Treatment is symptomatic; short term infusions of betalipoprotein have not affected red cell morphology or intestinal absorption. Steatorrhoea can be controlled by a reduction in dietary fat, and large doses of the fat-soluble vitamins A and E should be given.

(b) *Primary Hypobetalipoproteinaemia*[69]

Primary hypobetalipoproteinaemia, in which betalipoprotein is reduced to about half the normal concentration is inherited as an autosomal dominant,[71,71a] and is a different condition from abetalipoproteinaemia. Any of the clinical manifestations of abetalipoproteinaemia may occur, but not all have been found in the same patient, and some individuals have no abnormal clinical features.

(c) *Familial Alphalipoprotein Deficiency*[69,70] *(Tangier Disease)*

This disorder is inherited as an autosomal recessive, and about 13 cases have been reported. The gene defect concerns the synthesis of alphalipoprotein; a small amount of an abnormal high-density lipoprotein is formed. The clinical manifestations include large yellow-orange tonsils, lymphadenopathy, hepatosplenomegaly and corneal opacities. The abnormalities in the reticuloendothelial system are due to accumulation of cholesterol ester, and similar deposition has also been found in the bone marrow and rectal mucosa. The pathogenesis of the neuropathy is unknown, though deposition of cholesterol ester within a peripheral nerve has been demonstrated in one patient. The relationship between the accumulation of cholesterol ester in tissues and the absence of normal alphalipoprotein in plasma is not understood. It has been suggested that alphalipoprotein may inhibit the uptake of lipid by phagocytic cells, particularly those of the reticuloendothelial system. Heterozygotes show no clinical manifestations.

The plasma is turbid in the fasting state. Electrophoresis demonstrates that most of the lipid is in the pre-betalipoprotein fraction and that alphalipoprotein appears absent. Ultracentrifugation shows an increase in very low-density lipoprotein, and a reduction in low-density lipoprotein; a very small quantity of high-density lipoprotein can be detected. All the lipoproteins have an abnormal lipid composition with an excess of triglyceride. Plasma lipids show increased levels of triglyceride to about 300–400 mg/100 ml, and decreased levels of cholesterol (50–110 mg/100 ml) and of phospholipid (90–140 mg/100 ml). The proportion of cholesterol that is esterified is normal. The distribution of the individual phospholipids is abnormal with a relative increase in the proportion of phosphatidyl choline and decrease in sphingomyelin.

The diagnosis should be suspected if plasma triglyceride is raised and cholesterol reduced. Electrophoretic, ultracentrifugal, and immunochemical analyses are required for confirmation. Reduced levels of alphalipoprotein are found in most of the heterozygotes, particularly in males, and small quantities of the abnormal high-density lipoprotein have also been detected.

No specific treatment is available. The hypertriglyceridaemia has been shown to respond to a reduction in dietary carbohydrate.

(d) *Secondary Hypolipoproteinaemias* (Table 1.3)

Reduction in chylomicron formation and in plasma betalipoprotein concentration may result from impaired fat absorption due to any cause. Low levels of betalipoprotein occur in hyperthyroidism due to increased catabolism. In the terminal stages of hepatocellular failure all lipoproteins may be decreased. The reduction in betalipoprotein which may be found in some patients with chronic anaemia is unexplained.

Reduced levels of alphalipoprotein have been reported in familial lecithin-cholesterol acyl transferase deficiency.[73,73a]

REFERENCES

1. H. B. SALT and O. H. WOLFF, *Arch. dis. childh.*, **32** 404 (1957)
   The applications of serum lipoprotein electrophoresis in paediatric practice.
2. R. S. LEES and F. T. HATCH, *J. lab. clin. med.*, **61**, 518 (1963)
   Sharper separation of lipoprotein species by paper electrophoresis in albumin-containing buffer.
3. M. H. FLETCHER and M. H. STYLIOU, *Clin. chem.*, **16**, 362 (1970)
   A simple method for separating serum lipoproteins by electrophoresis on cellulose acetate.
4. R. M. IAMMARINO, M. HUMPHREY and P. ANTOLIK, *Clin. chem.*, **15**, 1218, (1969)
   Agar gel lipoprotein electrophoresis: a correlated study with ultra-centrifugation.
5. R. STRAUS and M. WURM, *Amer. J. clin. path.*, **29**, 581 (1958)
   A new staining procedure and a method for quantitation of serum lipoproteins separated by paper electrophoresis.

6. O. DE LALLA and J. W. GOFMAN, (ed., D. Glick),
Methods of biochemical analysis, vol 1, (Interscience, New York, 1954), p 459.
7. G. L. MILLS and P. A. WILKINSON, *Clin. chim. acta.*, **8**, 701 (1963)
The distribution of beta-lipoproteins in human plasma.
8. R. J. HAVEL, H. A. EDER and J. H. BRAGDON, *J. clin. invest.*, **34**, 1345, (1955)
The distribution and chemical composition of ultracentrifugally separated
lipoproteins in human serum.
9. D. G. CORNWELL, F. A. KRUGER, G. J. HAMWI and W. B. BROWN, *Amer. J.
clin. nutr.*, **9**, 24 (1961)
Studies on the characterisation of human serum lipoproteins separated by
ultracentrifugation in a density gradient.
10. R. S. LEES *Science*, **169**, 493 (1970)
Immunoassay of plasma low-density lipoproteins.
11. P. GRABAR and C. A. WILLIAMS, *Biochim. biophys. acta*, **17**, 67 (1955)
Méthode immunoèlectrophoretique d'analyse de mèlanges de substances
antigèniques.
12. C. B. LAURELL. *Anal. biochem.*, **15**, 45 (1966)
Quantitative estimation of proteins by electrophoresis in agarose gel containing antibodies.
13. D. S. FREDRICKSON, R. I. LEVY and R. S. LEES, *New Engl. J. Med.*, **276**, 32, 94,
148, 215, 273 (1967)
Fat transport in lipoproteins—an integrated approach to mechanisms and
disorders.
13a. D. S. FREDRICKSON and R. I. LEVY, (eds., J. B. Stanbury, J. B. Wyngaarden and
D. S. Fredrickson),
Familial hyperlipoproteinaemia in The Metabolic Basis of Inherited Disease,
(McGraw-Hill, New York, 3rd ed. 1972) p. 545.
14. K. W. WALTON and P. J. SCOTT, *J. clin. path.*, **17**, 627 (1964)
Estimation of the low-density (beta) lipoproteins of serum in health and
disease using large molecular weight dextran sulphate.
15. M. BURSTEIN, H. R. SCHOLNICK and R. MORFIN, *J. lip. res.*, **11**, 583 (1970)
Rapid method for the isolation of lipoproteins from human serum by precipitation with polyanions.
16. C. FRANZINI. *Clin. chim. acta.*, **14**, 576 (1966)
Gel filtration of human serum lipoproteins.
17. M. C. STONE and J. M. THORP, *Clin. chim. acta.*, **14**, 812 (1966)
A new technique for the investigation of the low-density lipoproteins in health
and disease.
17a. A. SCANU, (ed., R. M. S. Smellie),
Human plasma high density lipoproteins, in Plasma Lipoproteins, Biochemical
Society Symposia: no. 33. (Academic Press, London and New York, 1971) p. 34.
17b. G. KOSTNER and P. ALAUPOVIC, *FEBS Letters*, **15**, 370, (1971)
Studies of the composition and structure of plasma lipoproteins. C- and N-
terminal amino acids of the two non-identical polypeptides of plasma apolipo-
protein A.
18. D. B. ZILVERSMIT, (eds., E. Tria and A. M. Scanu),
Structural and functional aspects of lipoproteins in living systems, (Academic
Press, London and New York, 1969), pp. 329–365.
19. A. GUSTAFSON, P. ALAUPOVIC and R. H. FURMAN, *Biochemistry*, **4**, 596 (1965)
Studies of the composition and structure of serum lipoproteins: isolation,
purification and characterisation of very low density lipoproteins of human
serum.
20. V. P. DOLE and J. T. HAMLIN, *Physiol. Rev.*, **42**, 674, (1962)
Particulate fat in lymph and blood.
21. M. RODBELL and D. S. FREDRICKSON, *J. biol. Chem.*, **234**, 562 (1959)
The nature of the proteins associated with dog and human chylomicrons.

22. A. Scanu and I. H. Page, *J. exp. Med.*, **109,** 239 (1959)
    Separation and characterisation of human serum chylomicrons.
23. S. Margolis, (eds., E. Tria and A. M. Scanu), Structural and functional aspects
    of lipoproteins in living systems, (Academic Press, London and New York,
    1969), pp. 370–415.
24. J. L. Oncley, (eds., J. Folchi-Pi and H. Bauer),
    Brain lipids and lipoproteins, and the leucodystrophies (Elsevier, Amsterdam,
    1963), pp. 1–17.
25. V. P. Skipsky, M. Barclay, R. K. Barclay, V. A. Fetzer, J. J. Good and
    F. M. Archibald. *Biochem. J.*, **104,** 340 (1967)
    Lipid composition of human serum lipoproteins.
26. F. T. Lindgren and A. V. Nichols, (ed., F. W. Putnam),
    The plasma proteins, (Academic Press, London and New York, 1960), Vol. 2.
    pp. 1–58.
27. S. Margolis, *J. lipid. Res.*, **8,** 501 (1967)
    Separation and size determination of human serum lipoproteins by agarose
    gel filtration.
28. J. L. Granda and A. Scanu, *Biochemistry (N.Y.)* **5,** 3301 (1966)
    Solubilisation and properties of the apoproteins of the very low- and low-
    density lipoproteins of human serum.
29. R. S. Levy, A. C. Lynch, E. D. McGee and J. W. Mehl, *J. Lipid. Res.*, **8,** 463
    (1967)
    Amino acid composition of the proteins from chylomicrons and human serum
    lipoproteins.
30. M. Rodbell, *Science (N.Y.)* **127,** 701 (1958)
    N-terminal amino acid and lipid composition of lipoproteins from chyle and
    plasma.
31. P. Herbert, R. I. Levy and D. S. Fredrickson, *J. Biol. Chem.*, **246,** 7068 (1971)
    Correction of COOH-terminal amino acids of human plasma very low density
    apoliproteins.
32. A. Gustafson, P. Alaupovic and R. H. Furman, *Biochemistry (Wash.)*, **5,**
    632 (1966)
    Studies of the composition and structure of serum lipoproteins. Separation and
    characterisation of phospholipid-protein residues obtained by partial delipidiza-
    tion of very low density lipoproteins of human serum.
33. R. I. Levy and D. S. Fredrickson, *J. Clin. Invest.*, **44,** 426 (1965)
    Heterogeneity of plasma high density lipoproteins.
34. A. M. Scanu, (eds., E. Tria and A. M. Scanu),
    Structural and functional aspects of lipoproteins in living systems, (Academic
    Press, London and New York, 1969), pp. 425–444.
34a. J. I. Kessler, J. Stein, D. Dannacker and P. Narcessian, *J. Biol. Chem.*, **245,**
    5281 (1970)
    Biosynthesis of low density lipoprotein by cell-free preparations of rat intestinal
    mucosa.
35. V. Shore and B. Shore, *Biochemistry*, **7,** 3396 (1968)
    Some physical and chemical studies on two polypeptide components of high-
    density lipoproteins of human serum.
36. J. B. Marsh, (eds., E. Tria and A. M. Scanu),
    Structural and functional aspects of lipoproteins in living systems, (Academic
    Press, London and New York, 1969), pp. 447–462.
37. P. S. Roheim, L. Miller and H. A. Eder, *J. biol. Chem.*, **240,** 2994 (1965)
    Formation of plasma lipoproteins from apoprotein in plasma.
37a. J. B. Marsh, (ed., R. M. S. Smellie),
    Biosynthesis of plasma lipoproteins in Plasma Lipoproteins, Biochemical
    Society Symposia: no. 33. (Academic Press, London and New York, 1971)
    p. 89.

37b. D. W BILHEIMER, S. EISENBERG and R. I. LEVY, *Biochim. Biophys. Acta*, **L19**, 212 (1972).
The metabolism of very low density lipoprotein proteins I. Preliminary in vitro and in vivo observations.

38. D. GITLIN, D. G. CORNWELL, D. NAKASATO, J. L. ONCLEY, W. L. HUGHES JR. and C. A. JANEWAY, *J. clin. Invest.*, **36**, 172 (1957)
Studies on the metabolism of plasma proteins in the nephrotic syndrome. II The lipoproteins.

39. D. S. ROBINSON and A. SEAKINS, (ed., A. C. Frazer),
Biochemical problems of lipids. (Elsevier, Amsterdam, 1963), pp. 359–365.

40. C. GALLI, R. SIRTORI and R. PAOLETTI, (eds., E. Tria and A. M. Scanu),
Structural and functional aspects of lipoproteins in living systems, (Academic Press, London and New York, 1969), pp. 518–540.

41. D. E. HAFT, P. S. ROHEIM, A. WHITE and H. A. EDER, *J. clin. Invest.*, **41**, 842 (1962)
Plasma lipoprotein metabolism in perfused rat livers. 1. Protein synthesis and entry into plasma.

42. R. K. OCKNER and A. L. JONES, *J. lipid. Res.*, **11**, 284 (1970)
An electronmicroscopic and functional study of very low density lipoproteins in intestinal lymph.

43. M. I. GURR, W. F. R. POVER, J. N. HAWTHORNE and A. C. FRAZER, *Nature*, **197**, 79 (1963)
Phospholipid composition and turnover in rat intestinal mucosa during fat absorption.

44. S. M. SABESIN and K. J. ISSELBACHER, *Science (N.Y.)*, **147**, 1149 (1965)
Protein synthesis inhibition: mechanism for the production of impaired fat absorption.

45. E. GORDIS, *Proc. Soc. exp. Biol. (N.Y.)*, **110**, 657 (1962)
Demonstration of two kinds of fat particles in alimentary lipemia with PVP gradient columns.

46. P. J. NESTEL, *J. clin. Invest.*, **43**, 943 (1964)
Relationship between plasma triglycerides and removal of chylomicrons.

47. D. S. FREDRICKSON, K. ONO and L. L. DAVIS, *J. lipid. Res.*, **4**, 24 (1963)
Lipolytic activity of post-heparin plasma in hyperglyceridemia.

48. J. A. GLOMSET, *J. lipid. Res.*, **9**, 155 (1968)
The plasma lecithin: cholesterol acyltransferase reaction.

49. K. W. WALTON, P. J. SCOTT, P. W. DYKES and J. W. L. DAVIES, *Clin. Sci.*, **29**, 217 (1965)
Alterations of metabolism and turnover of I[131] low density lipoprotein in myxoedema and thyrotoxicosis.

50. R. H. FURMAN, P. ALAUPOVIC and R. P. HOWARD, *Prog. Biochem. Pharmacol.*, **2**, 215 (1967)
Effects of androgens and estrogens on serum lipids and the composition and concentration of serum lipoproteins in normolipemic and hyperlipidemic states.

51. J. SLACK, *Lancet*, **2**, 1380 (1969)
Risks of ischaemic heart disease in familial hyperlipo-proteinaemic states.

52. T. LANGER, W. STROBER and R. I. LEVY, *J. Clin. Invest.*, **48**, 49a (1969)
Familial type II hyperlipoproteinemia: a defect of beta lipoprotein apoprotein catabolism?

52a. J. SLACK and G. L. MILLS, *Clin. Chim. Acta*, **29**, 15 (1970)
Anomalous low density lipoproteins in familial hyperbetalipoproteinaemia.

52b. T. F. BAGNALL, *Clin. Chim. Acta*, **42**, 229 (1972)
Composition of low density lipoprotein in children with familial hyperbetalipoproteinaemia and the effect of treatment.

52c. D. S. FREDRICKSON, *Brit. Med. J.*, **2**, 187 (1971)
Mutants, hyperlipoproteinaemia, and coronary artery disease.

53. C. J. GLUECK, F. HECKMAN, M. SCHONFELD, P. STEINER and W. PEARCE, *Metabolism*, **20**, 597 (1971)
Neonatal familial type II hyperlipoproteinemia: cord blood cholesterol in 1,800 births.

54. J. M. DARMADY, A. S. FOSBROOKE and J. K. LLOYD, *Brit. Med. J.*, (1972) (in press)
Diagnosis of familial hypercholesterolaemia in the first year of life; a prospective study of serum cholesterol concentrations.

55. M. M. SEGALL, A. S. FOSBROOKE, J. K. LLOYD and O. H. WOLF, *Lancet*, **1**, 641 (1970)
Treatment of familial hypercholesterolaemia in children.

56. T. A. MIETTINEN, (ed., R. J. Jones), Atherosclerosis: Proceedings of the second international symposium. (Springer-Verlag, Berlin, Heidelberg and New York, 1970), pp. 508–515

57. W. R. HAZARD, F. T. LINDGREN and E. L. BIERMAN, *Biochim. Biophys. Acta*, **202**, 517 (1970)
Very low density lipoprotein subfractions in a subject with broad beta disease (type III hyperlipoproteinemia) and a subject with endogenous lipemia (type IV): chemical composition and electrophoretic mobility.

58. R. ZELIS, D. T. MASON, E. BRAUNWALD and R. I. LEVY, *J. clin. Invest.*, **49**, 1007 (1970)
Effects of hyperlipoproteinemias and their treatment on the peripheral circulation.

59. D. S. FREDRICKSON and R. S. LEES, (eds., J. B. Stanbury, J. B. Wyngaarden and D. S. Fredrickson), The metabolic basis of inherited disease, (McGraw-Hill Co., New York, Toronto, Sydney and London. 2nd ed. 1966), pp. 429–485

60. J. D. BAGDADE, D. PORTE, JR. and E. L. BIERMAN, *Diabetes*, **17**, 127 (1968)
Acute insulin withdrawal and the regulation of plasma triglyceride in diabetic subjects.

61. J. FERNANDES and N. A. PIKAAR, *Amer. J. clin. Nutr.*, **22**, 617 (1969)
Hyperlipemia in children with liver glycogen disease.

62. D. D. O'HARA, D. PORTE, JR. and R. H. WILLIAMS, *Metabolism*, **15**, 123 (1966)
The effect of diet and thyroxin on plasma lipids in myxedema.

63. I. F. C. MCKENZIE and P. J. NESTEL, *J. clin. Invest.*, **47**, 1685 (1968)
Studies on turnover of triglyceride and esterified cholesterol in subjects with the nephrotic syndrome.

64. J. S. CHOPRA, N. P. MALLICK and M. C. STONE, *Lancet*, **1**, 317 (1971)
Hyperlipoproteinaemias in nephrotic syndrome.

65. M. YAMADA and I. MATSUDA, *Clin. chim. acta.* **30**, 787 (1970)
Lipoprotein lipase in clincial and experimental nephrosis.

66. D. SEIDEL, P. ALAUPOVIC and R. H. FURMAN, (ed., R. J. Jones), Atherosclerosis: Proceedings of the second international symposium, (Springer-Verlag, Berlin, Heidelberg and New York, 1970), pp. 161–166

66a. D. SEIDEL, B. AGOSTINI and P. MULLER, *Biochim. Biophys. Acta*, **260**, 146 (1972)
Structure of an abnormal plasma lipoprotein (LP-X) characterising obstructive jaundice.

67. J. L. BEAUMOUNT, (ed., R. J. Jones), Atherosclerosis: Proceedings of the second international symposium, (Springer-Verlag, Berlin, Heidelberg and New York, 1970), pp. 166–178

68. W. R. HAZARD, M. J. SPIGER, J. D. BAGDADE and E. L. BIERMAN, *New. Engl. J. Med.*, **280**, 471 (1969)
Studies on the mechanism of increased plasma triglyceride levels induced by oral contraceptives.

69. D. S. FREDRICKSON, A. M. GOTTO and R. I. LEVY, (eds., J. B. Stanbury, J. B. Wyngaarden and D. S. Fredrickson), Familial lipoprotein deficiency in The Metabolic Basis of Inherited Disease (McGraw-Hill Co., New York, Toronto, Sydney and London, 3rd ed. 1972), pp. 493–530

70. J. K. LLOYD and O. H. WOLFF, (ed., F. Linneweh), Handbuch der Inneren Medizin, (Springer-Verlag, Berlin, Heidelberg and New York, in press)
71. H. MARS, L. A. LEWIS, A. L. ROBERTSON, A. BUTKUSS and G. H. WILLIAMS, *Amer. J. Med.*, **46,** 886 (1969)
    Familial hypo-betalipoproteinemia: a genetic disorder of lipid metabolism with nervous system involvement.
71a. A. S. FOSBROOKE, S. CHOKSEY and B. A. WHARTON, *Arch. Dis. Childh.* **48,** 729 (1973)
    Familial hypo-β-lipoproteinaemia.
73. J. A. GLOMSET, K. R. NORUM and W. KING, *J. Clin. Invest.*, **49,** 1827 (1970)
    Plasma lipoproteins in familial lecithin: cholesterol acyltransferase deficiency: lipid composition and reactivity in vitro.
73a. K. R. NORUM, J. A. GLOMSET and E. GJONE, (eds., J. B. Stanbury, J. B. Wyngaarden and D. S. Fredrickson), Familial lecithin—cholesterol acyl transferase deficiency in The Metabolic Basis of Inherited Disease, (McGraw-Hill, New York, 3rd ed. 1972) p. 531

Chapter 2

Hemopexin, The Heme-Binding Serum β-Glycoprotein*

Ursula Muller-Eberhard

and

Ham Heng Liem

*Scripps Clinic and Research Foundation,
La Jolla, California 92037, U.S.A.*

2.1. INTRODUCTION

Hemopexin, a heme-binding serum protein, was first described by Neale *et al* in 1958.[1] The high carbohydrate content and immuno-electrophoretic identity of this macromolecule led to its designation by Schultze *et al.*[2] as β_{1B}-glycoprotein. The name "hemopexin", given by Grabar *et al.*,[3] reflects the ability of the protein to bind heme† in an equimolar ratio.[4,5] The other serum protein which binds metallo-porphyrins, although with lower affinity,[6] is albumin. In several of the following sections, the two proteins will be compared.

The presence of heme-albumin (methemalbumin) in the sera of patients with severe hemolytic disease was recognized in the late thirties.[7] Heme-hemopexin complexes in these patients, however, were not discovered until twenty years later.[1] It subsequently became clear that, in general, the serum hemopexin concentration is lowered after the depletion of haptoglobin which specifically binds hemoglobin.[8,9] The determining factor for serum hemopexin depletion appears to be the accumulation of circulating heme, for an inverse correlation is found for hemopexin and heme levels.[10,11] Moreover, intravascular injection of heme leads to a decrease in hemopexin level[12] and an accelerated catabolism of the protein.[13] Removal of the heme-hemopexin

* From the Department of Biochemistry, Scripps Clinic and Research Foundation, and the Division of Pediatric Hematology, Department of Pediatrics, University of California at San Diego, La Jolla, Ca. 92037.

Supported by a Career Development Award (5-K3-AM-16,923) from the National Institute of Arthritis and Metabolic Diseases, and by research grants from the National Heart Institute (HE-08660) and the Institute of Child Health and Human Development (HD-04445), as well as from the San Diego County Heart Association (No. 83).

† heme = ferriprotoporphyrin IX.

complex, like that of hemoglobin-haptoglobin, is accomplished by the parenchymal cells of the liver.[14,15] Monitoring serum hemopexin levels has now become an invaluable, simple screening test for assessing the severity of hemolysis, especially in cardiac disease.[16] Whether hemopexin also plays a role in the disposal of porphyrins not containing iron,[13,17,18] which it binds *in vitro*,[6,19–21] is presently under investigation.

Recent studies have centered on the relationship of hemopexin to drug-metabolizing hepatic enzymes, some of which share physicochemical characteristics with hemopexin. A major point of comparison is the inducibility of certain hepatic microsomal enzymes and hemopexin by the same compounds. In this review we will present current knowledge on the molecular and biological properties of hemopexin.

2.2. PHYSICOCHEMICAL CHARACTERISTICS OF THE HEMOPEXIN MOLECULE

2.2.1. PHYSICOCHEMICAL PARAMETERS

Human hemopexin was first purified by the research team of Schultze et al.,[2] who reported its amino acid and carbohydrate composition and estimated its molecular weight as 80 000 daltons. A comparison of the human hemopexin to that of the rabbit, purified either by the technique described by Schultze et al.[2] or by a modification thereof,[22] shows them to be very similar.[5] Both were found to have a molecular weight of approximately 57 000 daltons,[23] to consist of a single polypeptide chain, and to possess no free sulfhydryl groups.[5] The assayable half-cysteines can, therefore, be assumed to form intramolecular disulfide bridges rather than to take part in porphyrin binding. The complex of heme-hemopexin with CN^- forms readily, whereas that of reduced heme-hemopexin with CO and O_2 does not.[33] Further evidence that the complex has no enzymatic activity is derived from the following studies. Rabbit heme-hemopexin complexes were thoroughly investigated for peroxidase and catalase activity by Dr. N. M. Alexander, University of California at San Diego, Ca., and Mr. F. W. Selby, Scripps Clinic and Research Foundation, La Jolla, Ca. No enzymatic activity was observed under varying conditions and protein concentrations.

The molecular weight and other physicochemical characteristics are listed in Table 2.1. The partial specific volume of rabbit hemopexin is 0.702;[23] this relatively low value for a protein is probably caused by the high carbohydrate content. As in most other serum proteins, the saccharides are mannose, galactose, glucosamine, and sialic acid.[5] The last comprises four percent of the total carbohydrate content. Carbohydrate is attached to the protein by N-glycosidic linkage.*

---

* This was verified by the kind cooperation of the late Dr. R. J. Winzler.

TABLE 2.1

Summary of physicochemical parameters of the hemopexins of human and rabbit

Molecular weight	57 000 $\pm$ 2000 daltons
Partial specific volume	0.702 $\pm$ 0.003 ml/gm
Carbohydrate content	20–22 %
(in N-glycosidic linkage)	
N-terminal	Valine (R); serine (H)
Extinction coefficient ($E_{1cm}^{1\%}$)	
apoprotein	280 nm = 19.7
equimolar heme:protein	$\begin{cases} 280 \text{ nm} = 21.8 \\ 414 \text{ nm} = 19.2 \end{cases}$
Absorption, Mössbauer and electron spin resonance spectra	{low-spin Fe
Binding of porphyrins (estimated K_d values)	heme $\sim$ mesoheme $\sim$ deuteroheme $\leq 10^{-8}$ M; Co-D $\leq 10^{-7}$ M; D $\simeq 10^{-6}$ M

Whether there are one or more sites of attachment of the carbohydrate moiety has not been determined. The amino-terminal amino acid residue of the rabbit molecule is valine, and that of the human is probably serine. The NH_2-terminal amino acid sequences* (available for up to 31 amino acids for the rabbit and up to 24 for the human) show extensive similarities. However, their fingerprint patterns differ, and the antigenic cross-reactivity is only about 30 per cent using sheep or goat antibody.

The extinction coefficient ($E_{1cm}^{1\%}$) for the apo-protein from rabbit is 19.7 at 280 nm. At this wavelength, the heme-hemopexin complex has an extinction coefficient of 21.8; at 414 nm it is 19.2.[23] The extinction coefficients of the human counterpart are nearly identical.[5]

The absorption spectrum of the heme-hemopexin complex exhibits ill-defined bands in the visible region which upon reduction resolve into distinct α and β bands at 556 and 526 nm, respectively.[5] These absorption properties of heme-hemopexin are similar to those of cytochrome b_5, observed by Strittmatter and Velick,[24] but not to hemoglobin or myoglobin, Smith and Williams,[25] and indicate the iron to be in low-spin form. This has been confirmed by detailed studies with electron spin resonance and Mössbauer spectroscopy. The iron of heme-hemopexin is not readily converted to the high-spin form with fluoride indicating stability of the low-spin complex.[26]

2.2.2. INTERACTION WITH PORPHYRINS

Like many hemoproteins, hemopexin also binds porphyrins other than heme. The binding of deuteroporphyrin and some of its derivatives

* In collaboration with Drs. J. D. Capra and J. M. Kehoe.

by hemopexin has been studied extensively by spectrophotometry, equilibrium dialysis and fluorimetry.[20,6] The apparent K_d values for the interaction of rabbit-hemopexin with protoheme, deuteroheme, and mesoheme are approximately 10^{-8} M, that of cobalt-deuteroporphyrin 10^{-7} M, and that of deuteroporphyrin is yet a magnitude lower at 10^{-6} M. Because of the inherent tendency of most porphyrins to aggregate, as already pointed out by Lemberg and Legge in 1949,[27] K_d values are difficult to determine accurately. The complexing of rabbit hemopexin with metal-free porphyrins and with certain metalloporphyrins produces only slight changes in extinction. However, these porphyrin hemopexin interactions can be conveniently observed by other techniques. The protein-bound porphyrin can be separated from the free porphyrin by polyacrylamide gel electrophoresis or gel filtration and can then be studied. Examined by this technique, the metalloporphyrins zinc- and nickel-deuteroporphyrin associated with hemopexin, as did the synthetic 2,4 disulfonic acid substituted deuteroporphyrin,[21] and coproporphyrin isomers I and III. The latter finding was first reported by Finnish investigators[19] who also observed binding of isomer III, but not I of uroporphyrin. Utilizing isotope-labeled compounds and polyacrylamide gel electrophoresis, we found that hemopexin formed complexes with cobinamide but not with other vitamin B_{12} analogues.* Bilirubin and biliverdin, gifts of Drs. R. Schmid and D. C. Mauzerall, did not associate with hemopexin. Work is in progress to establish with what affinity porphyrins containing more than two carboxyl groups and other non-metal containing porphyrins are complexed and whether they are bound in the heme crevice. The evidence thus far obtained suggests that the presence of a metal in the center of the porphyrin ring is of greater consequence for binding than are minor substitutions in the periphery of the ring (positions 2 and 4).[20]

Primarily two classes of protein-heme interaction are encountered; one is the iron-ligand bond and the other involves the amino acids which form the immediate environment of the porphyrin ring structure. Among the amino acid residues, which can theoretically coordinate with heme iron to form a low-spin heme-hemopexin complex, are a histidine and a methionine or two of each.[28] That the ligands in the case of hemopexin are two histidines is suggested by: (a) the similarity of the absorption and electron paramagnetic resonance spectra of heme-hemopexin and cytochrome b_5, a protein known to have two histidine residues coordinated to the heme iron;[29] (b) the similarity of the magnetic circular dichroism spectra of heme-hemopexin, cytochrome b_5, and dihistidyl-heme model compounds;[30] (c) the absence of a charge-transfer absorption band at 695 nm in the spectrum of heme-hemopexin[31] which in cytochrome c is produced by the methionine-iron coordination; and (d) chemical modification studies. In the last case, photo-oxidation

---

* These compounds were synthesized by Dr. D. W. Jacobsen.

of histidine residues with rose bengal or modification of histidine residues with diethylpyrocarbonate results in loss of the heme-binding activity of rabbit apo-hemopexin. Treatment of the heme-rabbit hemopexin complex by these means showed that bound heme protected histidine residues required for liganding.[32]

Conformational changes effected by the protein combining with heme have been observed by means of optical rotatory dispersion,[33] circular dichroism[20] and an immunological method.[34] The spectral observations in the far-ultraviolet region point to slight changes in secondary conformation, but those in the near-ultraviolet region to considerable changes in tertiary conformation. In addition, the effect of pH on the interaction of deuteroheme and rabbit hemopexin, monitored by changes in absorbance and circular dichroism, indicates particular conformation and ionization states to be essential for the binding process.[20]

Several observations suggest that certain tryptophan residues are essential for the interaction of heme with hemopexin. First, the band of positive ellipticity centered at 231 nm of hemopexin increases by 50 percent when an equimolar amount of deuteroheme is added to rabbit hemopexin.[20] This band is characteristic of poly-L-tryptophan[35] and is abolished by treatment with the tryptophan modification agents N-bromosuccinimide or dimethyl(2-hydroxy-5-nitrobenzyl)sulfonium bromide. The presence of bound heme does not greatly reduce the action of these agents.[36] These results imply that trytophan-tryptophan interactions are affected when heme is bound by hemopexin and in part account for changes in the tertiary structure noted above. Other observations indicate that tryptophan residues form part of the heme-binding site of hemopexin. When one mole of heme is bound by hemopexin, the fluorescence of tryptophan residues is quenched,[6] and its absorption spectrum is perturbed.[31]

2.2.3. STABILITY OF THE PROTEIN

The molecule is resistant to precipitation by low concentrations of sulfosalicylic or perchloric acid, allowing the latter to be used in its purification.[2] However, aggregation of hemopexin monomers occurs during the purification procedure,[22] and lowers the heme-binding affinity. The apoprotein polymerizes more readily than the hemeprotein, but is stable in highly purified form at 4° for up to three days in dilute solutions and indefinitely when lyophilized or kept in liquid nitrogen. Aggregation is also observed when the protein is heated, with or without heme. Polyacrylamide gel electrophoretic analysis shows polymer formation after the hemeprotein is exposed to 60° for 5 minutes or 85° for 1 minute, and this is accompanied by release of some heme. In serum, hemopexin remains stable for at least two weeks at 4°,[37] and longer if a preservative such as merthiolate is added. Even at room temperature, hemopexin in serum shows no tendency to polymerize for several days.

2.2.4. POLYMORPHISM AND IMMUNOLOGICAL CROSS-REACTIVITY

Genetic polymorphism of hemopexin has been reported for the rabbit[38] and the pig.[39] The electrophoretic mobility patterns of the human molecule have not shown genetic polymorphism in man.[40] However, a detailed analysis of the antigenic properties of the hemopexin may reveal genetic variability. This remains to be explored. The antigenic properties are shared to varying degrees by all eutherian mammals, but not by non-eutherian mammals and lower vertebrates.[41] Nevertheless, the non-eutherian mammals, as well as phylogenetically older vertebrates possess heme-binding protein(s) with an electrophoretic mobility distinctly different from that of albumin. Besides the hemopexin of human and rabbit, only the mouse[42,43] and the rat[44] analogues have been studied.

2.2.5. HEMOPEXIN ASSAY

Due to the stability of the protein in serum or plasma, as discussed in section 2.2.3, samples can be transported easily. The hemopexin levels can be measured by radial immunodiffusion according to Mancini.[45] Alternatively, the concentration can be determined by "rocket" or two-dimensional immunoelectrophoretic methods.[46,47] Immuno-plates for determination of the levels of human hemopexin are commercially available.* The loss of antigenicity which ensues upon the interaction with heme occurs only in the soluble antigen-antibody complexes[34] and does not interfere with the accuracy of the immunodiffusion techniques which are dependent upon precipitable complexes.

2.3. SYNTHESIS AND CATABOLISM

The liver was found to be the primary site of hemopexin formation. Synthesis was studied *in vitro* using various tissues from man,[48] the Rhesus monkey and the rabbit.[49] In the latter study, isotope-labeled amino acids were added to culture media containing cells from several organs of the adult Rhesus monkey and rabbit. Incorporation of the isotope-labeled amino acid into the hemopexin molecule occurred exclusively in liver cell cultures. In similar experiments, tissue fragments of the human fetal liver but not of the placenta, thymus or colon, were capable of forming hemopexin.[50] The results demonstrated a very active metabolism of this molecule in the fetus.

The relative ratios of albumin:hemopexin production in these culture systems for human fetal liver as well as for adult Rhesus monkey and rabbit are *ca.* 10:1. Values for the ratios of serum albumin:hemopexin,

* Behring Diagnostics, Division of American Hoechst Corp., Somerville, New Jersey, U.S.A. Behringwerke, Marburg, Germany. Orion Diagnostic, Division of Orion Pharmaceutical Co., Helsinki, Finland, and LKB-Produker Ab, S-16125 Bromma 1, Sweden.

on the other hand, are in all species approximately 100:1, a magnitude higher than that found in the culture system. This difference could be explained by delayed release of the glycoprotein hemopexin which raises the possibility of an intracellular function as a porphyrin carrier. Synthesis can also be shown in single cell cultures of fetal rat liver by a method recently described by Leffert and Paul.[51] The effect of inducers on rat hemopexin synthesis is presently being studied.*

Heme catabolism occurs in the hepatocytes rather than in the cells of the reticuloendothelial sytem of the liver. The parenchymal cells take up iodinated hemopexin and tritiated heme injected intravenously into rabbits.[14] Quantitative differences exist between the hepatic uptake of ^{125}I-hemopexin and that of ^{3}H-heme. One hour after administration, 60% of labeled heme, but only 10% of the labeled hemopexin, is taken up, while most of the remaining radioactive hemopexin continues to circulate.[52] This finding may indicate that hemopexin is reutilized for transporting heme or may reflect equilibration with endogenous protein.

The actual site of hemopexin catabolism is also as yet unknown. Turnover studies on the iodinated hemopexin molecule have been initiated in humans and in rabbits. The human serum protein declined with a half clearance time (T/2) of 7.10 $\pm$ 1.05 (1SD) days in ten control subjects. After intravenous injection of heme, this was reduced to 0.8 days in a healthy individual; the T/2 was shortened to 4.1 and 5.4 days in two patients with sickle cell disease. A shorter T/2 was also found in patients who suffered from aberrations of porphyrin metabolism. The T/2 of five patients with erythropoietic protoporphyria was 6.2 $\pm$ 0.8 days, and in the one patient with porphyria cutanea tarda who was available for study, it was 4.3 days.

The rate of synthesis in the control subjects was 6.90 $\pm$ 1.23 mg protein/kg/day. The synthetic rates in patients with the hemolytic or porphyric diseases investigated, fell within the range of normal, indicating that the reduced serum concentrations were not a result of defective synthesis but instead could be accounted for by accelerated breakdown.[13]

In rabbits, the T/2 of hemopexin, approximately one day, is shorter than we expected[53] and varies greatly from one animal to another. It is not due to differences between batches of protein or to the iodine-labeling,[54] but instead may be related to the genetic polymorphism of hemopexin in this species.[38] Animals receiving large amounts of heme (in solution or in the form of methemoglobin), well above the saturation point of hemopexin, also show a greatly reduced T/2.[54,55] Desialidation of hemopexin causes rapid clearance of the molecule from the circulation, but does not affect hemebinding.[31] Interestingly, the T/2 of rose bengal-inactivated protein (see 2.2.2) is not shortened.

* In collaboration with Drs. H. L. Leffert of the Salk Institute, La Jolla, Ca., and F. F. Becker of New York University, New York, N.Y.

Currently, we are studying hemopexin synthesis in the perfused isolated rat liver, employing an immunological technique. An astonishingly high rate of hemopexin synthesis of about 5 mg/hr/100 g liver, one fourth that of albumin, was found. Pretreatment of the Sprague-Dawley rats with sodium pentobarbital for 2 to 4 days did not increase hemopexin or albumin synthesis, although it caused greater removal of heme added to the perfusate which was not associated with an increased bilirubin excretion.

2.4. INDUCTION OF HEMOPEXIN IN RELATION TO HEPATIC PROTEINS

2.4.1. INDUCING AGENTS

A variety of substances are potent inducers of hemopexin synthesis. The first evidence of increased production came from experiments in rabbits, which showed that small amounts of heme given intravenously raised hemopexin levels up to two-fold.[56] Subsequently, the carcinogens, 3-methylcholanthrene (3-MC) and 3,4-benzpyrene, and the porphyrinogens, allylisopropylacetamide (AIA) and lead acetate, were recognized as inducers.[57,58] One of the classical agents, pentobarbital, also has an inducing effect in rabbits.*[37] Peak hemopexin levels in these experiments were achieved *in vivo* between 48 and 72 hours after administration of the inducer and rose to as high as 400 percent above baseline concentrations. Even higher levels were reached in mice after multiple injections with a streptococcal vaccine.[43] Griseofulvin, a porphyrinogen, is also an inducer of mouse hemopexin synthesis. The levels correlated directly with the concentrations of coproporphyrin and protoporphyrin in erythrocytes and liver tissue when the drug was offered in the food for several weeks.†[59]

2.4.2. INHIBITORS

Cycloheximide, an inhibitor of protein synthesis, interrupted the induction process,[37] whereas D,L-ethionine, an inhibitor of adenosine

* Unpublished work of Dr. J. M. Chandler (presently at the University of Oklahoma Medical Center, Oklahoma City, Oklahoma) corroborates these data. Experiments with Sprague-Dawley rats showed no evidence for induction of acute phase reactants associated with the proliferation of the smooth endoplasmic reticulum (SER) induced by phenobarbital. In short, phenobarbital was administered repeatedly to six animals, i.p., and their sleeping times were recorded. The periods of sleep were shortened in these rats in comparison to those of six rats the received saline injections. After an i.p. dose of ^{14}C-leucine, incorporation of label into circulating albumin, seromucoid and fractions containing hemopexin and haptoglobin was followed. No difference was observed in the radioactivity of these protein fractions between the treated animals and their controls, yet the former showed pronounced SER proliferation.

† Certain steroids that induce drug metabolism also have an effect on serum hemopexin levels in Fisher rats. Cortisone administration reduces and adrenalectomy may increase hemopexin concentrations (in collaboration with Dr. J. T. Wilson).

triphosphatase, abolished it in rabbits.[58] Chloramphenicol and amino-triazole, inhibitors of protein synthesis and of δ-aminolevulinic acid dehydrase, respectively, had no inhibitory action *in vivo*. SKF-525-A,* 2-diethylaminoethyl-3,3-diphenyl-propylacetate, does not inhibit the stimulatory effect of 3-MC on hemopexin synthesis. Administered by itself, we found that SKF-525-A induces rather than inhibits hemopexin synthesis, as does ethanol.

2.4.3. BINDING

An interaction of the inducing drugs and steroids with hemopexin is unlikely to be part of the induction process, as we have found no evidence for this in a study of a variety of compounds. Radioactively-labeled steroids (pregnenolone, cholesterol, progesterone and andro-stenedione,† DDC,‡ sodium phenobarbital and benzpyrene) were admixed with rabbit apo-hemopexin, and electrophoresed on polyacryl-amide gel. The hemopexin band was excised and found to contain no radioactivity. Non-radioactive AIA, DDC, lead acetate, benzpyrene and 3-MC were incubated with rabbit hemopexin; an electrophoretic mobility change was sought, but not found. In addition, no change in light absorption spectrum of rabbit heme-hemopexin was detected when it was incubated with either sodium phenobarbital or benzpyrene.

2.4.4. GENETICS

We have made a very interesting observation regarding the induction of hemopexin in mice. A pool of serum from groups of three to six animals from six different strains was compared with that from mice of the same strains 48 hours after injection of 3-MC. The following strains were used: AKR/J, C_{57}BL/6J, SWR/J, C_3H.Q/Sf, B_{10}.A/SqSn and A/J, and the results are summarized in Table 2.2. The baseline hemopexin levels differed remarkably from one strain to another, and the increase of the hemopexin concentrations after injection of 3-MC ranged from 60 to 500 per cent. The animals with low pre-experimental values showed the greatest increase in hemopexin levels, whereas those with initially high values responded least to the inducer. Thus, the maximum serum values after treatment with this inducing agent varied little among the strains.

Other sets of data bear on this subject. Hemopexin in an individual rabbit rises to a maximum induction level of 2–4 fold, regardless of whether 3-MC is injected in a dose of 4 mg/kg or 10 mg/kg. If either amount is adminstered again after a two week interval, the induction effect is similar, although some rabbits are no longer inducible.[37] The serum hemopexin concentrations of monozygotic human twins differ

---

* Kindly donated by Smith, Kline, and French Laboratories, Inc., Philadelphia, Pa.
† Gifts of Dr. K. J. Ryan, University of California at San Diego, La Jolla, Ca.
‡ DDC = diethyl 1,4-dihydro-2,4,6-trimethyl pyridine-3,5-dicarboxylate. Gener-ously supplied by Dr. G. S. Marks, Dept. Pharmacology, Queen's University, Kingston, Ontario, Canada.

TABLE 2.2

Hemopexin levels (u/ml) in several genetic strains of mice before and after a single injection i.p. of 3-methylcholanthrene (100 mg/kg)

Mice (pool 3–6 animals)	Before 3-MC	After 3-MC	% Increase
AKR/J	72	239	+233%
C$_{57}$BL/6J	110	176.5	+61%
SWR/J	34	229	+573%
C$_3$H.Q/Sf	47	237	+404%
B$_{10}$A/SqSn	78	246	+215%
A/J	75	218.5	+191%

little whereas those of dizygotic twins equal that of a random population (a collaborative effort with Dr. L. Wetterberg). These observations in man, rabbits and mice may imply genetic factors govern not only the baseline values, but also the response to environmental factors.

2.4.5. LINKS WITH CYTOCHROME P-450

A striking similarity exists in the response of certain inducers and inhibitors of hemopexin and certain hepatic enzymes, in particular cytochrome P-450. Earlier studies on heme-exchange between cytochrome P-450$_{cam}$ of *Pseudomonas putida* and hemopexin yielded information on the nature of the prosthetic group of this cytochrome.[60] Upon incubation with apo-hemopexin, the enzyme lost heme as well as hydroxylase activity and the latter was readily restored when heme was added. However, heme was not transferred to rabbit hemopexin from either "CO-binding particles" or solubilized cytochrome P-450 of the rat. After steapsin treatment, which converts cytochrome P-450 to P-420, heme was transferred to apo-hemopexin.[61] If cytochrome P-420 is a naturally occurring intermediate in the degradation of cytochrome P-450,[62] one may consider hemopexin as a carrier transporting heme derived from cytochrome P-450 to heme oxygenase.

The results of experiments conducted thus far on the induction of hemopexin synthesis suggest a connection between the metabolism of both hemopexin and mammalian cytochrome P-450. Recent genetic studies*[1] showed hemopexin induction by 3-MC in B$_6$ mice; aryl hydrocarbon hydroxylase, a cytochrome P-450, is also readily induced in this particular strain.[63] Neither protein was induced in the unresponsive strain, NZW. Cytochromes of the P-450 type are induced by phenobarbital and carcinogens,[64] but not by AIA.[65] The lack of induction of cytochrome P-450 by AIA may be explained by a rapid loss of the integrity of the heme moiety.[66] The induction of drug metabolizing enzymes, like that of hemopexin, is inhibited by cycloheximide[67] and

* In collaboration with Drs. D. W. Nebert and I. S. Owens.

D,L-ethionine.[68] However, the induction of cytochrome P-450, unlike that of hemopexin, is inhibited by SKF-525-A[69] and aminotriazole.[70,71] It, therefore, seems that induction of cytochrome P-450 does not necessarily accompany induction of hemopexin. Many more comparative studies are needed to establish the mechanism and sequence of events leading to the induction of hemopexin and cytochromes of the P-450 type.

2.4.6. LINK WITH HEME BIOSYNTHESIS

Whether there is a connection between the metabolism of δ-aminolevulinic acid synthetase, the first and rate-limiting enzyme of heme biosynthesis,[72] and hemopexin is a current subject of our research. Purified rabbit apo-hemopexin counteracts the inhibitory effect of heme on δ-aminolevulinic acid synthetase of *Rhodopseudomonas spheroides*,[73] and on that of rabbit bone marrow stimulated by erythropoietin.[74] It remains to be established whether hemopexin also exerts such an effect *in vivo*, thus being involved in the regulation of both heme and drug metabolism.

2.5. HEMOPEXIN IN HEALTH AND DISEASE

2.5.1. NORMAL SUBJECTS

Hemopexin concentrations in the sera of healthy human adults range from 500 to 1000 μg/ml.[8] In the fetus, hemopexin has been detected as early as the 29th day of gestation.[48] The concentrations range from 10 to 50 μg/ml,[75] and in general, rise as the fetus matures.[48] In the adult, low levels of hemopexin are found in amniotic,[76] synovial and cerebrospinal fluids,[44] but in the latter hemopexin could be detected only after the fluid had been concentrated. Premature babies have somewhat lower levels than those born at term.[77] Nevertheless, within six months* adult levels are attained by healthy offspring.[79] No difference in hemopexin levels are found between the sexes during the reproductive years. Individuals over 60 years of age generally have somewhat lower levels than younger adults.

2.5.2. HEMOLYTIC DISEASE

The part played by hemopexin as a disposer of heme in patients with hemolytic disease has been discussed extensively[8,80,81] Although a mild degree of intravascular red cell destruction with ensuing hemoglobinemia depletes serum haptoglobin, hemopexin is decreased only when

---

* Similarly, the hemopexin levels of rabbit neonates are lower than those of their mothers, reaching adult levels at 30–60 days of life. Another phenomenon, the postparturition drop in hemopexin levels in rabbit does[78] is striking and warrants investigation in other mammals. Hemopexin concentrations in rats are also low in the immediate post-natal period and increase during the first weeks post-partum (unpublished results obtained in collaboration with Dr. J. T. Wilson).

heme levels in the plasma are significantly elevated.[10,11] Hemopexin concentrations, therefore, reflect the degree of hemolysis, and are especially useful indicators for the cardiac surgeon. Surgical intervention is required for patients with severe cardiac hemolysis due to malfunction of grafted valves, as this process can lead to acute renal failure;[82] and even if the patient's condition is not acute, severe iron deficiency is inevitable. Other methods of assessing the severity of hemolysis have disadvantages. Measurement of the carbon monoxide content of plasma or exhaled air is costly, and test results depend on a variety of factors apart from hemolysis.[83] Counting the number of schistocytes (deformed red cells), measuring serum lactic dehydrogenase or urinary hemosiderin excretion, or following the half-life of isotope-labeled red cells are tedious methods and require the assistance of skilled technical personnel. The determination of plasma hemoglobin requires scrupulous care with blood sampling, which becomes unnecessary with hemopexin. Plasma heme levels,[10,11] schistocyte count, and hemosiderin excretion[16] correlate inversely with hemopexin levels that can be assessed simply and inexpensively by radial immunodiffusion.

Erythroblastosis fetalis is another hemolytic disease in which hemopexin determinations have proven to be of diagnostic value. With advancing gestation, the concentration of hemopexin in the amniotic fluid rises more rapidly than that of albumin, *i.e.* the hemopexin: albumin ratio increases. In the presence of significant hemolysis, this ratio levels off or falls. Consequently, in combination with the ratio of bilirubin:albumin, it provides a sensitive index of the severity of fetal hemolysis.[76]

2.5.3. OTHER DISEASES

To confirm that the hemopexin concentration is a dependable indicator of hemolysis, other factors that may influence hemopexin levels have been evaluated. Hemopexin does not appear to be an "acute phase reactant". Levels determined on samples drawn every 8 hours remained stable in young adult males who suffered fractures of major long bones and were under observation for fat embolism. While concentrations of the acute phase reactants, haptoglobin and C-reactive protein, invariably rose for at least four days after injury, hemopexin values were unchanged.[84] In severe infections, hemopexin levels usually remain stable,[80,84,85] indicating a rigid homeostatic control of this protein.

Elevated hemopexin levels, although rarely more than twice normal, occur in diabetes mellitus,[86] possibly in muscular dystrophies,[87] and definitely in various forms of cancer.[80,88–90] Whether this elevation is an early or late manifestation of cancer has not been ascertained. Very high hemopexin levels were encountered in patients with fast-growing melanomas.[89] In this context, it is interesting to recall the induced synthesis of rabbit and mouse hemopexin by carcinogenic agents discussed in sections 2.4.1 and 2.4.4 of this review.

Low levels of hemopexin may be associated with pathologic conditions other than hemolytic diseases. They occur in hemorrhagic pancreatitis[10,10a] and in kwashiorkor* (in collaboration with Drs. R. E. Olson and R. Suskind, Mai Thai, Thailand). In some cases of Dengue hemorrhagic fever, studied during an epidemic in Bangkok, Thailand, the serum levels of hemopexin were decreased to as little as 10 per cent of the normal values. As activation of the complement system has been implicated as the pathobiologic mechanism of this viral disease,[92] hemopexin levels were compared to those of complement proteins. However, no relationship between the levels of hemopexin and any of the complement components was encountered.†

In kidney disease of any etiology, hemopexin is excreted into the urine, together with proteins of similar molecular weight. During the course of severe liver disease, hemopexin levels decline presumably because of decreased synthesis rather than hemolysis, for haptoglobin levels are normal or elevated.[93]

2.5.4. PORPHYRIAS

Hemopexin was also reported to be undetectable by immunoelectrophoresis in sera of a significant number of patients with porphyria cutanea tarda.[94] We observed moderately low levels in only half of 64 patients with this disease, as well as in half of 37 patients with erythropoietic protoporphyria. On the other hand, in 22 of 26 patients with acute intermittent porphyria, the levels fell into the normal range. It is important to note that the patients with porphyria cutanea tarda also had diminished concentrations of albumin. In acute intermittent porphyria and porphyria cutanea tarda, but not in erythropoietic protoporphyria, the levels of hemopexin and albumin were directly correlated. In the patients with erythropoietic protoporphyria, hemopexin levels were inversely proportional to protoporphyrin levels in erythrocytes and feces. The amount of coproporphyrin in the urine related inversely to serum hemopexin levels in patients with porphyria cutanea tarda. The probability thus emerges that hemopexin is instrumental not only in the disposal of heme, but also of other porphyrins.[95]

2.6. CURRENT CONCEPT OF HEMOPEXIN FUNCTION(S)

The fates of hemoglobin and heme in nonimmune hemolytic states are depicted in the Fig. 2.1. The site of destruction of senescent erythrocytes is as yet unknown. Deformed cells are taken up by sinusoidal cells

---

* Worth mentioning is the fall in hemopexin levels of rabbits three to four days after commencing starvation.[37]

† Dr. N. R. Cooper kindly verified that addition of human hemopexin to serum did not influence complement activity as measured by hemolytic titration, nor did anti-hemopexin antibody react with any of the complement proteins.

HEMOGLOBIN AND HEME DISPOSAL IN NON-IMMUNE STATES

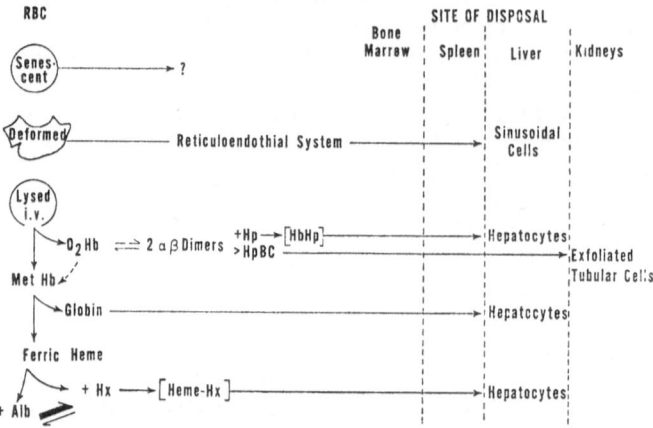

FIG. 2.1. Current concept of hemoglobin and heme disposal in non-immune states. The fate of senescent red blood corpuscles (RBC) is unknown. Deformed erythrocytes are filtered by cells of the reticulo-endothelial system, especially the Kupffer cells, *i.e.* sinusoidal cells of the liver. By contrast, hemoglobin (Hb) released from the red cells intravascularly, dissociates into $\alpha\beta$ dimers which are bound by serum haptoglobin (Hp) and enter the liver parenchymal cells. Dimers in excess of the haptoglobin binding capacity (HpBC) are excreted into the urine and in part reabsorbed by the renal tubular cells which are eventually exfoliated. This gives rise to hemosiderinuria (iron in the urine). Hemoglobin in excess of the absorptive capacity of the tubular cells, passes directly into the urine (hemoglobinuria). Varying quantities of hemoglobin circulate as methemoglobin which readily releases its heme moiety. The heme is taken up simultaneously by serum albumin (Alb) and hemopexin (Hx). Heme-albumin probably serves only as a plasma heme reservoir, whereas the hepatocytes engulf the heme-Hx complex.

of the liver.[96] It should be kept in mind that each hemoglobin molecule contains four molecules of heme; but hemopexin, which has a similar molecular weight, binds only one mole of heme. Although the amount of circulating hemopexin in a healthy adult is approximately 2.5 gm, only 25 mg of heme can be carried at a given time. In the event of liberation of all the heme from the hemoglobin molecules, intravascular lysis of a few milliliters of red blood corpuscles would lead to hemopexin depletion. Here haptoglobin comes into the picture. Hemoglobin released intravascularly from the erythrocytes dissociates into $\alpha\beta$ dimers that are bound by haptoglobin.[97,98] Subsequently, the hemoglobin-haptoglobin complex is removed by the liver. Recent evidence reinforces the previous indications that this complex is taken up by hepatocytes rather than by cells of the reticuloendothelial system.[15,99–101] When the quantity of circulating hemoglobin exceeds the haptoglobin binding capacity, the subunits pass into the glomerular filtrate where they are reabsorbed until the renal excretory cells are saturated. Amounts of hemoglobin in excess of the renal absorptive threshold appear unchanged in the urine (hemoglobinuria).[102] The hemoglobin-containing tubular cells are eventually exfoliated and may be determined in the urine as hemosiderin. The subject was reviewed by Pimstone in 1972.[103]

An uncertain amount, as yet, of hemoglobin from the red cells or in the circulation is oxidized to methemoglobin; heme is then released from the globin. The circulating heme is taken up by both hemopexin and albumin, the latter naturally comprising the greater reservoir. Nevertheless, the hemopexin possesses a much greater affinity for heme and apparently has the appropriate conformation to bring it into the parenchymal cells.[14,15]

The T/2 of heme complexed with hemopexin is shorter than that of the heme-albumin complex.[54] This was shown by Sears,[105] studying heme disposal in men previously depleted of hemopexin.

In vitro studies with ^{59}Fe-heme and purified hemopexin and albumin, which are easily separated on polyacrylamide gel electrophoresis, showed that fifty per cent of heme bound to albumin (1:1) prior to incubation with hemopexin was transferred to hemopexin within ten minutes. Equilibrium was reached after three to four hours, at which time only 4 to 11 per cent of the heme remained attached to albumin. Under the experimental conditions employed, nearly identical results were obtained for the uptake of heme by hemopexin and albumin from humans or rabbits.[104]

Earlier experiments in rabbits yielded similar results in vivo. Heme from intravenously administered heme-albumin complexes was found in the β-region on plasma electrophoresis within a few minutes.[56] By contrast, depletion of serum hemopexin in the rabbit by heme injections is difficult; this is probably owing to a rapid synthesis rate of the protein. Repeated injections of heme lower the hemopexin concentration by 30 to 40 per cent for the first few days after which the level stabilizes usually at a higher value. Subcutaneous injections of the redox agent phenylhydrazine, however, do cause a pronounced fall in hemopexin concentration in rabbits. Administration of a smaller dose for more than three days as well as intravenous injection of very large amounts of heme cause shock and death.[56]

Whereas ahaptoglobinemia has no consequence for the organism and even occurs congenitally, ahemopexinemia, unrelated to severe hemolysis or an extraordinary heme load, has not been reported in man. We have always been able to detect hemopexin, although sometimes only after concentrating the serum. Besides hemolysis following cardiac surgery, extremely low concentrations are encountered almost exclusively in sera from patients with thalassemia major. Their sera are more depleted of hemopexin than sera from patients with sickle cell disease, which is associated with a greater degree of hemolysis.[106,107] This fact suggests that hemopexin may be needed for the disposal of pigments deposited in tissues originating in the frequent blood transfusions that these patients need.

Many questions still remain to be answered in examining the function(s) of hemopexin. A few of these are: (1) Does heme enter the hepatocytes only when complexed to hemopexin, as indicated by autoradiographic studies? (2) Is heme brought to the liver as the hemehemopexin complex a good substrate for microsomal heme oxygenase?[108] (3) Is the heme of heme-albumin directly available for degradation or does it circulate until hemopexin becomes available? (4) Has hemopexin a part in intracellular heme transport, for rapid assembly and function of hemeproteins, and/or for control of their biosynthesis? (5) What factors control hemopexin synthesis?

In conclusion, now that about 15 years have elapsed since the discovery of hemopexin, more questions can be asked than answers given. The scavenger function of hemopexin for heme is well-established, and ground is being gained in our knowledge of its involvement in the disposal of other porphyrins. In the near future, we should learn whether or not hemopexin recycles or functions as a suicidal protein, like haptoglobin. The induction of hemopexin synthesis concomitant with that of hepatic enzymes engaged in the metabolism of drugs remains a most challenging phenomenon to be explored. Information emerging from this work may lead to new approaches for the treatment of drug toxicity.

ACKNOWLEDGEMENT

The authors are most grateful for the valuable and generous advice of Drs. T. P. Conway, E. F. Johnson, W. T. Morgan, I. M. Murray-Lyon, V. L. Seery, as well as Drs. S. Granick and T. E. Hugli. We also appreciate the expert assistance of K. H. Cox and D. Montoya.

REFERENCES

1. F. C. NEALE, G. M. ABER and B. E. NORTHAM, *J. Clin. Path.*, **11**, 206 (1958)
2. H. E. SCHULTZE, K. HEIDE and H. HAUPT, *Naturwissenschaften*, **48**, 696 (1961)
3. P. GRABAR, C. DEVAUX ST-CYR and H. CLEVE, *Bull. Soc. Chim. Biol. (Paris)*, **42**, 853 (1960)
4. K. HEIDE, H. HAUPT, K. STÖRIKO and H. E. SCHULTZE, *Clin. Chim. Acta*, **10**, 460 (1964)
5. Z. HRKAL and U. MULLER-EBERHARD, *Biochemistry*, **10**, 1746 (1971)
6. V. L. SEERY and U. MULLER-EBERHARD, *J. Biol. Chem.*, **248**, 3796 (1973)
7. N. H. FAIRLEY, *Nature (London)*, **142**, 1156 (1938)
8. U. MULLER-EBERHARD, *New Engl. J. Med.*, **283**, 1090 (1970)

9. H. J. Braun, *Klin. Wochenschrift*, **49**, 445 (1971)
10. D. A. Sears, *J. Lab. Clin. Med.*, **71**, 484 (1968)
10a. H. J. Braun, Verhandlungen der Deutschen Gesellschaft für innere Medizin, **78**, 1426 (1972)
11. U. Muller-Eberhard, J. Javid and H. H. Liem, *Blood*, **32**, 811 (1968)
12. D. A. Sears, *Proc. Soc. Exper. Biol. Med.*, **131**, 371 (1969)
13. R. D. Wochner, I. Spilberg, A. Lio, H. H. Liem and U. Muller-Eberhard, New Engl. J. Med. (April, 1974)
14. U. Muller-Eberhard, C. Bosman and H. H. Liem, *J. Lab. Clin. Med.*, **76**, 426 (1970)
15. C. Hershko, J. D. Cook and C. A. Finch, *J. Lab. Clin. Med.*, **80**, 624 (1972)
16. M. E. Eyster, T. S. Edgington, H. H. Liem and U. Muller-Eberhard, *J. Lab. Clin. Med.*, **80**, 112 (1972)
17. U. Muller-Eberhard, *Amer. Soc. Clin. Invest.—65th Ann. Meeting, 1973 Program*, p. 59a, Abstract #214 (1973)
18. U. Muller-Eberhard, H. H. Liem, M. M. Mathews-Roth and J. H. Epstein, Serum levels of hemopexin and albumin in disorders of porphyrin metabolism (in preparation).
19. P. Koskelo, I. Toivonen and P. Rintola, *Clin. Chim. Acta*, **29**, 559 (1970)
20. W. T. Morgan and U. Muller-Eberhard, *J. Biol. Chem.*, **247**, 7181 (1972)
21. T. P. Conway and U. Muller-Eberhard, *Federation Proceedings* **32**, 469 (Abstract #1386) (1973)
22. U. Muller-Eberhard and E. C. English, *J. Lab. Clin. Med.*, **70**, 619 (1967)
23. V. L. Seery, G. Hathaway and U. Muller-Eberhard, *Arch. Biochem. Biophys.*, **150**, 269 (1972)
24. P. Strittmatter and S. F. Velick, *J. Biol. Chem.*, **221**, 253 (1956)
25. D. W. Smith and R. J. P. Williams, (eds., R. Hemmerick *et al.*), Structure and Bonding (Springer-Verlag, New York 1970), pp. 2–45.
26. A. J. Bearden and U. Muller-Eberhard, Heme complexes of rabbit and human hemopexin and human serum albumin: Electron spin resonance and Mössbauer spectroscopic studies (in preparation).
27. R. Lemberg and J. W. Legge, Hematin Compounds and Bile Pigments (Interscience Publishers, Inc., New York, 1949).
28. C. E. Castro, *J. Theoretical Biology*, **33**, 475 (1971)
29. F. S. Mathews, M. Levine and P. Argos, *J. Molecular Biology*, **64**, 449 (1972)
30. L. E. Vickery and W. T. Morgan, Magnetic circular dichroism studies of hemopexin-porphyrin complexes (in preparation).
31. W. T. Morgan, personal communication.
32. W. T. Morgan and U. Muller-Eberhard, *Ninth International Congress of Biochemistry*, Stockholm, *1973 p. 113 (Abstract #267)* (1973)
33. U. Muller-Eberhard and K. Grizzuti, *Biochemistry*, **10**, 2062 (1971)
34. J. Javid, M. H. Fuhrman, H. H. Liem, A. Northway and U. Muller-Eberhard, *J. Lab. Clin. Med.*, **80**, 817 (1972)
35. E. Peggion, A. Cosani, A. S. Verdini, A. Del Pra and M. Mammi, *Biopolymers*, **6**, 1477 (1968)
36. W. T. Morgan and U. Muller-Eberhard, *Enzymes* (in press)
37. E., Smibert, H. H. Liem and U. Muller-Eberhard, *Biochemical Pharmacology*, **21**, 1753 (1972)
38. A. A. Grunder, *Genetics*, **54**, 1085 (1966).
39. L. N. Baker, *Vox Sang.*, **12**, 397 (1967)
40. R. E. Stewart and E. W. Lovrien, *Ann. Hum. Genet.*, **35**, 19 (1971)
41. S. B. Wormsley, N. A. Northway, K. H. Cox and U. Muller-Eberhard, Immunological cross-reactions between heterologous hemopexins (in preparation).
42. I. Witz and J. Gross, *Proc. Soc. Exper. Biol. Med.*, **118**, 1003 (1965)
43. D. G. Klapper, M. A. Cuchens and L. W. Clem, *Lab. Invest.*, **26**, 731 (1972)

44. U. MULLER-EBERHARD and K. H. COX, *Biochim. Biophys. Acta*, (in press)
45. G. MANCINI, A. O. CARBONARA and J. F. HEREMANS, *Immunochemistry*, **2**, 235 (1965)
46. C. B. LAURELL, *Scand. J. Clin. Lab. Invest.*, **29** *(Suppl. 124)*, 21 (1972)
47. P. O. GANROT, *Scand. J. Clin. Lab. Invest.*, **29** *(Suppl. 124)*, 39 (1972)
48. D. GITLIN and A. BIASUCCI, *J. Clin. Invest.*, **48**, 1433 (1969)
49. G. J. THORBECKE, H. H. LIEM, S. KNIGHT, K. COX and U. MULLER-EBERHARD, *J. Clin. Invest.*, **52**, 725 (1973)
50. U. MULLER-EBERHARD, *Pediatric Research*, **7**, 86 (Abstract) (1973).
51. H. L. LEFFERT and D. PAUL, *J. Cell. Biol.*, **52**, 559 (1972)
52. H. H. LIEM, *Biochim. Biophys. Acta*, (in press)
53. R. S. LANE, D. M. RANGELEY, H. H. LIEM, S. B. WORMSLEY and U. MULLER-EBERHARD, *J. Lab. Clin. Med.*, **79**, 935 (1972)
54. H. H. LIEM, J. I. SPECTOR and U. MULLER-EBERHARD, (in preparation).
55. R. S. LANE, D. M. RANGELEY, H. H. LIEM, S. B. WORMSLEY and U. MULLER-EBERHARD, *Brit. J. Haematol.* **25**, 533 (1973)
56. U. MULLER-EBERHARD, H. H. LIEM, A. HANSTEIN and P. A. SAARINEN, *J. Lab. Clin. Med.*, **73**, 210 (1969)
57. J. D. ROSS and U. MULLER-EBERHARD, *J. Lab. Clin. Med.*, **75**, 694 (1970)
58. J. D. ROSS, and U. MULLER-EBERHARD, *Biochem. Biophys. Res. Commun.*, **41**, 1486 (1970)
59. D. J. CRIPPS, H. H. LIEM and U. MULLER-EBERHARD, (in preparation)
60. U. MULLER-EBERHARD, H. H. LIEM, C. A. YU and I. C. GUNSALUS, *Biochem. Biophys. Res. Commun.*, **35**, 229 (1969)
61. M. D. MAINES, M. W. ANDERS and U. MULLER-EBERHARD, Molecular Pharmacology, **10**, 204 (1974)
62. M. D. MAINES and M. W. ANDERS, *Molecular Pharmacology*, **9**, 219 (1973)
63. J. E. GIELEN, F. M. GOUJON and D. W. NEBERT, *J. Biol. Chem.*, **247**, 1125 (1972)
64. J. R. GILLETTE, D. C. DAVIS and H. A. SASAME, *Ann. Rev. Pharmacol.*, **12**, 57 (1972)
65. R. BARNES, M. S. JONES, O. T. G. JONES and R. J. PORRA, *Biochem. J.*, **124**, 633 (1971)
66. F. DE MATTEIS, *Biochem. J.*, **124**, 767 (1971)
67. W. R. JONDORF, D. C. SIMON and M. AVNIMELECH, *Molecular Pharmacology*, **2**, 506 (1966)
68. A. H. CONNEY, *Pharmacological Revues*, **19**, 317 (1967)
69. N. E. SLADEK, and G. J. MANNERING, *Molecular Pharmacology*, **5**, 186 (1969)
70. J. BARON and T. R. TEPHLY, *Molecular Pharmacology*, **5**, 10 (1969)
71. R. KATO, *Jap. J. Pharmacol.*, **17**, 56 (1967)
72. S. GRANICK, and S. SASSA, *Metabolic Regulation*, **5**, 77 (1971)
73. G. R. WARNICK and B. F. BURNHAM, *J. Biol. Chem.*, **246**, 6880 (1971)
74. S. S. BOTTOMLEY, G. A. SMITHEE, S. B. WORMSLEY and U. MULLER-EBERHARD, *Blood*, **38**, 796 (Abstract #30) (1971)
75. U. MULLER-EBERHARD, H. H. LIEM and K. H. COX, (in preparation).
76. U. MULLER-EBERHARD and R. BASHORE, *New Engl. J. Med.*, **282**, 1163 (1970)
77. B. LUNDH, F. A. OSKI and F. H. GARDNER, *Acta Paediat. Scand.*, **59**, 121 (1970)
78. J. D. ROSS, H. H. LIEM and U. MULLER-EBERHARD, *Proc. Soc. Exper. Biol. Med.*, **136**, 127 (1971)
79. A. HANSTEIN and U. MULLER-EBERHARD, *J. Lab. Clin. Med.*, **71**, 232 (1968)
80. H. J. BRAUN and F. W. ALY, *Klin. Wschr.*, **49**, 451 (1971)
81. H. F. BUNN, *Seminars in Hematology*, **9**, 3 (1972)
82. G. W. MARSH and S. M. LEWIS, *Seminars in Hematology*, **6**, 133 (1969)
83. S. R. LYNCH and A. L. MOEDE, *J. Lab. Clin. Med.*, **79**, 85 (1972)
84. I. KUSHNER, T. S. EDGINGTON, C. TRIMBLE, H. H. LIEM and U. MULLER-EBERHARD, *J. Lab. Clin. Med.* **80**, 18 (1972)
85. H. G. M. CLARKE, T. FREEMAN and W. PRYSE-PHILLIPS, *Clin. Sci.*, **40**, 337 (1971)

86. H. CLEVE, K. ALEXANDER, H. J. MITZKAT, P. NISSEN and I. SALZMANN, *Diabetologia*, **4**, 48 (1968)
87. W. ASKANSAS, *Life Sciences*, **5**, 1767 (1966)
88. J. KOŘÍNEK, *Cas Lek Cesk*, **108**, 919 (1969)
89. Y. MANUEL, M. C. DE FONTAINE, J. J. BOURGOIN, M. DARGENT and J. M. SONNECK, *Clin. Chim. Acta*, **31**, 485 (1971)
90. S. SNYDER and G. ASHWELL, *Clin. Chim. Acta*, **34**, 449 (1971)
91. H. J. BRAUN and F. W. ALY, *Klin. Wschr.*, **48**, 760 (1970)
92. V. A. BOKISCH, F. H. TOP, Jr., P. K. RUSSELL, H. J. MÜLLER-EBERHARD, and F. J. DIXON, *New Engl. J. Med.*, **289**, 996 (1973)
93. I. M. MURRAY-LYON, Studies on plasma proteins in liver disease using quantitative immunoelectrophoresis. Dissertation at University of Edinburgh, Scotland (1973)
94. H. G. MEIERS and H. IPPEN, *Klin. Wschr,*. **46**, 560 (1968)
95. U. MULLER-EBERHARD, H. H. LIEM, M. M. MATHEWS-ROTH and J. H. EPSTEIN, (in preparation)
96. J. H. JANDL, A. R. JONES and W. B. CASTLE, *J. Clin. Invest.*, **36**, 1428 (1957)
97. E. CHIANCONE, A. ALFSEN, C. IOPPOLO, P. VECCHINI, A. F. AGRÓ, J. WYMAN and E. ANTONINI, *J. Molec. Biol.*, **34**, 347 (1968)
98. R. L. NAGEL and Q. H. GIBSON, *J. Biol. Chem.*, **246**, 69 (1971)
99. S. KORNFELD, B. CHIPMAN and E. B. BROWN, *J. Lab. Clin. Med.*, **73**, 181 (1969)
100. S., GOLDFISCHER, A. B. NOVIKOFF, A. ALBALA and L. BIEMPICA, *J. Cell. Biol.*, **44**, 513 (1970)
101. D. M. BISSELL, L. HAMMAKER and R. SCHMID, *Blood*, **40**, 812 (1972)
102. H. F. BUNN and J. H. JANDL, *J. Exp. Med.*, **129**, 925 (1969)
103. N. R. PIMSTONE, *Seminars in Hematology*, **9**, 31 (1972)
104. U. MULLER-EBERHARD and H. H. LIEM, *Federation Proceedings*, **27**, 723 (Abstract #2850) (1968)
105. D. A. SEARS, *J. Clin. Invest.*, **49**, 5 (1970)
106. M. E. ERLANDSON, I. SCHULMAN, G. STERN and C. H. SMITH, *Pediatrics*, **22**, 910 (1958)
107. M. E. ERLANDSON, I. SCHULMAN and C. H. SMITH, *Pediatrics*, **25**, 629 (1960)
108. R. TENHUNEN, H. S. MARVER and R. SCHMID, *J. Biol. Chem.*, **244**, 6388 (1969)

Chapter 3

Haptoglobin

E. Giblett

*King County Central Blood Bank,
Seattle, Washington, U.S.A.*

3.1. INTRODUCTION

Haptoglobin (Hp) is an α-2 globulin with several unusual properties. For example, it has the unique capability of combining with the globin moiety of hemoglobin (which accounts for its name), thereby providing the major determinant of the renal threshold for hemoglobin. Also, it is the chief component of the "acute phase" proteins, its concentration increasing up to several-fold during inflammatory reactions. Finally, Hp is not a single structural entity; the genetic polymorphism of one of its polypeptide chains represents a very important form of molecular evolution. These properties have made Hp the subject of a large variety of laboratory and clinical investigations which can be only partially covered in this chapter. Other reviews[1-7] may be consulted for more complete details.

3.2. HAPTOGLOBIN STRUCTURE

3.2.1. POLYPEPTIDE CHAINS

The molecular structure of Hp resembles that of the immunoglobulins, in that its basic unit consists of two identical heavy polypeptides (called β chains) and two identical light polypeptides (called α chains).[8] Two separate chromosomal loci, Hp α and Hp β, control the structure of these chains. Genetic variation at the Hp β locus appears to be very rare. However, the Hp α locus has three common alleles: Hp_α^{1F}, Hp_α^{1S} and Hp_α^2; their corresponding polypeptides are hp α^{1F}, hp α^{1S} and hp α^2.

(a) *The Alpha Chains*

When purified Hp is subjected to reductive cleavage and electrophoresis in acid-urea starch gels, the polypeptide chains can be distinguished. As shown in Fig. 3.1, the hp β chain remains near the origin; the hp α^{1F} chain migrates slightly faster than hp α^{1S}, while hp α^2 has a considerably slower mobility. Hp chain differentiation has also been achieved by immunological methods.[10]

3

FIG. 3.1. Diagram showing electrophoretic migration of the haptoglobin polypeptide chains in acid-urea starch gel.

The hp α^{1F} and hp α^{1S} chains have a molecular weight of 9100. The sequence of their respective 84 amino acids differs at only one point: a substitution of glutamic acid in hp α^{1S} for lysine in hp α^{1F} at position 54 of the chain.[8] However, the hp α^2 chain is nearly twice as long as either hp α^{1F} or hp α^{1S}, and its molecular weight is 16 000. Furthermore, the sequence of its 143 amino acids consists of the first 71 of one hp α^1 chain (F or S) and the last 72 of the other hp α^1 chain (S or F).[11,85]

Cysteine is located in the hp α^1 chains at positions 21, 35, 69 and 73. The inter-chain disulfide linkage is made at position 21, while the next two cysteines take part in an intra-chain disulfide loop. Two such intra-chain loops are reported in the hp α^2 chain. In intact molecules of Hp, each 73α cysteine is linked to a β chain cysteine.[11–13,85–87]

There appears to be a homology between residues 22 to 38 in the α chain of Hp and residues 73 to 89 in the κ and λ (light) chains of the immunoglobulins. In a comparison of these two sequences of 17 amino acids,[11] five were identical, including a cysteine at position 35 in Hp and 86 in the κ and λ chains. When the RNA codons were compared, base sequence homologies were noted for six of the other twelve amino acids. These findings suggest that the light chains of both Hp and immunoglobulin molecules may have had a common evolutionary origin.[11]

(b) *The Beta Chain*

The β chain of Hp has a molecular weight of approximately 40 000, of which about one-fifth is carbohydrate.[14,15] (Note the analogy to the immunoglobulin heavy chains, which also carry the carbohydrate moiety). Of the 275 to 300 amino acids of hp β, five are cysteine residues. Neither the hp α nor hp β chains have detectable free sulfhydryl groups; thus all of the cysteines of the intact molecule are disulfide bonded.[12] There are no β-β disulfide bonds.[87]

The amino acid sequence has so far been determined for only the first 18 and the last eight positions of the hp β chain.[16,17] Quite surprisingly, six of the eight C-terminal and eight of the 18 N-terminal residues are identical in sequence to the C and N terminal sequences, respectively, of either bovine chymotrypsin A or bovine trypsin. A common ancestral gene for these serine proteinases and either one or both of the polypeptide chains of Hp has therefore been proposed.[16]

The large carbohydrate moiety (18%) of the Hp molecule has not been completely analysed. Its constituent sugars are galactose (28%), glucosamine (28%), sialic acid (28%), mannose (12%) and fucose (1%).[18] It has been suggested that each hp β chain carries four branched carbohydrate chains which probably do not have identical sequences.[18,19]

3.2.2. SUBUNIT INTERACTION

In individuals who inherit one of the Hp_α^1 genes from both parents, the genotype may be $Hp_\alpha^{1F}Hp_\alpha^{1F}$, $Hp_\alpha^{1S}Hp_\alpha^{1S}$ or $Hp_\alpha^{1F}Hp_\alpha^{1S}$. Although the products of the two Hp_α^1 genes can be distinguished after purification by reductive cleavage and electrophoresis in 8M urea gel at a low pH, they present the same pattern, called Hp 1, on conventional electrophoresis in starch or acrylamide gel at alkaline pH[19]. Its single electrophoretic component, shown in Fig. 3.2, is a tetrad[21] with the formula (hp α^1)$_2$(hp β)$_2$.

On the other hand, the electrophoretic pattern of Hp_α^2 homozygotes (phenotype Hp 2) consists of a series of bands[22] which represents "multimers" of the (hp α^2)$_2$(hp β)$_2$ tetrad, each held together by disulfide bonds.[23] Thus, in both Hp 1 and Hp 2, the proportion of α and β chains is equimolar.[24] (see Addendum)

The pattern of $Hp_\alpha^2Hp_\alpha^1$ heterozygotes consists of another series of bands,[22] of which the fastest-moving coincides with the single (hp α^1)$_2$-(hp β)$_2$ tetrad of Hp 1. The remaining bands migrate somewhat faster than those of Hp 2; they consist of another multimeric series[25-27] in which the basic subunit is the tetrad: hp α^1hp α^2 (hp β)$_2$. Again the hp α and β chains are equimolar,[14] but there may be a variable ratio of hp α^1 to hp α^2 chains. As a result, both the number of bands and their staining intensity are shifted toward the cathode or the anode, depending on whether the hp α^2 or hp α^1 chains, respectively, predominate.[28]

Molecules of Hp 1 have a uniform weight of about 100 000, while the
multimeric components of Hp 2 and Hp 2-1 have a wide range in molecular weight. Because of the filtration effect of starch or acrylamide

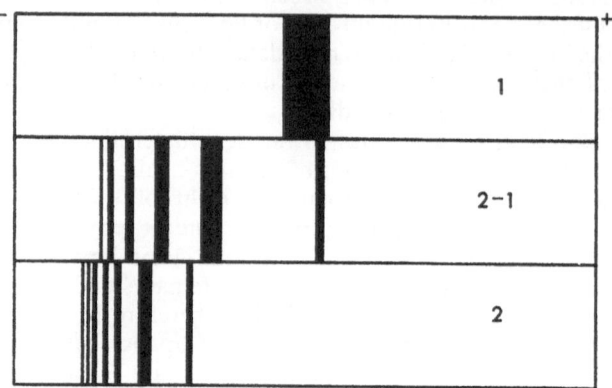

FIG. 3.2 Diagram showing electrophoretic patterns of the three common
phenotypes of haptoglobin.

gels, the heaviest components have the slowest migration rates. Presumably, in spite of their large size, these components can bind hemoglobin on the same equimolar basis as the smaller sized Hp 1 molecules.[29]

3.3. THE HAPTOGLOBIN—HEMOGLOBIN COMPLEX

3.3.1. GENERAL PROPERTIES

Haptoglobin combines stoichiometrically with the globin moiety
of hemoglobin and not with heme or any other heme-containing compounds, including myglobin.[7] Several lines of evidence (reviewed in
reference 20) indicate that Hp does not combine with the intact tetramer of hemoglobin, but rather, with its $\alpha\beta$ dimers. The globin-binding
sites of Hp appear to be on the hp β chains.[31] Four such binding sites
have been found on the Hp 1 molecule; in the reaction, the α globin
chains are preferentially bound.[32,33] Binding of the β globin chains is
apparently minimal until the Hp molecule undergoes a critical configurational change secondary to α globin chain binding.[30]

Although Hp forms a highly stable complex with oxyhemoglobin,
carbon monoxyhemoglobin, methemoglobin and cyanmethemoglobin,
it reacts very poorly with deoxyhemoglobin,[34] and sub-optimally with
hemoglobin treated with the sulfhydryl reagent, bis-(N-maleimido-
methyl) ether.[35] This low reactivity is apparently due to impaired
ability of the globin chains in these forms of hemoglobin to undergo

reversible dissociation. The fact that neither the β chain tetramer, Hb H, nor the γ chain tetramer, Hb Barts, can combine with Hp may also be due to reduced dissociability of globin chains.[36] However, an equally plausible explanation rests on the very poor affinity of Hp for the β and γ globin chains, as compared with the α globin chains.[32]

The linkage between Hp and hemoglobin does not involve covalent bonds,[35,37] nor does the sialic acid content of Hp influence the binding.[38] Furthermore, treatment of hemoglobin with ethyl acetimidate to block its ϵ-amino groups (in lysine) has no deterrant effect on the combination with Hp.[39] All of the lysine side chains are on the exterior of the hemoglobin molecule. This and other evidence suggests that the Hp-binding site is located on the $\alpha\beta$ dimer interface of hemoglobin.[39]

3.3.2. INTERMEDIATE COMPLEXES

When Hp 1 is only partially saturated with hemoglobin, two complexes are formed,[7] as shown in Fig. 3.3. The slower-moving (saturated)

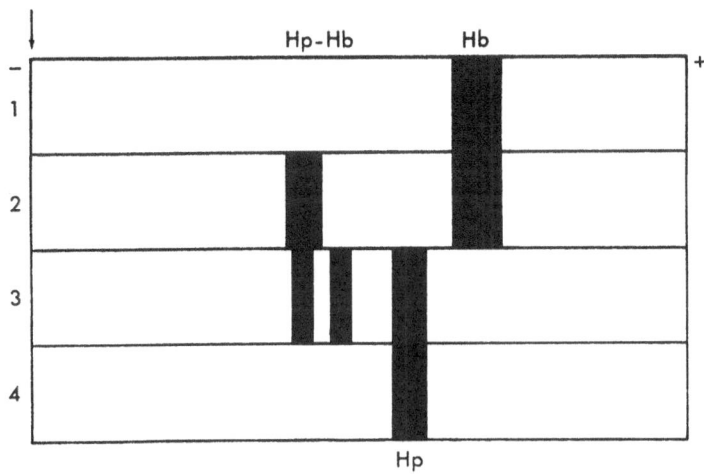

FIG. 3.3 Diagram showing electrophoretic migration of (1) free hemoglobin; (2) the fully saturated complex formed when hemoglobin is in excess; (3) the intermediate and saturated complexes formed when haptoglobin is in excess; (4) free haptoglobin.

complex consists of one Hp molecule and two $\alpha\beta$ dimers of hemoglobin, while the partially saturated (intermediate) complex consists of one Hp molecule and one $\alpha\beta$ dimer.[29,40] The multiple components of Hp 2 and Hp 2-1 can combine with progressively larger numbers of hemoglobin dimers. Thus, in each phenotype, the heavier the individual component, the greater its number of intermediate HpHb complexes, depending on the degree of saturation with hemoglobin.[1]

3.3.3. EFFECT ON HEMOGLOBIN FUNCTION

The presence of heme has little effect on the interaction of globin and Hp.[7] However, several of the properties of heme are altered in the complex. For example, peroxidase activity is enhanced,[41] denaturation by guaiacol is blocked,[42] and the exchange of heme with other molecules is prevented.[43,44]

The affinity of Hp-bound hemoglobin for oxygen and other ligands is greatly increased, resulting in a hyperbolic oxygen-dissociation curve.[34] This loss of heme-binding interaction is apparently due to configurational constraint on the complexed hemoglobin, as reflected by changes in its circular dichroism and electron magnetic resonance measurements.[29,42] These changes resemble those which occur during deoxygenation, suggesting that in the Hp–Hb complex, the F helix of the β globin

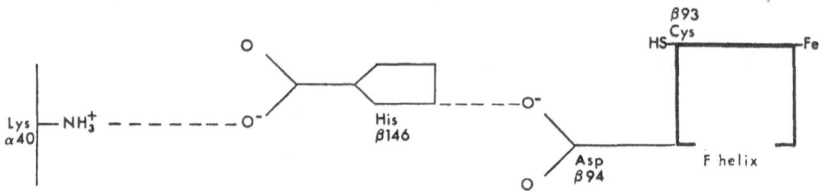

FIG. 3.4. Diagram showing immobilization of the C-terminal histidine (β146) in deoxyhemoglobin due to formation of salt bridges with aspartic acid at β94 and with lysine at position 40 on the adjacent α chain (adapted from Perutz; reference 45).

chain assumes a deoxy-like conformation independent of the oxidation and ligand state of the heme iron,[42] which is directly linked to the β 92 histidine on the F helix.

The work of Perutz[45] has shown that one of the important characteristics of deoxyhemoglobin is immobility of the C-terminal histidine in both β globin chains. This immobility is due to the formation of salt bridges between the C-terminal (β146) histidine and two other residues: one is the aspartic acid at position 94 (FG4), just beyond the F helix of the same beta chain, while the other is the lysine at position 40 (C5) on the adjacent α chain (See Fig. 3.4). This terminal histidine fixation is reversible on oxygenation, and is thought to be partially responsible for the alkaline Bohr effect.[45] Irreversible fixation of the terminal histidines on the Hp-complexed hemoglobin could account for its greatly reduced Bohr effect.[34,42] It could also explain why enzymatic removal of the terminal histidines, which prevents hemoglobin from assuming the deoxy configuration, permits it to combine with Hp regardless of the oxygenation state.[46]

3.4. HAPTOGLOBIN PHYSIOLOGY

3.4.1. METABOLISM

The plasma level of Hp is fairly constant during health in any given person, but it varies considerably among normal individuals, from as low as 30 mg % to as high as 300 mg %. One reason for this wide range is the fact that the Hp level reflects not only the rates of Hp synthesis and degradation, but also the removal of circulating Hp which accompanies intravascular release of hemoglobin. These factors will be briefly considered.

(a) *Synthesis: Normal and Abnormal*

The liver is the major site of Hp synthesis,[48] although other organs, such as the spleen and lymph nodes, may make a minor contribution.[49] After liver transplantation, Hp of the donor's genetic type becomes established in the patient, while Hp of his own type disappears.[50] Thus, if there is extrahepatic synthesis of Hp, it must be of a low order.

Only 10 to 15 per cent of newborn infants have sufficient plasma Hp to be detectable by the usual methods,[51] which depend upon the hemoglobin-binding capacity of the protein. Those infants in whom there is a measurable amount of Hp often have a different genetic type from that of their mothers; thus, the fetus is capable of synthesizing its own Hp protein. However, it has not been conclusively shown whether the common anhaptoglobinemia of the newborn is due to limited synthesis of Hp or to its rapid removal by hemoglobin released from fetal red cells unusually susceptible to destruction.

The Hp level increases during the first several months, but tends to be lower in children and young adults than in older people.[1,52] There is a correlation of Hp phenotype with Hp plasma concentration,[53] so that individuals of type 1 generally have higher levels than those of type 2. Sex is not an important determinant, although it has been reported that androgens increase, and estrogens depress the Hp level, presumably by affecting its synthesis.[53]

Under the stimulus of inflammation, neoplasia or severe stress, there may be as much as a 10-fold increase in Hp.[6] A similar rise is seen in experimental animals given injections of irritant or pyrogenic substances.[54] Such increases in Hp are usually accompanied by elevated levels of several other plasma proteins, particularly orosomucoid and fibrinogen.

The Hp concentration is depressed in patients with severe hepatocellular damage, consistent with the hepatic origin of the protein. However, since there is often associated red cell destruction, some of the effect is due to increased catabolism rather than decreased synthesis of the protein.

(b) *Catabolism: Normal and Abnormal*

Hemoglobin released within the circulating blood combines immediately with haptoglobin, and the complex is removed at a rate dependent on its concentration.[43,55,56] At the low physiological concentration of 0.3 mg of hemoglobin in 100 ml of plasma, removal of the complex is very rapid, with a half-time of about nine minutes.[57] Nevertheless, the amount of hemoglobin normally catabolized via complex formation (about one gram per day) is only a sixth of the total hemoglobin turnover. Similarly, only a portion (probably less than 50 per cent) of the Hp produced is removed in this way. The remainder is catabolized independently, with a half-time of about four days.[58]

In most patients with a hemolytic process, there is an increase in the amount of hemoglobin released into the plasma. Under these circumstances, the amount of Hp synthesized is insufficient to compensate for its continuous removal as the HpHb complex. Thus, hypohaptoglobinemia or anhaptoglobinemia is a common finding in hemolytic states,[43,59] including those with ineffective erythropoiesis (such as thalassemia and pernicious anemia), where intra-marrow hemoglobin turnover is excessive.

It was previously noted that when the amount of Hp greatly exceeds that of hemoglobin in vitro, most of the HpHb complex exists as an intermediate form in which the two components are not equimolar. Since the normal plasma level of hemoglobin is far below the binding capacity of Hp, it is likely that the intermediate complex is more prevalent *in vivo* than the saturated complex. On the other hand, with brisk hemolysis, the binding capacity may be reached or exceeded, resulting in a predominance of the saturated complex. Since HpHb removal appears to be inversely related to the level of hemoglobin in the plasma, it may be that the intermediate and saturated forms have different clearance rates (*i.e.* the more saturated the complex, the slower its removal).[7] Furthermore, differences in Hp phenotype might be expected to influence the rate of HpHb removal; however assessments of this possible role of Hp polymorphism have yielded conflicting results.

The liver is the major site of HpHb removal, with some uptake by the bone marrow and spleen.[60] Most investigators have inferred from their studies that this activity is a function of the reticuloendothelial cells. However, there is recent evidence that in the liver, the hepatocyte, rather than the Kupffer cell, is primarily involved.[61]

Accelerated removal of Hp seems to have only a moderate effect on its synthesis, so that if hemolysis is suddenly terminated, several days are usually required for restoration of the normal Hp level, unless there is a concomitant stimulus such as inflammation.[59]

A small proportion (about 4 per cent) of adult American Negroes and a higher proportion of Negro children have apparent anhapto-globinemia without an associated increase in hemolysis.[62] This subject is discussed in a later section on genetic variation, but it is mentioned here as a reminder that the absence of demonstrable Hp does not necessarily imply the presence of liver damage or accelerated red cell breakdown. For this reason, and because the normal Hp level varies so widely, it is impossible to interpret the clinical significance of a laboratory value obtained from a single specimen of blood, particularly when the patient is very young or is of African ancestry. Even when serial specimens are obtained, the results may be misleading. However, in conjunction with other tests, the Hp level has some utility for monitoring the clinical course in hepatic or hemolytic disease.

3.4.2. FUNCTION

A true carrier protein, such as transferrin, reversibly combines with a transported substance so that it can be taken up by specific receptor cells. The carrier molecule is thereby freed to pick up and release additional loads before being itself metabolized. The behavior of Hp is quite different, both because its bond with hemoglobin is not physiologically reversible without intramolecular disruption, and because the binding of hemoglobin to Hp has little influence on the site of its subsequent deposition.[60] Thus, in animals depleted of Hp, injected hemoglobin is taken up by the liver, bone marrow and spleen, just as it would be if Hp were present. However, there is also considerable renal tubular uptake, which does not occur when Hp is present, because the HpHb complex is too large to pass the glomerular membrane.

Due to this filtration effect, the renal threshold of hemoglobin is determined to a large extent by the amount of Hp in the plasma.[55] Once the binding capacity of Hp is exceeded, molecules of hemoglobin become more vulnerable to loss of heme and to dissociation into $\alpha\beta$ dimers, in which form hemoglobin is apparently excreted.[63]

The amount of Hp normally present in the total plasma volume is sufficient to bind about 3 grams of Hb, which is equivalent to about 10 ml of red cells or 12 mg of iron. Under normal circumstances, this large amount of hemoglobin (half the total daily turnover) is never released into the plasma. Thus, Hp serves as a continual blockade to iron loss.[64]

It is not clear to what extent this iron-conservation property of Hp is critical for the maintenance of life. Individuals with inherited forms of hemolytic anemia usually have no detectable plasma Hp, because Hp is continually being removed by the hemoglobin released from red cells destroyed within the circulation. Nevertheless, these patients rarely become iron deficient, presumably because there is sufficient renal tubular reabsorption of heme to prevent large losses of iron,[65] rather than because the rate of Hp synthesis is increased to the extent necessary to accommodate the large load of free hemoglobin. However, it is impossible to obtain an accurate assessment of Hp turnover during active hemolysis. Furthermore, if hemolysis is prevented by replacing the defective red cells with transfused normal cells, thereby suppressing the bone marrow, there is no longer any need for the patient to produce increased amounts of Hp, and turnover measurement has little meaning.

Since the Hp level rises markedly in the presence of inflammation, it is tempting to believe that the protein plays some important role in combating disease. According to one study,[66] Hp is a strong non-competitive inhibitor of cathepsin B, a proteolytic enzyme released in association with tissue injury. Another study[67] indicated that Hp can inhibit the agglutination of chicken erythrocytes by influenza virus. However, whether or not these properties have any physiological significance remains to be determined.

3.5. Hp GENETICS

3.5.1. ALPHA CHAIN DUPLICATION

It was noted in an earlier section that the heavy (β) and light (α) chains of Hp are under the control of genes at two separate loci, called Hp α and Hp β, and that the three common alleles at the Hp α locus are called Hp_α^{1F}, Hp_α^{1S} and Hp_α^2. The two Hp_α^1 genes produce polypeptide chains of 84 amino acids, with a single substitution at position 54 accounting for their different electrophoretic properties. This amino acid exchange of lysine in hp α^{1F} for glutamic acid in hp α^{1S} can be readily explained by a single step mutation in their respective codons.[8] However, point mutation cannot be invoked to explain the structure of hp α^2, which consists of the first 71 amino acids of one hp α^1 chain, and the last 72 of the other hp α^1 chain[11]. Clearly, the Hp_α^2 gene must have been formed as the result of some kind of duplicative mechanism in an early human ancestor who was heterozygous for the two Hp_α^1 alleles.[8]

The nature of the genetic mechanism is not known, but it is likely to have been unequal crossing over between two dissimilar regions of Hp[1] genes on opposite chromosomes. Thus, if misalignment occurred in a $Hp_\alpha^{1S}Hp_\alpha^{1F}$ heterozygote during meiosis (See Fig. 3.5), that portion of one Hp_α^1 gene corresponding to the C terminus of its polypeptide product probably overlapped the N terminal regional of the other Hp_α^1 gene, on the opposing chromosome.[8] Then, crossing over in the region of the seventy-first codon of one gene and the twelfth codon on the other resulted in the formation of a nearly duplicated gene, Hp_α^2. (Whether its sequence is 1S1F or 1F1S is not known).

An objection to this theory[68] arose from the fact that the amino acids of the hp α polypeptides in the two regions of the postulated crossing over show very little homology. Thus, if it is necessary that two opposing DNA sequences be homologous in order for pairing and subsequent crossing-over to occur, it would seem that some other mechanism of chromosome breakage and joining would have to be invoked. However, a comparison of the actual codon sequence corresponding to the amino acids at positions 9–17 and 67–75 showed many identical bases, suggesting that pairing could take place, with subsequent unequal crossing-over.[12]

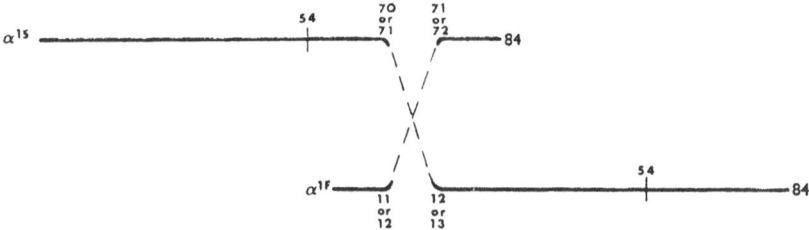

F$_{\text{IG}}$. 3.5. Diagram showing the effect on the Hp polypeptide of unequal, non-homologous crossing over in an individual heterozygous for Hp_α^{1S} and Hp_α^{1F} genes. The resulting long segment, consisting of 143 amino acids, represents the hp^2 chain. The short segment has no known polypeptide counterpart. Position 54 in both chains is the site of amino acid substitution which differentiates hp α^{1S} from hp α^{1F}.

Regardless of the mechanism involved, it seems very likely that the genetic event giving rise to Hp_α^2 was unique, both because of its unusual duplicated structure and because there is no evidence for the existence of other kinds of hp α^2 chains with a different point of junction between the hp α^1 chains. Such chains could theoretically escape detection by electrophoresis. However, since the hp α^2 chain which has been analyzed consists of nearly equal numbers of amino acids from hp α^{1F} and hp α^{1S}, other hp α^2 chains would be expected to have different lengths and/or to consist of unequal contributions from the two chains. Either of these kinds of alteration should be detected by electrophoresis, particularly in a filtrative medium such as starch or acrylamide gel.

Because the Hp_α^1 genes are only about half as long as Hp_α^2, it has been proposed that variation in their alignment during meiosis in Hp 2-1 heterozygotes might be a fairly common event. As a result, new kinds of Hp_α^2 alleles, corresponding to 1F1F and 1S1S, might be generated.[8,68] There is indirect evidence for the occurrence of such chains, based on electrophoretic mobility differences.[69] However, these variants appear to be uncommon.

3.5.2. GENETIC LINKAGE

From studies of families with inherited chromosomal anomalies, it has been virtually established that the Hp α locus is on the sixteenth chromosome.[70] Reported linkage with the HL-A locus[71–73] has been withdrawn. Another possible linkage is with the locus controlling catalase molecular structure.[74] Although linkage data on the Hp α and Hp β loci are limited, the currently available information indicates that they are not closely linked.[75]

3.5.3. QUANTITATIVE VARIATION

As noted earlier, the plasma Hp level is subject to considerable variation under the influence of a number of diseases. In addition, there are hereditary factors influencing the amount of Hp produced.[53] For example, individuals of Hp type 2 tend to have lower levels than those of type 1. Other kinds of hereditary influences on the Hp level are also apparent. As a result, the electrophoretic pattern varies from complete loss of detectable Hp to alterations in the relative quantities of the individual polymers (*i.e.* "multimers") of Hp 2 and Hp 2-1 phenotypes.

(a) *Hp 2-1 Variation*

In the typical Hp 2-1 electrophoretic pattern, there are six to eight visible multimers with a gradual increase in staining intensity accompanying mobility rate. The fastest-moving band, which corresponds to the Hp 1 tetrad, is often very weak. However, there is a wide spectrum of Hp 2-1 phenotypes (See Fig. 3.6), with anodal bands predominating at one end of the spectrum and cathodal bands at the other end. Since the assembly of Hp polypeptide chains occurs within the cell, the most likely explanation for this range of phenotypic expression is a difference in the relative quantity of the Hp_α^1 and Hp_α^2 gene products.[76]

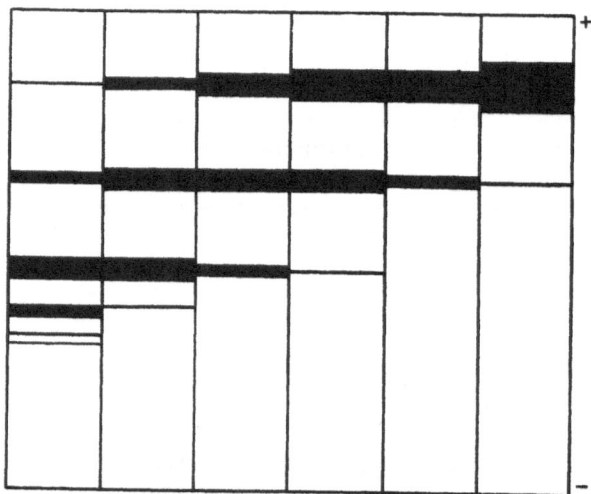

FIG. 3.6. Diagram showing a spectrum of Hp 2-1 electrophoretic patterns, with successive shifts from left to right of the predominant components toward the anode.

Best studied are the variants collectively named Hp 2-1 (mod), in which hp α^1 is predominant over hp α^2. The result is heavy banding of the fastest-moving components and loss of the slowest bands. The effect is apparently due to relative deficiency of hp α^2 chain production. Inheritance of this phenotype has been well established,[77] and there is also evidence for heritable differences in the degree of hp α^2 deficiency.[78] The genetic mechanism responsible for this quantitative variability is unknown. According to one theory, there are control genes which have a differential effect on the activity of Hp_α^1 and Hp_α^2 alleles.[79] However, studies of families with members of Hp 2-1 (mod) type have provided no evidence for independent segregation of the hypothetical controller effect from the Hp_α^2 allele itself. Thus, if control genes are involved, they must be very closely linked to the Hp α locus.

The so-called Carlberg type[51] resembles a mixture of Hp 2 and Hp 2-1. Its inheritance is consistent with the existence of an allele at the Hp α locus which has a product indistinguishable from hp α^1 except for its diminished quantity.[76] Presumably, when hp a^1 chains are deficient, there is a surplus of hp α^2 chains which, with hp β chains, form multimers typical of the Hp 2 phenotype. Thus, the phenotype consists of both Hp 2 and Hp 2-1 multimers. It has not been conclusively shown, with either the Hp 2-1 (mod) or Carlberg phenotypes, that the peptide "fingerprints" or amino acid sequences of the deficient chains are identical to those of the more common hp α^1 and hp α^2 chains. However, the possibility of structural variation seems remote, because the electrophoretic patterns of the separated chains of both Hp 2-1 (mod) and Hp Carlberg differ in staining intensity, but not in mobility, from the common hp α chains.[74]

(a) *Anhaptoglobinemia*

Hypohaptoglobinemia or anhaptoglobinemia, designated phenotypically Hp 0, is characteristic of most newborn infants and accompanies a variety of diseases in which there is either hepatocellular damage or increased intravascular hemolysis. However, there are genetic factors which lead to the same phenotype. As noted above, individuals of Hp type 2 tend to have lower levels than those of type 1. Thus, a healthy person with little or no demonstrable Hp in his serum is more likely to carry two Hp^2 genes than two Hp^1 genes. In fact, studies of appropriate families have shown that hypohaptoglobinemia rarely occurs in subjects who have inherited Hp_α^1 from both parents. However, the $Hp_\alpha^2 Hp_\alpha^1$ genotype may be expressed as Hp 0, particularly when one or the other parent has the Hp 2-1 (mod) phenotype, thus indicating an inherited decrease in hp α^2 chain production.[77,78] Since Hp 2-1 (mod) occurs most frequently in people of African origin (about 10 per cent of American Negroes), it is not surprising that these people have the highest frequencies of Hp 0 (among American Negroes, about 4 per cent of adults and 12 per cent of children).[62]

There is no evidence that the cases of Hp 0 observed in the Afro-Americans are due to the presence of a Hp_α allele which has no recognizable polypeptide product. However, there have been a few reports of families which contained some members with Hp 0 and others with aberrant inheritance (*e.g.* an Hp 1 child with an Hp 2 parent, or vice versa).[79–81] Determining the subtypes of Hp 1 in these families provided evidence strongly favoring a "silent" Hp^0 gene which, in the heterozygous state, was associated with the phenotypes Hp 0, Hp 1 or Hp 2, depending on whether or not the Hp α allele on the opposite chromosome was expressed sufficiently to be detected.[2]

Because of the very low frequency of Hp_α^0 in all of the populations tested, the likelihood of finding an individual homozygous for this allele is very small. Nevertheless, it would be important to know if the complete absence of haptoglobin is compatible with life. It appears almost certain that most individuals with Hp 0 phenotype due to deficient production of hp α chains are capable of synthesizing some Hp, but the amount in their plasma is too small to be detected unless the individual is subjected to some stress such as that of inflammatory disease. On the other hand, one would expect that an individual homozygous for a functionally "silent" Hp allele could not respond to disease in this manner, and might thereby be severely handicapped, depending upon whether or not Hp is an essential component of the body's defense mechanisms.

3.5.4. STRUCTURAL VARIANTS

Hp variants which contain electrophoretic components not found in the three common phenotypes are very rare. The variant most often observed is the so-called "Johnson" type (Fig. 3.7), isolated examples having been reported from most of the laboratories throughout the world where large-scale Hp testing is performed. On subtyping (after reductive cleavage and electrophoresis in an acid-urea gel), this Hp consists of either hp α^{1F} or hp α^{1S} and a second polypeptide which migrates much more slowly than hp α^2 and appears to have a greater molecular weight.[8] This polypeptide, called hp α^J, is the product of an Hp_α^J allele, which may have arisen by homologous but unequal crossing-over in an individual homozygous for Hp_α^2. In this case, mispairing of the alleles and subsequent crossing over presumably produced a partially triplicated gene. The fact that there is evidence for more than one kind of Hp_α^J can be readily explained by assuming different points of crossing-over.[8]

Other structural variants due to mutations at the Hp α locus are extremely rare, and only a few have been described. While individual instances of hp β chain variation are also highly unusual, several phenotypes have been described which are either known or assumed to represent examples of Hp_β gene mutation. When carefully studied by chemical and immunological methods, these variants have shown some degree of defective hemoglobin binding,[75,82,83] consistent with the acknowledged importance of hp β chains in forming the HpHb complex. For example, one of the antigenic determinants of the Hp molecule is ordinarily blocked from combining with its specific antibody when hemoglobin is present. However, in the Marburg and Bellevue hp β chain variants, this antigenic site remains free to react with its antibody after HpHb complex formation has occurred.

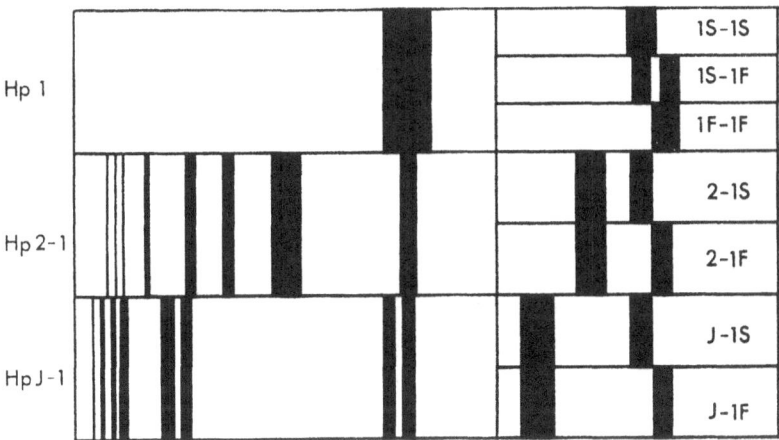

FIG. 3.7. Diagram showing electrophoretic patterns of three Hp phenotypes (on the left) and their respective sub-types (on the right).

A comparison of the Marburg beta chain (hp β^{Mb}) with hp β chains derived from the common Hp types showed the presence of either three or four additional tryptic peptides in digests of the aberrant chain.[84] Thus, although the two kinds of hp β chains did not differ significantly in their amino acid composition, the multiple differences in their peptide "fingerprints" suggested that some mutational event other than single amino acid substitution was responsible for the Hp_β^{Mb} gene.

3.5.5. Hp POPULATION GENETICS

A large body of data on the geographic distribution of the three major Hp phenotypes is available, showing that a Hp polymorphism exists in all of the many populations tested.[2,4] If one accepts the suggestion that the common Hp_α^2 gene arose by partial duplication of two Hp_α^1 genes on only one occasion, then it is necessary to conclude that the duplicative event occurred in a very early ancestor. Since the haptoglobin in primates other than man does not have a multiple-banded pattern, Hp α^2 must have arisen after separation of the human evolutionary line but before the population became widely dispersed. Another indication of its relative recency is the retention in hp α^2 of the same amino acid sequences as those of its two parent hp α^1 chains. However, they, too have remained the same, suggesting that the point mutation which differentiated hp α^{1S} from hp α^{1F} (or vice versa) also took place after the separation of human from other evolutionary lines.

The rarity of Hp_α^2 genes having the 1F1F or 1S1S sequence is unexpected, if unequal but homologous crossing over in Hp 2-1 heterozygotes occurs fairly often. On the other hand, it is possible that selective factors maintain the common 1F1S (or 1S1F) Hp_α^2 allele.

The Hp gene frequencies vary considerably among different ethnic groups. The Hp_α^1 genes (usually not further differentiated by Hp typing) have their highest frequencies in various parts of Africa, South and Central America, and several of the Pacific Islands. Low frequencies of Hp_α^1 are most notable among the peoples of Asia, where, of course the highest frequencies of Hp_α^2 occur. Most European populations have a somewhat higher frequency of Hp_α^2 than Hp_α^1. There have been attempts to associate Hp type with certain diseases, but so far, these suggested associations have not been confirmed.

Since Hp subtyping is a fairly difficult procedure, less is known about the relative distribution of Hp_α^{1F} and Hp_α^{1S}. Nevertheless, it appears that in most populations with a high frequency of Hp_α^2, the Hp_α^{1F} frequency is low. Conversely, when Hp_α^2 is relatively infrequent, Hp_α^{1F} is more prevalent. Although it seems almost certain that these associations reflect the action of selective factors, it has not yet been possible to determine what those factors are now or may have been before the onset of the modern era.

ADDENDUM

In a recent paper by Fuller et al. (Biochemistry, **12**, 253 1973), the multimers of Hp2 were shown to consist of successive additions of $\alpha^2\beta$ subunits rather than $\alpha_2^2\beta_2$ tetramers. Substantiating evidence for a similar polymeric model in Hp 2-1 was provided by Ogawa et al. (Proc. Jap. Acad., **46**, 814, 1970).

REFERENCES

1. H. E. SUTTON (eds., A. G. Steinberg and A. B. Bearn), The Haptoglobins, Progress in Medical Genetics, vol. 7, p. 163 (Grune & Stratton, New York, 1970)
2. E. R. GIBLETT, Genetic Markers in Human Blood (Blackwell, Oxford and F. A. Davis, Philadelphia, 1969)
3. J. JAVID, Sem. Hemat., **4**, 35 (1967)
4. R. L. KIRK, The Haptoglobin Groups in Man (S. Karger, Basel & New York, 1968)
5. H. E. SCHULTZE and J. F. HEREMANS, Molecular Biology of Human Proteins vol. 1. Nature and Metabolism of Extracellular Proteins (Elsevier, Amsterdam, 1966) p. 384

6. M. F. JAYLE and J. MORETTI (eds., C. V. Moore and E. B. Brown), Haptoglobin: biochemical, genetic and physiological aspects, *Progress in Hematology*, vol. 3, p. 342 (Grune & Stratton, New York, 1962)
7. C. B. LAURELL and C. GRONVALL (eds., H. Sobotka and C. P. Stewart), Haptoglobins, *Advances in Clinical Chemistry*, vol. 5, p. 135 (Academic Press, New York 1962)
8. O. SMITHIES, G. E. CONNELL and G. H. DIXON, *Nature*, **196**, 4851 (1962)
9. G. E. CONNELL, O. SMITHIES and G. H. DIXON, *J. Molec. Biol.*, **21**, 225 (1966)
10. C. EHNHOLM, Ph.D. Dissertation (University of Helsinki, 1969)
11. J. A. BLACK and G. H. DIXON, *Nature*, **218**, 736 (1968)
12. O. SMITHIES, G. E. CONNELL and G. H. DIXON, *J. molec. Biol.*, **21**, 213 (1966)
13. J. A. BLACK and G. H. DIXON, *Canad. j. Biochem.*, **48**, 133 (1970)
14. J. A. BLACK, G. F. Q. CHAN, C. L. HEW and G. H. DIXON, *Canad. j. Biochem.*, **48**, 123 (1970)
15. H. CLEVE, B. H. BOWMAN and S. GORDON, *Humangenetik*, **7**, 337 (1969)
16. D. R. BARNETT, T. H. LEE and B. H. BOWMAN, *Nature*, **225**, 938 (1970)
17. D. R. BARNETT, T. H. LEE and B. BOWMAN, *Fed. Proc.*, **30**, 1295 (1971)
18. R. I. CHEFTEL, L. CLOAREC, J. MORETTI and M. F. JAYLE, *Bull. Soc. chim. Biol.*, **47**, 385 (1965)
19. C. M. GERBECK, A. BEZKOROVAINY and M. E. RAFELSON, *Biochemistry*, **6**, 403 (1967)
20. O. SMITHIES, G. E. CONNELL and G. H. DIXON, *Am. J. hum. Genet*, **14**, 14 (1962)
21. B. S. SHIM and A. G. BEARN, *J. exp. Med.*, **120**, 611 (1964)
22. O. SMITHIES, *Biochem. J.*, **61**, 629 (1955)
23. O. SMITHIES, *Science*, **150**, 1595 (1965)
24. G. M. FULLER, M. L. MCCOMBS and D. R. BARNETT, *Fed. Proc.*, **30**, 1071 (1971)
25. O. SMITHIES, *Adv. Prot. Chem.*, **14**, 65 (1959)
26. A. C. ALLISON, *Nature*, **183**, 1312 (1959)
27. J. JAVID, *Proc. natn. Acad. Sci.*, **52**, 663 (1964)
28. E. R. GIBLETT, *Cold Spring Harbor Symposium on Quantitative Biology*, **29**, 321 (1964)
29. H. HAMAGUCHI, *Am. J. hum. Genet.*, **21**, 440 (1969)
30. R. L. NAGEL and Q. H. GIBSON, *J. biol. Chem.*, **246**, 69 (1971)
31. S. GORDON and A. G. BEARN, *Proc. Soc. exp. Biol.*, **121**, 846 (1966)
32. R. L. NAGEL and Q. H. GIBSON, *J. biol. Chem.*, **242**, 3428 (1967)
33. E. CHIANCONE, A. ALFSEN, C. IOPPOLO, P. VECCHINI, A. AGRO, J. WYMAN and E. ANTONINI, *J. mol. Biol.*, **34**, 347 (1968)
34. R. L. NAGEL, M. C. ROTHMAN, T. B. BRADLEY, and H. M. RANNEY, *J. biol. Chem.*, **240**, PC4543 (1965)
35. H. F. BUNN, *J. Lab. clin. Med.*, **70**, 606 (1967)
36. R. L. NAGEL and H. M. RANNEY, *Science*, **144**, 1014 (1964)
37. L. ROBERT, G. BOUSSIER and M. F. JAYLE, *Experientia*, **13**, 111 (1957)
38. B. S. BLUMBERG and L. WARREN, *Biochim. biophys. Acta*, **50**, 90 (1961)
39. W. L. LOCKHART and D. E. SMITH, *Canad. j. Biochem.*, **49**, 148 (1971)
40. A. C. PEACOCK, J. V. PASTEWKA, R. A. REED and A. T. NESS, *Biochemistry*, **9**, 2275 (1970)
41. M. POLONOVSKI and M. F. JAYLE, *C.r. Soc. Biol.*, **129**, 457 (1938)
42. M. W. MAKINEN and H. KON, *Biochemistry*, **10**, 43 (1971)
43. A. ALLISON and W. AP REES, *Brit. med. J.*, **ii**, 1137 (1957)
44. H. F. BUNN and J. H. JANDL, *Proc. natn. Acad. Sci.*, **56**, 974 (1966)
45. M. F. PERUTZ, *Nature*, **228**, 726 (1970)
46. E. CHIANCONE, J. B. WITTENBERG, B. A. WITTENBERG, E. ANTONINI and J. WYMAN, *Biochim. biophys. Acta*, **117**, 379 (1966)
47. B. MALCHY and G. DIXON, *Canad. j. Biochem.*, **47**, 1205 (1969)
48. C. A. ALPER, J. H. PETERS, A. E. BIRTCH and F. H. GARDNER, *J. clin. Invest.*, **44**, 574 (1965)

49. T. WADA, H. O'HARA, K. WATANABE, H. KINOSHITA and H. NISHIO, *J. retic. Soc.* **8**, 195 (1970)
50. N. KASHIWAGI, C. G. GROTH and T. E. STARZL, *Proc. Soc. exp. biol. Med.*, **128**, 247 (1968)
51. F. GALATIUS-JENSEN, *Acta genet.*, **8**, 248 (1958)
52. M. HAUGE, A. HEIKEN and C. HOGLUND, *Human hered.*, **20**, 557 (1970)
53. M. NYMAN, *Clin. chim. Acta*, **3**, 111 (1958)
54. R. K. MURRAY and G. E. CONNELL, *Nature*, **186**, 86 (1960)
55. C. B. LAURELL and M. NYMAN, *Blood*, **12**, 493, (1957)
56. L. GARBY and W. D. NOYES, *J. clin. Invest.*, **38**, 1479 (1959)
57. T. FREEMAN, *Protides Biol. Fluids*, **12**, 344 (1964)
58. W. D. NOYES and L. GARBY, *Scand. J. clin. lab. Invest.*, **20**, 33 (1967)
59. M. NYMAN, *Scand. J. clin. lab. Invest.*, **11**, suppl. 39 (1959)
60. W. R. KEENE and J. H. JANDL, *Blood*, **26**, 705 (1965)
61. C. HERSHKO and J. D. COOK and C. A. FINCH, *J. Lab. Clin. Med.*, **80**, 624 (1972)
62. E. R. GIBLETT, *Nature*, **183**, 192 (1959)
63. H. F. BUNN, W. T. ESHAM and R. W. BULL, *J. exper. Med.*, **129**, 904 (1969)
64. A. C. ALLISON, *Proc. roy. Soc. Med.*, **51**, 641 (1958)
65. H. F. BUNN and J. H. JANDL, *J. exper. Med.*, **129**, 925 (1969)
66. O. SNELLMAN and B. SYLVEN, *Nature*, **216**, 1033 (1967)
67. W. DOBRYSZYCKA and E. LISOWSKA, *Biochim. biophys. Acta*, **121**, 42 (1966)
68. O. SMITHIES, *Cold Spring Harbor Symposia on Quantitative Biology*, **29**, 309 (1964)
69. W. E. NANCE and O. SMITHIES, *Nature*, **198**, 869 (1963)
70. E. B. ROBSON, P. E. POLANI, S. J. DART, P. A. JACOBS and J. H. RENWICK, *Nature*, **223**, 1163 (1969)
71. W. B. BIAS, D. B. AMOS, F. E. WARD, O. C. YODER, J. H. RENWICK and V. A. McKUSICK, *Abstract, Amer. Soc. Hum. Genet.*, p. 40 (1970)
72. R. M. HARDISTY, M. M. TILL, S. D. LAWLER, P. T. KLOUDA, J. R. BATCHELOR, J. H. EDWARDS, J. STUART, P. J. L. COOK and E. B. ROBSON. *Ann. hum. Genet.*, **35**, 161 (1971)
73. W. F. BODMER, J. G. BODMER, A. COUKELL, H. CANN and B. VAN WEST, *Ann. hum. Genet.*, **35**, 167 (1971)
74. W. E. NANCE, J. E. EMPSON, T. W. BENNETT and L. LARSON, *Science*, **160**, 1230 (1968)
75. J. JAVID, *Proc. natl. acad. Sci.*, **57**, 920 (1967)
76. E. R. GIBLETT, *Cold Spring Harbor Symposia on Quantitative Biology*, **29**, 321 (1964)
77. E. R. GIBLETT and A. G. STEINBERG, *Amer. J. hum. Genet.*, **12**, 160 (1960)
78. H. E. SUTTON and G. W. KARP, *Amer. J. hum. Genet.*, **16**, 419 (1964)
79. W. C. PARKER and A. G. BEARN, *Amer. J. hum. Genet.*, **15**, 159 (1963)
80. H. HARRIS, E. B. ROBSON and M. SINISCALCO, *Nature*, **182**, 1324 (1958)
81. W. SCHWERD and I. SANDER, *Blut*, **15**, 99 (1967)
82. G. WEERTS, W. NIX and H. DEICHER, *Blut*, **12**, 65 (1965)
83. J. JAVID and W. YINGLING, *J. clin. Invest.*, **47**, 2297 (1968)
84. H. CLEAVE, B. H. BOWMAN and S. GORDON, *Humangenetik*, **7**, 337 (1969)
85. B. MALCHY and G. H. DIXON, *Canad. J. Biochem.*, **51**, 321 (1973)
86. B. MALCHY and G. H. DIXON, *Canad. J. Biochem.*, **51**, 249 (1973)
87. B. MALCHY, O. RORSTAD and G. H. DIXON, *Canad. J. Biochem.*, **51**, 265 (1973)

Chapter 4

Acute-Phase Reactants

Their Synthesis, Turnover and Biological Significance

A. Koj

Department of Animal Biochemistry, Institute of Molecular Biology, Jagiellonian University, Cracow, Poland

4.1. INTRODUCTION

The term "acute-phase reactants" (AP-reactants) is generally considered to refer to protein components of plasma whose concentration is significantly increased in the acute phase of inflammatory processes. This category includes several proteins with various physicochemical properties (Table 4.1) and biological functions, but at least two common features may be ascribed to them: almost all contain significant amounts of carbohydrates and all are synthesized in liver parenchymal cells. Hence AP-reactants may be defined as trauma-inducible liver-produced plasma glycoproteins.

When comparing physicochemical properties of proteins listed in Table 4.1 it may be concluded that their carbohydrate content decreases with the rise of isoelectric point and molecular weight. This list of AP-reactants is far from being complete, but there is no general consensus in respect of which other proteins should be included. For example some authors have reported that macroglobulins from rabbit serum[7,8] are increased during inflammatory reactions, and others[9] have observed increased synthesis of transferrin during perfusion of the liver from laparatomized or partially hepatectomized rats. On the other hand, Crockson *et al.*[10] and Werner and Odenthal[11] found no change or even decreased level of macroglobulin and transferrin in patients after surgical trauma. The situation is even more complicated by the demonstration of Heim[4] that a typical AP-reactant, slow α_2 globulin of the rat, is related to α_2 macroglobulin of other mammals. Confusion may also arise from the fact that the seromucoid fraction is heterogenous and includes some other AP-reactants listed in Table 4.1. Since at least 30 different glycoproteins are present in plasma it is likely

TABLE 4.1

Some properties of typical acute-phase reactants listed in alphabetic order

The data refer to human plasma except α_1 and α_2 acute-phase globulins from the rat and C_x-reactive protein from rabbit

Protein	Amount in normal plasma mg/100 ml	Molecular weight	Isoelectric point pI	Carbohydrate content %	Reference
α_1-acid glycoprotein	75–100	44 000	2.7	41.4	1
α_1-AP globulin (rat)	120	45 000	4.5	16	2
α_1-antitrypsin	210–287	45 000	4, 0	12.4	1, 3
α_2-AP globulin (rat)	0	950 000 approx.	?	?	4
Ceruloplasmin	27–63	160 000	4.4	8.0	1
C-reactive protein	0	138 000	?	?	5
C_x-reactive protein (rabbit)	0	120 000	?	?	5
Fibrinogen	200–600	341 000	5.8	2.5	1
Haptoglobin (Hp 1–1)	30–190	85 000	4.1	19.3	1
Seromucoid	61	Heterogenous fraction		25 approx.	6

that some of them will in the future be identified as other acute-phase reactants. It should be remembered that plasma contains also a great variety of trace protein components which are detectable only by virtue of their biological properties, such as clotting factors, enzymes and hormones. Some are glycoproteins of liver origin and show augmented level after injuries, but since they are not regarded as typical plasma proteins they will not be discussed here.

A considerable body of information on the conditions leading to increased concentrations of acute-phase ractants has already been obtained both in clinical studies and with experimental animals, mainly rats and rabbits. Necessarily the selected references are limited to the last few years, the earlier studies being exhaustively reviewed by Winzler,[6,12,13] Owen[14] and others. The typical conditions include: local inflammation[15–21] thermal or mechanical injury,[22–24] major surgery,[10,11,25] bacterial infection or endotoxin injection[12,13,16,18,26–28] and neoplastic growth.[12,13,16,29,30,32]

The response pattern of individual AP-reactants to various stresses and diseases may differ considerably (see also section 11 of this chapter) and it must be understood that any attempt to discuss all these proteins as one group is certainly an oversimplification. For this reason the main

AP-reactants will be reviewed separately. At the same time it should be emphasized that none of the proteins mentioned in Table 4.1 is associated with one particular type of trauma only; in other words no strictly specific reaction has been observed in this respect.

Increased concentration in the blood of any AP-reactant does not necessarily imply enhanced synthesis, and some authors still believe that part of the serum glycoproteins comprised in the seromucoid fraction arises from the ground substance of connective tissue; but there has been controversy whether tissue destruction, or healing and tissue proliferation cause it.[12] However, experiments with isotopic techniques and the isolated liver have proved beyond doubt that in response to trauma the proteins in question are synthesized at an increased rate in the liver,[18,33-35]—see also reference 12.

Acute-phase proteins are of considerable interest to practical and experimental medicine, biochemistry and biology. Changes in their plasma concentration are regarded as sensitive (although rather non-specific) tests used for diagnostic and prognostic assessments. Some of these proteins are associated with such disturbances of homeostasis as bleeding and blood clotting (fibrinogen), intravascular haemolysis (haptoglobin), or increased activity of some proteolytic enzymes (α_1-antitrypsin). All AP-reactants represent a useful model for investigating the regulation of protein synthesis in the intact animal.

The present review will be concerned mainly with the synthesis, turnover and biological functions of AP-reactants. The physico-chemical properties and molecular structure of individual proteins belonging to this group are in most cases well established and described at length in other chapters of this book, or in other studies concerning plasma proteins.[1,36] The metabolism of plasma proteins, including some acute-phase reactants, has been recently reviewed by Bocci.[37]

4.2. STUDIES OF THE DYNAMIC STATE OF ACUTE-PHASE REACTANTS

4.2.1. CHANGES IN PROTEIN CONCENTRATION IN BLOOD PLASMA

The easiest and most common method of investigating the response of AP-reactants to trauma depends on measuring changes in the blood level of these proteins with appropriate corrections, when necessary, for haemodilution or haemoconcentration occurring in some pathological states. Almost each AP-reactant can be individually determined with a sufficient degree of accuracy by specific methods which will be discussed briefly in appropriate sections. At the same time it is possible to estimate approximately changes in the blood level of all AP-reactants by taking advantage of the fact they are glycoproteins.

Protein-bound hexoses are usually measured either by the method employing orcinol,[5,38] or with the aid of the anthrone reagent.[39,40]

Hexosamines (glucosamine and galactosamine) may be determined by various modifications of Elson–Morgan reaction.[41,42] Sialic acids appear to occupy a non-reducing terminal position in glycoproteins and are easily split off by acid or neuraminidase. The liberated sialic acid can be quantitatively determined in reaction with resorcinol[43] or thiobarbituric acid.[44] Determinations of serum glycoproteins based on protein-bound hexose, hexosamine, fucose and sialic acid are described by Winzler.[5] This author reported the following values for normal human plasma: protein-bound hexose 121 $\pm$ 21 mg/100 ml, hexosamine 83 $\pm$ 4

TABLE 4.2

The contents of seromucoid, haptoglobin, protein-bound hexose and sialic acid in the serum of children with scarlet fever (after Wiedermann et al[26])

Mean values of 50 children in hospital aged 4 to 14 years,* denotes statistical difference from the control at P < 0.001.

				Protein-bound	
Day after admission	Phase	Seromucoid mg % tyr.	Haptoglobin mg %	Hexose mg %	Sialic acid mg %
1	Clinical activity	7.32*	150.9*	133.2*	95.3*
6	Onset of convalescence	6.10*	131.5*	126.9*	88.3*
21	Close of convalescence	4.35	91.8	110.4	73.9
90	Control	4.02	79.5	105.1	69.9

mg/100 ml and sialic acid 60 $\pm$ 3.1 mg/100 ml.[13] Critical evaluation of various methods for determining sugar moieties in glycoproteins may be found in such monographs and textbooks as those edited by Gottschalk,[45] Balazs and Jeanloz[46] and Schultze and Heremans.[1]

Several authors have reported the existence of significant correlations between protein-bound hexose, hexosamine or sialic acid in plasma and the level of typical AP-reactants—fibrinogen, haptoglobin or seromucoid fraction.[20,26,47,48] This can be illustrated by the data in Table 4.2 derived from the study of Wiedermann et al.[26] on the changes of haptoglobin and other AP-reactants in the blood of children with scarlet fever.

It should be remembered, however, that several other plasma proteins not belonging to AP-reactants, such as γ-globulins, contain significant amounts of covalently bound carbohydrates. They may contribute to the plasma level of protein-bound hexose or sialic acid, especially at later stages of infectious diseases. For this reason some authors prefer the estimation of total glycoproteins to be carried out after electrophoretic separation of serum. Protein-bound carbohydrates can be then stained with the periodic acid-Schiff procedure according to Köiv and Grönvall.[49] After paper electrophoresis at pH 8.6 most of the AP-reactants appear to be present in α_1 and α_2 fractions.

Recent advances of immunological methods have overcome some difficulties in simultaneous quantitation of several individual proteins. Radial immunodiffusion introduced by Mancini et al.[50] has been successfully applied to plasma glycoproteins.[11,51-53] The method requires protracted incubation to complete the diffusion of high-molecular weight antigens but shows remarkable accuracy and sensitivity. With regard to AP-reactants it seems to be particularly suitable for studying response patterns of several proteins after various injuries.[54]

4.2.2. TURNOVER MEASUREMENTS WITH IODINE-LABELLED PROTEINS

It is not always remembered that even the most accurate determination of any AP-reactant changes in plasma level does not provide exact data on the changes in its synthesis rate. This is mainly due to the fact

TABLE 4.3

Distribution and turnover of iodine-labelled albumin and some AP-reactants in healthy human subjects

Protein	I.V. pool as per cent of total body pool	Half-life in plasma $T_{1/2}$(days)	Per cent of body pool degraded per day	Daily turnover mg/kg body per day	Reference
Albumin	42	14.8	4.5	164–236	55
α_1-acid glycoprotein	?	5.2	12	?	56
α_1-antitrypsin	?	3.9*	?	?	57
Haptoglobin	50†	2–4	14–18	14–23	58,59,60,61
Ceruloplasmin	40	4.25	16	5.9–6.2	62, 63
Fibrinogen	75–84	3.1–3.4	20–24	31–46	64, 65

* Sialic acid removed by treatment with neuraminidase before injecting the labelled protein.
† According to Krauss[61] 60–75% of total body pool of Hp 2-2 is present in the I.V. pool.

that the plasma concentration of a protein is also influenced by its catabolism and diffusion to the extravascular space. The ratio of intravascular to interstitial pool depends on the type of protein and animal species; in human subjects some 20% of total body fibrinogen or 60% of total body albumin are present in the extravascular compartment (Table 4.3). For this reason the rise of any AP-reactant in the plasma after injury is delayed and attains lower values than corresponding changes in the synthesis rate measured directly with tracer methods (see also Fig. 4.1).

The turnover of plasma proteins may be determined *in vivo* by injecting iodine-labelled protein and following disappearance of the label from plasma, or its excretion into urine. The labelling of isolated protein, which should be free of denatured material, is usually accomplished either by iodine-monochloride method[66] or by chloramine T method.[67] The injected protein equilibrates with the extravascular pool during 36–48 hours and then is eliminated from the body with the half-life of 3–14 days in humans depending on the type of protein (Table 4.3). The catabolic rate can be calculated either from the plasma slope of protein-bound activity, or from the activity of free iodide deriving from decomposed proteins and excreted into urine. Practical details of the procedure and kinetic considerations are described in several original papers and textbooks.[1,55,66,68,69]

The method based on iodine-labelled proteins provided basic information on the turnover rates and absolute catabolic rates of several acute-phase reactants, as shown in Table 4.3. The rates of protein synthesis can be determined indirectly by measuring the disappearance of ^{131}I-labelled molecules, provided that the steady state prevails during the period of measurement. Such a method would obviously not be applicable to longer observations, nor to pathological situations in which synthetic and catabolic rates are unequal—as in the case of AP-reactants. Several authors[68–70] have attempted to calculate the difference between rates of absolute synthesis and absolute catabolism using ^{131}I-labelled proteins but the results are not wholly satisfactory and the direct biosynthetic approach is clearly the method of choice.

4.2.3. SYNTHESIS RATES *IN VIVO* DETERMINED WITH LABELLED PRECURSORS

The direct method of estimating the synthesis of plasma glycoproteins depends on the measurement of incorporation of precursors (such as amino acids or glucosamine) labelled with ^{3}H, ^{14}C or ^{35}S, the radioactivity incorporated being regarded as a direct proof of net protein synthesis. The labelled precursor is usually administered parenterally and 1–5 hours later the investigated proteins under investigation are isolated from a sample of plasma to determine their specific activities. Some authors[23,24,71] advocate taking several blood samples to obtain the exponential curve of disappearance of labelled protein from plasma, allowing the calculation of catabolic rates (as with iodine-labelled proteins) but the results are not fully reliable because of reutilization of the label. Some kinetic aspects of the catabolism of plasma glycoproteins labelled biosynthetically with ^{14}C in the N-acetylglucosamine and sialic acid moieties were recently discussed by Robinson.[72]

When the precursor specific activity in the site of synthesis is not known accurate synthetic rates cannot be calculated and in most cases only gross estimates based on comparison of relative protein activities are possible. However, because of its simplicity the method based on injection of labelled amino acids has been extensively used in studies of AP-reactants, both in experimental animals[17,73–77] and in human subjects.[23,24,71]

At present the only method of specified accuracy which enables measurements of absolute synthesis rates of liver-produced plasma proteins to be made in a short-term experiment *in vivo* is based on using plasma urea specific activity as the indicator of intracellular specific activity of arginine (labelled ^{14}C in the guanidine carbon). The ^{14}C-carbonate method proposed by McFarlane[78] and subsequently developed and improved[27,79] depends on injecting ^{14}C-carbonate and then comparing the synthesis rate of the protein with that of urea, the ratio being proportional to the total ^{14}C-radioactivities incorporated into the guanidine carbon of protein arginine and into urea carbon during the synthesis interval. For the highest accuracy of measurements the losses of labelled urea and protein from the plasma during the synthesis interval should be independently determined with injected ^{13}C-urea and iodine-labelled protein,[27] although a simplified procedure is also possible.[80] The ^{14}C-carbonate method has been successfully[80–84] used both in experimental animals and in clinical studies. With the reference to AP-reactants this method allowed the determination of increased synthesis rates of fibrinogen after injection of endotoxin, turpentine or vitamin A in rabbits,[27,85,86] or growth hormone in rats.[87]

The carbonate method is technically exacting, however, and cannot be easily repeated at short intervals in the same experimental subject. An alternative approach developed by Koj[85] depends on using another protein synthesized in the liver parenchymal cells as a reference standard for the incorporation of a labelled amino acid. Plasma albumin appears to be a useful reference protein in relation to studies of acute-phase reactants since its metabolism is only little affected by local inflammation or injury.[76,88,89] The relative method is based on consecutive injections of different labelled amino acids (or other precursors such as ^{14}C-carbonate) and isolation of plasma albumin and the protein in question to compare their specific activities; the synthesis rate relative to albumin can then be calculated. Fig. 4.1 shows the results of application of the relative method to study the kinetics of trauma-induced fibrinogen synthesis in a rabbit.

Synthesis of fibrinogen increases considerably within the first day after induction if inflammation while the rise of plasma fibrinogen is delayed and less dramatic. The mean relative synthesis rate of fibrinogen during 48 h after turpentine injection calculated by integrating the area under the relative synthesis rate curve was 440% of the initial value (before treatment) while the fibrinogen concentration was increased to 270% in the same interval. It should be noted that when estimations of fibrinogen synthesis are given as specific activities neglecting

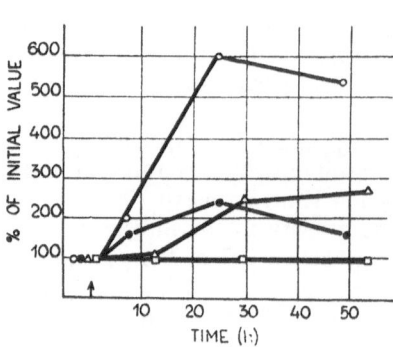

F<small>IG.</small> 4.1. Typical changes in relative synthesis rate and plasma fibrinogen in a rabbit after injected of turpentine. Results are expressed as percentage of initial values before injection of turpentine. o-relative synthesis rate of fibrinogen. Experimental points before injection of turpentine and after 7, 24 and 48 hours were obtained with using ³H-lysine, ¹⁴C-leucine, ¹⁴C-lysine and ¹⁴C-carbonate, respectively; ●-ratio of amino acid specific activities in fibrinogen and albumin (without corrections for changes in concentration of the two proteins in plasma); △ fibrinogen concentration in plasma; ☐ albumin concentration in plasma (after Koj[85]).

the augmented pool of this protein the results also markedly underestimate the real synthesis rate, especially at later periods of the experiment. Hence the expression of the data as specific activities alone (unfortunately still used by some authors) can be misleading as to the true rate of plasma protein synthesis under these conditions.

4.2.4. LIVER PERFUSION AND CELL-FREE SYSTEMS

Detailed studies on the mechanism of acute-phase reactant synthesis require a relatively simple system with controlled and easily varied conditions; hence the isolated perfused liver is a method of choice. The liver can be isolated at a specified time after injury and perfused with blood from normal or injured animals, with substitutes of blood enriched in hormones, metabolites or drugs. The rate of protein synthesis is usually determined by incorporation of labelled precursors, but direct measurement of net biosynthesis by immunological methods is also possible.[90] Most investigators used rat liver in such experiments, and details of the procedure can be found in papers by Miller,[33] Gordon and Darcy,[91] John and Miller,[90] and others.

Although liver slices and cultures of liver cells have also been used in studies of AP-reactants,[18,92,93] their application is rather limited will not be discussed here. On the other hand, cell-free preparations obtained by fractionation of liver homogenates show growing importance in investigations of synthesis of acute-phase reactants. Some of the results obtained with the perfused liver and cell-free systems will be discussed in section 4.10.

Here attention should be drawn to the fact that studies with subcellular components have provided an answer to an interesting question: how the carbohydrate moiety of glycoproteins is being synthesized. Results of investigations of Winzler and co-workers,[94-96] Sarcione et al.[97] Lawford and Schachter[98] and others (for detailed references see[37,99,100]) indicate that sugar components are being attached to the recently completed polypeptide molecules in the liver microsomal fraction. At first activated glucosamine or mannose are transferred by specific enzymes to a polypeptide receptor, perhaps even before its release from the ribosome occurs. Then, during transport of the protein along the rough and smooth endoplasmic reticulum, other sugars and amino sugars are added in consecutive reactions catalysed by specific transferases in a sequence characteristic for a given protein. Finally, after the terminal moiety of sialic acid is attached, the glycoprotein molecule can be released from the reticular system and secreted from the liver into the blood. The importance of sialic acid is emphasized by observations of Morell et al.[101] indicating that artificially desialylated glycoproteins are promptly removed from the circulation.

4.3. FIBRINOGEN

4.3.1. DETERMINATION AND ISOLATION OF PLASMA FIBRINOGEN

Most of the methods of fibrinogen determination are based on clot formation after adding calcium ions or thrombin to plasma. The amount of fibrin obtained can be measured by gravimetric procedure or employing colorimetric reactions for proteins or ammonia. Critical evaluation of various methods of fibrinogen estimation may be found in papers by Jacobsson[102] or Hirsch and Cattaneo.[103] A fast, although not very accurate procedure depends on measuring plasma turbidity directly after addition of thrombin.[104] An accurate isotrope dilution method has been developed by Atencio et al.[105] The following procedure suitable for small aliquots of plasma can be recommended by the author:[27] Blood samples are collected in test tubes containing solid EDTA-K_2 (in proportion 4 mg EDTA per 1 ml blood) and plasma separated by ordinary centrifugation. 0.1–0.2 ml of EDTA-treated plasma in a narrow test tube is clotted by adding 5 units of thrombin in 1 ml of sodium phosphate buffer pH 6.1.[102] After 2 h at room temperature the clot is wound on to a thin glass rod, washed with 0.9% NaCl solution (2×10 min) and dissolved by heating in 3 ml of 1 N NaOH at 80°C for 5 min. Then the amount of protein is measured by the method of Lowry et al. The standard curve should be prepared by clotting samples of plasma of known fibrinogen concentration (determined by gravimetric method). It should be mentioned that for dissolving of fresh fibrin clot alkaline urea solution may be used[105] and the amount of protein determined from O.D. at 280 nm.

Isolation of highly purified fibrinogen with preservation of its native state is required for metabolic studies with iodine-labelled proteins.[106] On the basis of a long experience Regoeczi[107] recommends the following procedure: Blood is collected into a 2% (w/v) solution of oxalates (40% potassium oxalate and 60% ammonium oxalate) using 1 vol. of the anticoagulant for 4 vol of blood. Prothrombin is removed by adsorption on $BaSO_4$ and the plasma is fractionated with a neutral solution of 2.05 M $(NH_4)_2SO_4$. To ensure optimal metabolic homogeneity, the fibrinogen fraction of low solubility is eliminated by preliminary precipitation at 18% $(NH_4)_2SO_4$ saturation, then the degree of saturation is increased to 23.8% and the precipitate collected by centrifugation at 600 g. The protein pellet is washed twice with 0.967 M $(NH_4)_2SO_4$ and redissolved in 0.9% NaCl containing 0.005 M trisodium citrate. After removing traces of ammonium sulphate by dialysis against the above solvent, the precipitation at 23.8% saturation is repeated and the pellet washed with 0.976 M $(NH_4)_2SO_4$ until the supernatant remains clear on mixing with an equal volume of 20% (w/v) trichloroacetic acid. The precipitate is dissolved in a minimal volume of 0.9% NaCl containing 0.005 M trisodium citrate and dialysed against the same solution overnight at $+4°C$. The purity of the preparation obtained is most accurately estimated by measuring the clottability of iodine-labelled fibrinogen.[108,109]

Isolation of fibrinogen from plasma in order to determine its radioactivity (in metabolic experiments with iodine-labelled fibrinogen or with labelled precursors) does not require such precautions and the main purification step usually includes clotting; either diluted plasma is clotted directly with thrombin[110] or crude fibrinogen is precipitated from plasma at 25% saturation with ammonium sulphate, clotted with thrombin, and the clot washed thoroughly with isotonic saline.[78]

4.3.2. TURNOVER OF FIBRINOGEN IN HEALTHY INDIVIDUALS

Fibrinogen is metabolized rather rapidly and its half-life in plasma ranges from 1.3 days in the rat to 2.8 days in rabbit and 3.4 days in man, cf.[1]

The fluorescent antibody technique demonstrated that hepatic parenchymal cells are the site of fibrinogen formation[111] and it is generally assumed that the liver is the only source of this protein, although some authors claim that blood platelets or megacaryocytes may also contribute to plasma fibrinogen (cf. Marcus and Zucker[112])

The liver does not contain significant amounts of stored fibrinogen[113] and the release of newly synthesized molecules occurs promptly. Alper *et al.*[75] and Smallwood *et al.*[114] demonstrated that fibrinogen (like other plasma proteins such as albumin or haptoglobin) passes directly from the liver into the intravascular compartment without appreciable prior mixing with the extravascular pool. Our results[27] indicate that labelled fibrinogen appears in the blood of a rabbit some 30 min after injecting ^{14}C-carbonate and the maximum specific activity of this protein is observed after 2 hours.

The absolute synthesis rate of fibrinogen determined by the ^{14}C-carbonate method is equal to 63.9 mg/kg body per day in the rat,[87] to 54.7 mg/kg body per day in the rabbit,[27] and to approximately 31 mg/kg body per day in man.[78] Atencio *et al.*[70] reported great daily fluctuations in the liver secretion rate of fibrinogen in healthy rabbits. This is in agreement with our observations[27] that the absolute synthesis rate of fibrinogen in these animals, determined in short-term experiments with ^{14}C-carbonate, ranges in normal conditions within $+50\%$ of the mean value, while the synthesis of albumin is more stable.

The fractional transfer rate of fibrinogen from plasma to the extravascular compartment is constant and does not depend on the pool size of this protein.[115] This means that at increased plasma concentrations of fibrinogen the absolute amounts of fibrinogn passing the capillary wall (transcapillary flux) are proportionally increased. The same is true of fibrinogen catabolism: quantities of fibrinogen catabolized daily are directly proportional to the pool size and correspond to first order kinetics.[66,70,107,115–117] This may indicate that the enzymatic system responsible for the normal breakdown rate of fibrinogen is not saturated even at high plasma fibrinogen levels. The site of fibrinogen catabolism, as with the majority of plasma proteins, is not known, but indirect evidence suggests that fibrinogen degradation, under physiological conditions, is confined to cells.[108] It is therefore reasonable to assume that the determining factor is the rate of uptake of fibrinogen by catabolic cells, the concentration-dependence of this process being explained by pinocytosis, with or without selective adsorption, as suggested by Regoeczi.[107] The normal breakdown of fibrinogen does not involve passage through fibrin since heparin and coumarin derivatives do not influence fibrinogen catabolism.[118] Also the inhibition of fibrinolysis by administration of epsilon-aminocaproic acid does not influence the half-life of ^{131}I-labelled fibrinogen in the dog and rabbit.[119,120] Hence it appears that fractional catabolic rates of fibrinogen are fixed by a fundamental biological function which is unrelated to the catabolic process itself, and that fibrinogen catabolism is probably not governed by any genuine control mechanism. The actual concentration of fibrinogen in the blood represents an equilibrium between the variable rate of synthesis and the constant fractional rates of catabolism or transfer of fibrinogen to the extravascular pool.

4.3.3. EFFECTS OF INJURY ON FIBRINOGEN TURNOVER

Fibrinogen is the most thoroughly investigated AP-reactant. It should be mentioned that already in 1897 Biernacki[121] observed the relationship between the plasma level of fibrinogen and increased erythrocyte sedimentation rate in infectious disease. Classical studies on the blood fibrinogen in various pathological states were accomplished by Foster and Whipple[122] and Ham and Curtis.[123] The latter authors proved that hyperthermia produced by heating of the body has no appreciable effect on fibrinogen concentration in distinction to fever in infectious diseases. The use of isotopes in metabolic studies during the last two decades elucidated many controversial problems of fibrinogen turnover after injuries and now rather a clear picture emerges.

Numerous data indicate that there is always a lag period of a few hours between the occurrence of injury and change in fibrinogen synthesis rate (cf. Fig. 4.1). This interval is probably essential for the development of local inflammation, release of active factors from the site of injury, their transport to the liver and the final effect on the synthesis of specific proteins. The maximum rate of fibrinogen synthesis is reached only after approximately 24-48 hours and then it declines slowly. Provided that other factors such as the nutritional state or hormonal balance of the injured subject are unchanged, the rising slope of fibrinogen concentration in plasma and the attained maximum are roughly proportional to the severity of damage while the descending slope shows the recovery from the effect of injury.

A basically similar picture is obtained with the perfused rat liver isolated at various time after subcutaneous injection of talc. In this case maximum synthesis rates are observed as soon as 9 hours after trauma (Fig. 4.2).

More severe injuries, such as extensive skin burns in man, lead to a prolonged increase of fibrinogen synthesis, and higher concentrations of fibrinogen in blood may persist for several weeks.[22,71] In certain diseases, such as rheumatoid arthritis, fibrinogen synthesis is turned on to three or more times its normal rate.[117] This may persist for many months and maintains fibrinogen concentrations 3 times or more the normal levels. A question arises of what the maximum synthesis rate of fibrinogen may be after trauma. The highest values observed in rabbits 24-48 hours after injection of endotoxin[27] or turpentine[85] were approximately seven times greater than those in control animals and reached the absolute synthesis rate of albumin. Even higher values obtained by an indirect method were reported by Reeve et al.[27] and Atencio et al.[113] This channelling of protein formation in the liver must be associated with profound changes in cellular organization and energy metabolism after trauma.

The mechanism of stimulation of the liver to produce more fibrinogen is still obscure and the problem will be discussed in detail in section 10, also in relation to other acute-phase reactants and the effect of various hormones. Some authors suggest the existence of a direct feed-back mechanism: products of fibrinogen degradation, as formed during fibrinolysis especially after defibrination produced by injection of thrombin, may stimulate fibrinogen synthesis in the liver.[27,111,125,126]

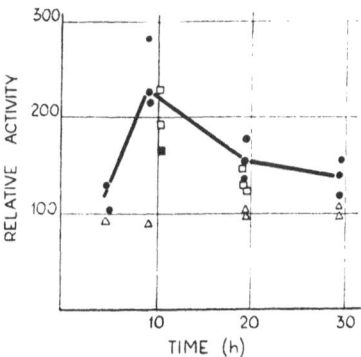

FIG. 4.2. Incorporation of ^{14}C-leucine into fibrinogen after 5-hour perfusion of livers from rats at different times after subcutaneous injection of talc. Radioactivity of total plasma fibrinogen expressed as a percentage of incorporation in control animals. ● livers from normal rats injected with talc; the line is an average of these values; □ livers from adrenalectomized rats after talc injection; ■ liver from an adrenalectomized rat after talc injection (perfused with blood from adrenalectomized rats); △ livers from normal rats, not injected with talc and perfused with blood from rats after talc injection (after Gordon and Koj[124]).

It should be remembered, however, that increased formation of fibrinogen may occur after even a mild stress, and the defibrination syndrome produced by thrombin infusion is certainly a stressful situation. Miller and John[127] demonstrated that livers isolated from rats 10 to 24 hours after defibrination produced by intraperitoneal injection of thrombin synthesized at the increased rate not only fibrinogen but also other acute-phase reactants. Thrombin had no direct effect on fibrinogen synthesis in the perfused liver. Nilehn and Ganrot[128] demonstrated that infusion of streptokinase resulted in 2–4-fold decrease of circulating fibrinogen without any sign of greatly increased synthesis. The return to normal values was slow and required 5–6 days.

Fractional rates of transfer of fibrinogen from plasma to the interstitial fluid seem to be unimpaired by trauma, even at high fibrinogen concentration; hence absolute amounts of fibrinogen diffusing through

A. KOJ

the capillary wall must be markedly increased.[129] A rather exceptional situation was observed in rabbits after injecting endotoxin[27] when it was found that the fraction of injected labelled fibrinogen leaving the plasma compartment during the 5 hours of the experiment was significantly reduced in comparison with the control animals. The ratio of intravascular to interstitial fibrinogen is not markedly altered in diseases associated with hyperfibrinogenemia[64,129] except in cases when all plasma proteins appear in larger quantities outside the vascular bed, e.g. after burns.[22]

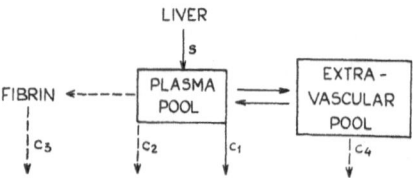

FIG. 4.3. Diagrammatic representation of fibrinogen turnover. s-synthesis of fibrinogen; c_1-"normal" breakdown of fibrinogen; c_2-fibrinogenolysis; c_3-fibrinolysis; c_4-sequestration at the site of injury (after Koj,[135] courtesy of Ciba Foundation).

The fractional catabolic rate of fibrinogen remains as a rule constant in various pathological states such as local inflammation, active tuberculosis, neoplastic disease, rheumatoid fever, glomerulonephritis[27,64,129,130] while absolute amounts of degraded fibrinogen are proportional to the total mass of circulating fibrinogen. However, the situation is complicated because of the fact that apart from "normal" fibrinogen breakdown by a mechanism common to all plasma proteins, in some cases fibrinogen may be catabolized in additional ways (Fig. 4.3). These include intravascular fibrinogenolysis by plasmin or other proteases, as occurs in certain diseases,[27,131] as well as a faster conversion into fibrin which is known to be deposited around some lesions, especially in kidney, metastases of cancer, joints, wounds and skin burns.[22,132-134]

4.4. HAPTOGLOBIN

4.4.1. DETERMINATION AND ISOLATION OF PLASMA HAPTOGLOBIN

Haptoglobin (Hp), discovered by Polonovsky and Jayle,[136] is an α_2-glycoprotein exhibiting the fundamental property of combining with haemoglobin (Hb) to give a complex (HpHb) which is characterized by peroxidase activity. This enzymatic property has been commonly used for quantitative haptoglobin determination.[137] In the convenient procedure described by Owen et al.[138] an excess of methaemoglobin is added to a sample of serum. 0.1 ml of the mixture is transferred to 5 ml of guaiacol solution pH 4.0, followed by 1 ml of 0.05 M H_2O_2. After 8 min incubation at 25°C optical density is measured at 470 nm. The results are usually expressed as mg of haemoglobin binding capacity per 100 ml of serum. In other methods the amount of haemoglobin-haptoglobin complex in the serum enriched with Hb is determined after electrophoretic separation.[139,140] A simple procedure described recently by Roy et al.[141] is based on the observation that haptoglobin protects haemoglobin from acid denaturation. Haptoglobin may be also estimated by immunological methods using either nephelometric measurement[142] or single radial diffusion technique of Mancini,[90] although our observations indicate that HpHb complex may not be precipitated quantitatively by the antihaptoglobin serum (Gordon and Koj, unpublished).

Isolation and purification of free haptoglobin from human, rabbit or rat-plasma was described among others by Connell and Shaw,[143] Lombart et al.[144] and Mouray.[17] The principal step in these procedures depends on chromatography on DEAE-cellulose, ceruloplasmin being the main impurity. John and Miller[90] employed additionally preparative acrylamide gel electrophoresis but still their preparation from rat serum was slightly contaminated with α_1-acid glycoprotein.

Isolation of HpHb complex offers certain advantages, especially in experiments with radioisotopes. Krauss and Sarcione[35] employed for this purpose starch block electrophoresis and column chromatography on DEAE-cellulose. Kluthe and Isliker[145] suggested precipitation of HpHb complex by zinc acetate with following chromatography on CM-cellulose. An elegant procedure has been devised by Dobryszycka et al.[146] for human, rat or rabbit serum. Diluted human serum is applied on the top of DEAE-cellulose column equilibrated with 0.01 M acetate buffer pH 4.7. In these conditions free haptoglobin is bound to cellulose while majority of serum proteins are recovered in the eluent. After additional washing of the column with the buffer of slightly higher ionic strength a calculated amount of haemoglobin is used for eluting HpHb complex which shows a smaller affinity to cellulose than free haptoglobin. In this way a high purification of haptoglobin in a single step is possible. Pure HpHb complex shows the following extinction coefficients[146] at concentration 1 mg per 1 ml: $E_{278} = 1.45$, $E_{406} = 2.90$.

4.4.2. CHARACTERISTIC FEATURES OF HAPTOGLOBIN TURNOVER

Mouray et al.,[74] Krauss and Sarcione,[35] Alper et al.[75] and Peters and Alper[147] have published good evidence that the liver is the sole site of haptoglobin synthesis. Normal values of haptoglobin plasma level in man vary in rather wide range around 1 mg/ml and are markedly above those found in other animals such as rabbits or rats.[144] The distribution

4

of haptoglobin in body compartments is probably related to the molecular weight of haptoglobin variants: for Hp 1-1 (m.w. 85 000) the
ratio E.V./I.V. is close to one[58] while in the case of Hp 2-2 (m.w. above
220 000) the extravascular pool constitutes only ¼ of plasma haptoglobin.[61] Krauss, Schrott and Sarcione[59] found a constant fractional
catabolic rate of haptoglobin despite great variations observed in Hp
level in plasma (cases of Hodgkin's disease 1.4–7 mg Hp/ml). Also

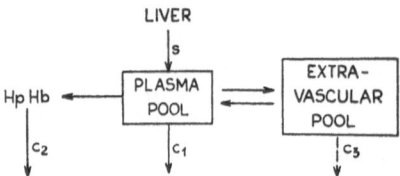

FIG. 4.4. Diagrammatic representation of haptoglobin turnover. s-
synthesis of haptoglobin; c_1 catabolism of free haptoglobin; c_2-cata-
bolism of haptoglobin-haemoglobin complex (HpHb); c_3-participation
of haptoglobin in tissue repair, (after Koj,[135] courtesy of Ciba Foundation).

Mouray[17] in rabbits with local inflammation (plasma haptoglobin
concentration exceeding in some cases normal values by 30 times)
estimated the fractional catabolic rate as constant and equal to 17–20%
of total body pool. On the other hand Freeman[58] concluded that the
fractional turnover rate of this protein is inversely proportional to its
concentration in plasma, since the absolute mass catabolized per day
remains approximately constant. The discrepancy may arise from the
fact that, according to Noyes and Garby,[148] in normal conditions at
least 50% of haptoglobin is eliminated after binding to haemoglobin
(Fig. 4.4), and this value may be further increased in haemolytic conditions.[61]

This complex (HpHb) is promptly removed from the circulation,[149–152] so that enhanced intravascular haemolysis will increase the
catabolism of haptoglobin. Engler, Moretti and Jayle[153] observed that
at high concentrations of HpHb in the blood of rabbits the elimination
rate is constant and amounts to 25 mg/hr. This may represent the effect
of saturation of the reticuloendothelial system which is believed to be
responsible for removing the complex. At lower concentrations HpHb
disappears from the plasma at an exponential rate with a half-life of
approximately 90 min, whereas the half-life of free haptoglobin is
around three days.

Noyes and Garby[148] infused intravenously healthy men with haemoglobin and attempted to estimate the rate of haptoglobin synthesis
from the speed of return to normal levels. Their estimates ranging

between 30 and 50 per cent of the intravascular pool synthesized per day indicate that haptoglobin belongs to plasma proteins with very fast turnover. The importance of the haemoglobin pathway in the turnover of haptoglobin is well illustrated by the fact that acceleration of red cell destruction, occurring in certain haemolytic diseases, may lead to ahaptoglobinaemia.[154] Noyes and Garby[148] emphasize that, in the absence of inflammation, plasma haptoglobin determinations are at least as sensitive as radiochromium-labelled red cell survival in detecting haemolysis.

Other interesting biochemical, genetic and physiological data concerning haptoglobin may be found in the relevant chapter of this book, as well as in monographs by Allison and Rees,[155] Nyman,[139] Jayle an Moretti,[156] Mouray,[17] Bocci[37] and Sutton.[157]

4.4.3. SYNTHESIS RATE OF HAPTOGLOBIN AFTER INJURIES

It is noteworthy that at the beginning of the inflammatory reaction a significant drop in the haptoglobin content in plasma occurs, as judged by determinations of peroxidase activity.[158,159] Engler and co-workers[160] demonstrated that this is not caused by migration of haptoglobin to the damaged site but rather is due to the appearance in the blood of a substance inhibiting formation of HpHb complex. The inhibitor derives most probably from the altered tissues and can be dissociated from haptoglobin on a DEAE-cellulose column.

The first signs of increased synthesis of haptoglobin after injury appear with a delay of 4–6 hours, and the rise of haptoglobin concentration is then very rapid. Typical time-response curves in acute inflammation in rats and rabbits are shown in Figs. 4.5a and 4.5b.

Fig. 4.5a indicates that rats injected with turpentine and simultaneously irradiated with X-rays develop lower haptoglobin level, so clearly the two types of injuries are not additive. On the other hand Mouray[17] found that in healthy rabbits the haptoglobin response was proportional to the dose of turpentine injected subcutaneously in the amount 0.2–2 ml per animal. Robert et al.[161] observed that connective tissue disorders induced by inflammatory granuloma, acute scurvy or intravenous injections of enzymes degrading connective tissue (elastase, hyaluronidase) lead to marked increases of haptoglobin level. Haptoglobin formation was also stimulated by injections of tissue extracts, but the effect was probably indirect since it could not be reproduced during liver perfusions.[162] On the other hand, local ulceration without great inflammatory reaction caused by subcutaneous injection of 2 N HCl, failed to produce marked elevation of haptoglobin.[15] The response of serum haptoglobin to inflammation is greatly impaired in the absence of the adrenals[163] and in general malnutrition.[17] Dobryszycka et al.[23] studied the incorporation of ^{14}C-glucosamine and ^{35}S-methionine into human and canine haptoglobin in inflammatory or traumatic conditions, and observed faster turnover rates of the carbohydrate moiety of hapto-globin in comparison with the polypeptide part of the molecule. These results are rather difficult to reconciliate with present views on the synthesis and breakdown of proteins as an irreversible process occur-ring in one direction only, and may perhaps be explained by various rates of reutilization of the ^{14}C and ^{35}S labelled precursors in the organism.

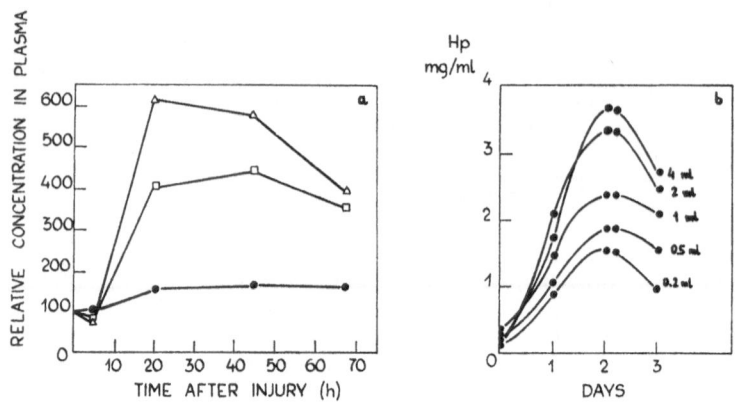

FIG. 4.5. Typical time-response curves of haptoglobin. 5a-relative increase in plasma haptoglobin in 3 groups of injured rats; ●—● rats irradiated with 800 R X-rays; △-turpentine injected rats; □ irradiated and turpentine injected rats (after Koj[159]). 5b-changes in the level of plasma haptoglobin in rabbits receiving at time 0 subcutaneous injections of turpentine oil in the amounts indicated (after Mouray[17]).

The haptoglobin concentration is usually elevated in the blood of patients with infectious diseases, cancer, leukaemia, as well as after surgery and burns,[23,26,54,156,164] and haptoglobin is regarded as a very sensitive indicator of disease persistence.

4.5. CERULOPLASMIN

4.5.1. DETERMINATION OF CERULOPLASMIN AND ITS ISOLATION

This α_2-globulin, accounting for the major portion of copper in plasma, shows oxidase activity against a number of substrates includ-ing p-phenylenediamine (PPD).[165] Several methods for quantitative assay of ceruloplasmin are based on this oxidase activity.[166-169] Bingley and Dick[170] drew attention to the importance of a constant pH in the incubation mixture, especially when unbuffered PPD.2HCl is used as substrate. These authors found that maximum activity is exhibited by human ceruloplasmin at pH 5.9, by that of rat at 6.2 and rabbit 6.4. Optimal conditions for ceruloplasmin estimations have been recently evaluated by Sunderman and Nomoto.[169] The reactions catalysed by ceruloplasmin are activated by certain ions and inhibited by metal chelators (for references see McDermott et al.[171]). Kinetic considerations of the reaction catalysed by ceruloplasmin may be found among others in a paper by Curzon.[172]

In the method of ceruloplasmin estimation according to Ravin[167] 0.1 ml of serum is pipetted to a test tube followed by 8 ml of 0.4 M acetate buffer pH 5.5. Then 1 ml of substrate (fresh 0.5% solution of PPD in water) is added and the mixture incubated at 37°C for 60 min. Finally the reaction is inhibited by adding 1 ml of 0.5% sodium azide and O.D. is determined at 530 nm against the appropriate blank. The amount of ceruloplasmin can be calculated from the standard curve obtained with pure ceruloplasmin.

The presence of ceruloplasmin in particular fractions of serum separated by disc electrophoresis or immunoelectrophoresis can be demonstrated either by incubation of the gel with PPD, or with ferrous sulphate and differential staining.[173]

Highly purified ceruloplasmin has been obtained by many authors.[174–178] Almost all the methods take advantage of the high affinity of ceruloplasmin to DEAE-cellulose or Sephadex. In the procedure described by Stokes[177] a sample of Cohn fraction IV-1 from human plasma is allowed to soak through a small DEAE-Sephadex column equilibrated with 0.02 M sodium pyrophosphate-acetic acid buffer pH 7.5. After washing out of inactive proteins the ceruloplasmin of approximately 65% purity is eluted by 0.25 M NaCl solution. The remaining impurities are removed by chromatography on a CM-Sephadex column in 0.02 M sodium acetate buffer pH 5.5, with a final yield of approximately 70%.

Because of the blue colour of ceruloplasmin solutions the purity of preparations can be easily estimated on the basis of optical density ratio at 610 and 280 nm equal to 0.042–0.047 for pure ceruloplasmin.[174,175,178]

4.5.2. PLASMA CONCENTRATION AND TURNOVER OF CERULOPLASMIN IN NORMAL INDIVIDUALS

The mean serum ceruloplasmin level of 32 ± 4.9 mg per cent was found in human adults by Ravin[167] by the enzymatic method. A very similar value (30.4 mg per cent) was reported by Cox,[174] but she observed age-dependent variations with the maximum in early childhood. The overall picture of ceruloplasmin is that of very low levels in the newborn, rising to adult levels in the first year and continuing to rise to a peak at 2 or 3 years, then declining to adult levels by 12 years of age. Adolescent males show a further slight decline and subsequent rise to adult levels. Ceruloplasmin levels appear to be influenced by genetic factors both in man and animals.[179,180] A pronounced inborn deficiency of this protein is characteristic of Wilson's disease (hepatolenticular degeneration).[181]

Ceruloplasmin has been detected in the plasma of many species of vertebrates.[168,170,182] When the values reported by Bingley and Dick[170] are expressed in relative terms (man = 100%) it appears that fowl blood shows only 43% of this level, rabbit 51%, sheep 182%, rat 205% and pig 292%.

Evans and co-workers[183–186] established that hormones of the pituitary-adrenal system markedly affect plasma copper and ceruloplasmin in the rat. Thus, one of the consequences of adrenalectomy or hypophysectomy is an increased ceruloplasmin, as measured by serum oxidase activity, with a restoration of normal levels when the animals are given corticosterone. On the basis of these results Evans and Wiederanders[183] have suggested that adrenalectomy removes an inhibitor, presumably a steroid, of ceruloplasmin biosynthesis. In contrast to these results with rats, the treatment of chicks with hydrocortisone increases the ceruloplasmin level from the very low normal values found in this species.[187] Recently Evans and his group[185] reported that removal of thyroid or injections of estradiol also produce a significant increase in serum ceruloplasmin. Loading of adrenalectomized rats with copper sulphate results in accumulation of Cu^{+2} ions in the liver[188] and this may directly induce ceruloplasmin synthesis.[186]

Turnover of iodine-labelled ceruloplasmin in man was studied by Kekki and Koskelo.[62,63] According to their observations the average daily catabolism of ceruloplasmin in healthy individuals amounts to approximately 6 mg per kg body weight. Distribution of ceruloplasmin in plasma and interstitial fluid, as well as its relative turnover rate, show some similarities with haptoglobin (see Table 4.3). Terminal sialic acid residues in the carbohydrate chains of ceruloplasmin may be important for the normal catabolism of ceruloplasmin, as suggested by elegant studies of Morell, Scheinberg, Ashwell and co-workers[176,189,190] Removal of sialic acid by neuraminidase from ceruloplasmin, labelled with ^{64}Cu or ^{3}H, resulted in the rapid disappearance of this protein from the circulation in rabbits and rats. This seems to be a rather general rule since desialylated orosomucoid, haptoglobin, fetuin, α_2-macroglobulin and some other plasma glycoproteins are also promptly removed from the blood and taken up by hepatocytes.[101]

4.5.3. INCREASED SYNTHESIS OF CERULOPLASMIN IN PATHOLOGICAL STATES

In 1955 Wintrobe and co-workers[191] demonstrated that oxidase activity of ceruloplasmin is elevated in sera of pregnant women, patients with infectious diseases and the nephrotic syndrome. Rice[192,193] found a good correlation between serum ceruloplasmin and other acute-phase reactants, especially C-reactive protein, in various pathological states. Sass-Kortsak[194] reviewed the effect of infection, disease, pregnancy and hormone administration on the ceruloplasmin level and copper metabolism. Increased level of ceruloplasmin in the blood of patients with rheumatoid arthritis was observed by Niedermayer[195] and Koskelo et al.,[196] while McCathie et al.,[25] Werner and Odenthal[11] and Werner[54] reported a similar phenomenon after surgery, myocardial infarction and in some other diseases. The stress reaction caused by injections of pyrogens over a period of time is known to increase ceruloplasmin and plasma copper concentration.[184,197] Evans et al.[198] demonstrated that inhibitors of RNA and protein synthesis (actinomycin D and cycloheximide) inhibit the plasma ceruloplasmin increase occurring in rats subjected to swimming-invoked stress. These results, however, cannot be interpreted in terms of de novo synthesis of ceruloplasmin since the experiments were of very short duration and it is known that in the rat at least 30 min are required for the liver-produced plasma proteins to appear in the blood.[77]

4.6. SEROMUCOID FRACTION AND ITS COMPONENTS

4.6.1. GENERAL CONSIDERATIONS

Approximately 1% of serum proteins (containing 10% of the total protein-bound hexose and hexosamine) is not precipitated from normal human serum by boiling at pH 4.5 or by sulphosalicyclic or perchloric acids. Due to the simplicity of isolation this "seromucoid" fraction has been investigated by many authors both in respect of composition and changes in concentration in various pathological states (for references see Winzler[6,12,13]). The spectrum of diseases showing elevated seromucoid levels is very broad and includes acute inflammation, various forms of trauma, infectious diseases, cancer etc.

The most commonly used procedure for the determination of seromucoid fraction in serum is that described by Winzler:[6] 0.5 ml of serum is mixed with 4.5 ml of 0.85% saline and then 2.5 ml of 1.8 M perchloric acid is added with shaking. Exactly 10 min later the precipitate is centrifuged or filtered, and the proteins soluble in perchloric acid are precipitated with 1 ml of 5% phosphotungstic acid in 2N HCl. After 10 min the sample is centrifuged, decanted, and the amount of seromucoid precipitate determined either from the protein content (by biuret or Lowry et al methods), from "tyrosine" content (with Folin-Ciocalteau reagent), or from hexose content (e.g. with orcinol). Normal values of seromucoid concentration in human serum as reported by Winzler[6] are: 61.2 mg% (as protein), 12.4 mg% (as hexose) and 3.38 mg% (as tyrosine).

It should be emphasized that the procedure does not give a quantitative measure of the seromucoid fraction components since some of them are co-precipitated by perchloric acid with other serum proteins.

Moreover, tracer experiments indicate that at higher seromucoid concentration the effect of "solubilization" of other plasma proteins in perchloric acid may occur as well.[199] For this reason the turnover or metabolic data concerning seromucoid should be treated with extreme caution.

The seromucoid fraction is very heterogenous and Robert et al.[200] demonstrated the presence of at least seven components by means of immunoelectrophoresis, while Doležalova et al.[201] separated the sulphosalicylic extract of human serum into eleven fractions by chromatography on DEAE-cellulose. A method for subfractionation of the $HClO_4$-soluble proteins by means of ammonium sulphate is described by Schultze and Heremans.[1] The main components of the seromucoid fraction are: α_1-acid glycoprotein, α_1-antitrypsin, haptoglobin, some other α and β globulins and small amount of albumin.

The relative amounts of proteins and carbohydrates in the serormucoid fraction are fairly constant in most normal and pathological sera. However, in certain conditions, such as liver or renal diseases and rheumatoid arthritis, the ratio of carbohydrate to protein in seromucoid may be distinctly elevated (cf. Winzler[6]), while in others, such as diabetes[202] or silicosis[203] this ratio may be reduced. The most likely explanation of this phenomenon is that in those conditions the proportion of carbohydrate-rich and carbohydrate-poor components in the seromucoid fraction is altered, although the possibility of synthesis of modified glycoprotein molecules cannot be entirely ruled out at present.

4.6.2. α_1-ACID GLYCOPROTEIN

α_1-acid glycoprotein (synonym = orosomucoid) is a well characterized constituent of normal human plasma.[204–207] Several authors obtained highly purified orosomucoid from human, rabbit or rat plasmas by utilizing ammonium sulphate fractionation and chromatography on DEAE-cellulose.[208–211] Physicochemical properties of this protein are described in detail by Jeanloz.[207]

Extensive studies of Schmid and associates showed that α_1-acid glycoprotein exhibits polymorphism although it appears homogenous as judged by classical criteria of purity.[212–216] After treatment with neuraminidase to cleave off sialyl residues, and after subsequent starch gel electrophoresis at pH 5, three different patterns are obtained in Caucasian patients (phenotypes I, II and III). Each variant has the same gross amino acid and carbohydrate composition and exhibits a sedimentation constant 2.7 S identical with native protein. The relative distribution of these patterns was found to be very similar in normal Caucasian adults and 44 patients with various diseases (including cancer, leukaemia, diabetes, rheumatoid arthritis). Moreover, it was shown that patients subjected to hypophysial stalk section operation or to uterectomy and irradiation (treatment of the cancer) preserved their type of α_1-acid glycoprotein variant despite considerable changes in the plasma level of this protein.

Several authors reported plasma concentrations of α_1-acid glycoprotein in healthy individuals around 75–100 mg/100 ml (*cf.* Schultze and Heremans[1]) but Schmid *et al.*[213,215] gave lower values (around 46 mg%). Surgery may double plasma level of orosomucoid[10,11,215] and a sustained elevation of this protein in burned patients was described by Zeineh and Kukral.[24] It is generally accepted that α_1-acid glycoprotein is responsible to a great extent for the increased concentration of seromucoid fraction observed in various pathological states.

The turnover rate of α_1-acid glycoprotein has been studied in normal man by Weisman *et al.*[56] and in burned patients by Zeineh and Kukral.[24] The picture is far from being complete but it appears that the half-life of iodine-labelled orosomucoid, 5–6 days, is not altered significantly by parenchymal liver disease (low level of α_1-acid glycoprotein) or inflammation (increased level of this protein). By the analysis of results obtained with ^{14}C-glucosamine incorporated into α_1-acid glycoprotein *in vivo*, Zeineh and Kukral[24] came to the conclusion that the half-life of this protein in burned patients is only around two days and increases during healing to 3.5 days. These authors observed, moreover, that some of the burned patients excreted with urine considerable amounts of orosomucoid, especially during the first days after injury. The ratio between intravascular and extravascular pool of α_1-acid glycoprotein seems to be unchanged by acute inflammation in the dog.[217]

4.6.3. α_1-AP GLOBULIN (RAT)

Apart from orosomucoid rat serum contains another α_1-glycoprotein behaving as typical AP-reactant and investigated in detail by Darcy.[218–223] Its relation to the seromucoid fraction is not quite clear but Gordon and Louis[2] differentiated it from α_1-acid glycoprotein on the basis of isoelectric point and amino acid composition. This α_1-AP globulin contains approximately 15% of carbohydrate and in isoelectric fractionation shows heterogeneity which disappears after treatment with neuraminidase. Its level in normal rat blood, estimated as approximately 1.2 mg/ml,[124] is greatly increased by turpentine injection, ischaemic necrosis or tumour implantation.[219] The 48 hr response of α_1-AP globulin is proportional to log turpentine dose and to the granuloma weight.[223] The liver is the sole site of α_1-AP globulin synthesis.[90,91,124,224] The trauma-induced synthesis of this protein is inhibited by puromycin and actinomycin D,[221] but increased by adrenalectomy.[124,219]

4.7. ALPHA-1-ANTITRYPSIN

4.7.1. TRYPSIN INHIBITORS IN BLOOD PLASMA

It has long been recognized that mammalian serum has the capacity to inhibit the enzymatic activity of various proteases. Trypsin can be

4A

inhibited by three or four, probably different plasma proteins desig-
nated as α_1-antitrypsin (α_1-AT), α_2-macroglobulin, the inter-α-trypsin
inhibitor and α_2-antitrypsin.[225–228] α_1-antitrypsin is responsible for
approximately 90% of total trypsin inhibiting capacity of human serum
and its reaction with trypsin is stoichiometric. There is a good evidence
that α_1-AT shows affinity to various proteases and inhibits not only
trypsin and chymotrypsin but also proteolytic enzymes from leucocytes[229,230,231] and Aspergillus oryzae,[232] pancreatic and skin elas-
tases[233,234] plasmin[235,230] and thrombin.[235] On the other hand some
reports indicate that it has no effect on hyaluronidase[236] or bacterial
proteases from Staphyloccus aureus or Proteus vulgaris,[229] and its
action on plasmin is also questioned.[102,236] Recently it has been shown
that the serum of patients with α_1-AT deficiency possesses less binding
capacity for plasmin and thrombin.[237] However, Bieth et al[227] could
not demonstrate any competition between plasmin and trypsin or
chymotrypsin for inhibitors of proteolysis from human serum. This may
indicate that α_1-AT has different binding sites for plasmin and for
trypsin, but the problem requires further study.

4.7.2. DETERMINATION AND ISOLATION
OF α_1-ANTITRYPSIN

Alpha-1-antitrypsin (synonyms: α_1-trypsin inhibitor, α_1-3.5 glycoprotein) is usually estimated either by radial immunodiffusion[3,51,238]
or by measuring the trypsin inhibiting capacity of serum. In the latter
test the residual activity of trypsin (with proteins or synthetic peptides
as substrates) is measured after the enzyme had been incubated for a
constant period of time with a dilution of the serum to be tested.[227,239]
This method is simple but it cannot differentiate between various inhibitors present in human serum. Moreover, it should be remembered
that depending on the substrate serum may not only inhibit trypsin but
also protect it from the effect of other inhibitors.[227]

The α_1-AT concentration in serum varies over a rather broad range.
The mean values for normal human subjects determined by radial
immunodiffusion are between 212 and 287 mg/100 ml,[3,51,225] although
recently Makino and Reed[238] reported higher figures based on a large
number of analyses: 310 mg% in men and 336 mg% in women.

Isolation of α_1-AT from serum is a difficult procedure and the results
are not always satisfactory. Bundy and Mehl[240] achieved 67-fold purification (with 4% yield) by employing ammonium sulphate precipitation,
chromatography on Dowex-1 and zone electrophoresis on starch gel,
while Schultze et al. (cf. Schultze and Heremans[1]) used Rivanol as a
precipitant. Wu and Laskowski[241] obtained crystalline acid-labile
trypsin inhibitor from bovine plasma by a four-step procedure. Column
chromatography on DEAE-Sephadex and preparative electrophoresis
on starch gel were used by Shamash and Rimon[242] for purification of an
α_1-globulin inhibiting trypsin and plasmin, probably identical with

α_1-AT. Kueppers[3] isolated apparently pure α_1-AT by electrophoresis on a Pevikon block with subsequent filtration on a Sephadex G-200 column, but one of the steps included removal of sialic acid by neuraminidase. Such desialylated α_1-AT preparation preserved its antigenic properties and after labelling with ^{125}I behaved *in vivo* as a native protein.[57] This is rather an exceptional situation as indicated by recent studies of Morell *et al.*[101] on desialylated plasma glycoproteins.

4.7.3. INBORN DEFICIENCY OF α_1-AT

Laurell, Eriksson and Kueppers were the first to describe the hereditary α_1-AT deficiency and its association with obstructive lung disease.[243–245] In recent years the syndrome has been recognized with increasing frequency and investigated in detail.[3,57,229,238,244] The members of families with α_1-AT deficiency show a trimodal distribution of α_1-AT concentration: normal individuals—220 mg/100 ml, heterozygotes—120 mg/100 ml, and deficient homozygotes—25 mg/100 ml[3]. Fagerhol and co-workers[247,248] suggested the presence of different hereditary variants of α_1-AT detectable during starch gel electrophoresis at acidic pH. This method led to the detection of several genotypes and construction of a genetic model termed P_i (proteinase inhibitor system). In this nomenclature apart from the most common or normal gene P_i^M there exist two deficiency genes P_i^Z and P^S, responsible for the production of electrophoretically distinguishable proteins.[228]

It has been well established that deficient individuals are prone to develop severe pulmonary emphysema beginning at an early age. The pathogenic mechanism is not entirely elucidated but it seems that proteolytic enzymes derived from leucocytes, tissues or bacteria may become overoperative *in vivo* when the serum inhibitory capacity is reduced. The failure of inhibition of proteolytic enzymes may lead to increased breakdown of alveolar connective tissue. It has been reported that human lung tissue, unlike bovine lung, does not possess a specific protease inhibitor. The constant physical stress of the lung during ventilation precipitates gross structural defects presenting as emphysema. Similar defects have not been detected until now in any other organ of patients with α_1-AT deficiency. The physiology, genetics and pathology of α_1-AT have been thoroughly reviewed by Kueppers[228] and in the chapter by Davies in this book.

4.7.4. EFFECT OF INJURIES ON α_1-AT

Several authors demonstrated that the trypsin-inhibiting property of α_1-globulin fraction in serum, or the level of α_1-AT, considerably increases in a variety of pathologic and physiologic conditions such as local inflammation, surgery, malignant tumours, injection of bacterial endotoxin, nephritis, pregnancy, and even during administration of contraceptive hormones.[10,11,54,102,193,226,249–521] The rise of α_1-AT concentration following trauma is quite steep and maximum values (twice the normal level) are reached in human subjects on the third day after gastrectomy[11] or after injection of typhoid vaccine.[226,249] Other serum trypsin inhibitors do not change appreciably in this period. Heterozygotes for the deficiency gene have an elevation of α_1-AT but it reaches barely 50 per cent of the response in normal individuals (Fig. 6.6).

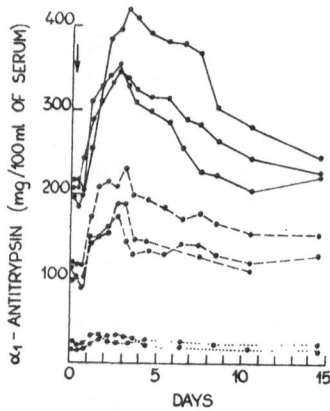

Fig. 4.6. Changes in α_1-antitrypsin levels of serum following the intravenous injection of 0.2 ml of typhoid-paratyphoid vaccine (arrow) in the genetically different individuals. Homozygotes for common gene: solid line, heterozygotes for α_1-AT deficiency gene: dashed line, homozygotes for deficiency gene: dotted line (after Kueppers[226]).

The turnover of α_1-AT has not been investigated in detail. Kueppers and Fallat[57] injected ^{125}I-labelled α_1-AT into 3 normal and 2 deficient subjects and observed that plasma half-lives of the protein were equal to 3.9 and 6 days, respectively. The results in healthy men compare fairly well with those for haptoglobin or ceruloplasmin, although further studies are required to demonstrate whether the difference between normal and deficient subjects is significant. The experiments convincingly proved that the deficiency of α_1-AT is due to the primary defect of synthesis and not catabolism. Very similar results on the elimination of α_1-AT from plasma were obtained by Makino and Reed.[238] The rate of disappearance of α_1-AT infused into deficient subjects correspond to a half-life between 5 and 6 days. Small amounts of exogenous α_1AT were demonstrated in the patient's sputum.

4.8. C-REACTIVE PROTEIN

Human C-reactive protein (CRP) and its rabbit analogue, C_x-reactive protein (C_xRP), are acute-phase components which are normally undetectable but appear in serum in response to injuries. Tillet and

Francis[252] were the first to observe that the serum of patients with acute inflammatory reaction contains a protein yielding precipitate with the C polysaccharide from the pneumococcus. A simple method of CRP estimation described by Anderson and McCarthy,[253] and used for a long time in clinical studies, depends on measuring the amount of precipitate formed after mixing an examined serum with the serum of animal immunized against the C polysaccharide. Nilsson[254] demonstrated that the capillary precipitation test is far less sensitive than the radial diffusion procedure now commonly employed for CRP determination. Concentration of CRP in the blood is usually expressed in arbitrary units but Crockson et al.[10] reported 0.5 mg CRP/ml plasma as maximum level of this protein in patients after surgical trauma. Early clinical and experimental studies on C-reactive protein were exhaustively reviewed by Hedlund.[255]

CRP is detected in blood in all febrile inflammatory diseases and its correlation with increased levels of haptoglobin, α_1-acid glycoprotein or ceruloplasmin has been firmly established.[193,256] The effect of surgical trauma on CRP was studied in recent years among others by Crockson et al.[10] and by Werner and Odenthal.[11] CRP was detected in all cases within 24 hr after operation and reached highest concentration usually on the second day. In the absence of postoperative complications it disappeared from the blood after 10–14 days.

The CRP test is not specific in that it accompanies any acute phase of inflammation independently of its aetiology. On the other hand, the positive CRP test occurs earlier than conspicuous increases of other AP-reactants or erythrocyte sedimentation rate;[10] hence it may be employed in early diagnosis, e.g. in heart infarct.

The classical method for the isolation of human CRP or rabbit C_xRP depends on their ability to form insoluble complexes in the presence of calcium ions with the somatic C or C_x polysaccharide of the pneumococcus. Hokama and Riley[257] satisfactorily purified human CRP by chromatography, while Ganrot and Kindmark[258] described a two-step procedure consisting of an initial adsorption of CRP onto $BaSO_4$ and elution with citrate followed by a second adsorption to, and elution from agar. Salting out with ammonium sulphate and subsequent filtration on Sephadex G-200 were employed by Croxatto et al.[259] Gotschlich and Edelman[260] obtained crystalline CRP from pleural and peritoneal fluids. The principle of affinity chromatography has been employed by Davis et al.[261] and Riley and Coleman[262] to isolate C-reactive proteins of man, monkey, rabbit and dog, although the yield was rather low.

Physicochemical properties of C-reactive proteins were studied among others by Gotschlich and Edelman[260] and Kushner and Sommerville.[4] The approximate molecular weight of CRP as it exists in serum lies between 135 and 140 000. The molecule is probably composed of 6 identical subunits of m.w. equal to 23 000 each. The presence in serum of a pentameric form as well was suggested by an asymmetric elution pattern on gel filtration. Riley et al.[263] demonstrated that CRP may dissociate during isolation and purification. The approximate m.w. of C_xRP is between 114 000 and 123,000 with no direct evidence for multiple polymeric forms. On the other hand Hokama et al.[264] observed in some cases, in rabbit serum subjected to immunoelectrophoresis two precipitin bands, in the β and γ regions, which might indicate heterogeneity of this protein.

The liver is known to be the site of synthesis of C-reactive protein. Hurliman et al.[18] using radioimmunoelectrophoretic techniques demonstrated in vitro incorporation of ^{14}C-labelled amino acids into CRP by cultures of rabbit, monkey and human liver tissue. The animals were stimulated to produce the acute-phase protein by injection of turpentine, croton oil, endotoxin, paratyphoid vaccine or pneumococci. Only livers taken 16–24 hours after these stimuli were active. Similar conclusions on the site of CRP formation were recently reached by Riley et al.[265] who employed a different method.

In a series of publications Hokama and associates[264,266] drew attention to the relationship between the synthesis of CRP and haemoproteins, such as catalase, in the liver. They demonstrated the occurrence of common antigens between CRP and human hepatic catalase, as well as similarities of the pattern obtained in disc polyacrylamide electrophoresis and tryptic peptide maps of these two proteins. When hematin was administered to rabbits a rise in both liver and kidney catalases ensued with a concomitant decrease in serum C_xRP. Purified CRP interacted with hematin to give variable peroxidatic activity. These authors suggest that C-reactive protein may represent a precursor for the formation of haemoproteins.

4.9. ALPHA-2-ACUTE PHASE GLOBULIN OF THE RAT

In the years 1961–65 several authors demonstrated independently that the serum of adult healthy rats is devoid of a specific α_2-glycoprotein which can be detected in foetal, neonatal, pregnant, tumour-bearing and injured rats. Beaton et al.[267] showed the protein to have the mobility of an α_2-globulin in paper electrophoresis, but because it migrates more slowly than β globulin in vertical starch gel electrophoresis it has been designated as slow α_2-globulin. The same term was used by Heim[268-270] while other authors employed different names for this apparently identical protein: "abnormal" serum globulin— Lawford,[271] α_2-glycoprotein (GP)—Bogden et al.[272] and Stanislawski et al.,[273] α_2-acute-phase globulin (α_2-AP)—Weimer and Benjamin,[16] α_2-macroglobulin—Boffa et al.,[274] and foetal globulin—van Gool and Ladiges.[275] Relatively simple methods of detection of α_2-AP globulin in rat serum are based on starch gel electrophoresis[276] or immunoelectrophoresis,[16] while quantitative determinations require radial immunodiffusion.[19,277] Purified α_2-AP globulin is not essential for the preparation of the antiserum and probably for this reason rather few attempts were made to isolate α_2-AP globulin by typical biochemical methods.[16,278] Recently Menninger et al.[279] obtained homogenous and pure α_2-AP globulin from the serum of croton oil-treated rats by a three-step procedure including salting out at 1.75 M ammonium sulphate concentration, gel filtration on a column of Sephadex G-200 and preparative starch block electrophoresis. Physicochemical properties of this protein were not investigated in detail but Heim[4] estimated its molecular weight as being around $0.9 - 1.0 \times 10^6$ and demonstrated its immunological similarity to α_2-macroglobulin of various mammalian species.

Branceni and Gonin[277] measured the level of α_2-AP globulin in rats at different ages. Assuming the concentration of α_2-AP globulin in newborn rats as equal to 100 they found it to decrease by the 14th day of life to 5 % and later below 1 %, except in pathological states. Sarcione and Bohne[280] established the immunological identity of α_2-AP globulin from foetal, neonatal, pregnant and injured rats. According to some authors[273,275] α_2-AP globulin represents a larger group of foetal proteins which are normally absent in adult life but may appear in certain conditions, especially in primary liver carcinoma.

Overwhelming evidence indicates that the appearance of α_2-AP globulin in the blood of injured rats is accompanied by a similar rise of other acute-phase reactants. Branceni and Gonin[277] reported that in adjuvant arthritis the level of α_2-AP globulin is the best humoral indicator of the inflammatory process. On the other hand, Heim[281] observed that rats irradiated with 600 r of gamma rays did not respond with formation of α_2-AP globulin. This was confirmed by John and Miller[282] who reported that rat liver isolated from animals subjected to ionizing radiation synthesized other AP-reactants but not α_2-AP globulin. In general, small injuries are insufficient to stimulate α_2-AP globulin formation;[268] van Gool and Ladiges[275] found that skin incisions were without effect in contrast to deeper injuries involving muscles.

By using perfusion technique[280,283] or tissue cultures[93] it was demonstrated that the liver is the site of α_2-AP globulin synthesis in injured adult rats. The synthesis in the stimulated liver occurs promptly and α_2-AP globulin can be detected in intact rats as early as 8 hr after injury.[16] The kinetics of formation of α_2-AP globulin in the perfused liver was studied among others by John and Miller[90] and Sarcione.[284,285] A pronounced inhibition of the α_2-AP globulin response to trauma has been observed in adrenalectomized rats.[16,31,48,272,286] The effect of adrenalectomy may be reversed by administration of corticosterone,[31] but large doses of glucocorticoids also inhibit synthesis.[48] John and Miller[90] were able to induce formation of α_2-AP globulin in the isolated normal rat liver during prolonged perfusion in the presence of insulin, cortisol and high concentrations of amino acids.

All this indicates that synthesis of α_2-AP globulin in injured rats represents the reversion to some foetal biosynthetic pathway rather than formation of a new abnormal protein.

4.10. MECHANISM OF INDUCED SYNTHESIS OF ACUTE-PHASE REACTANTS

4.10.1. LOCAL AND SYSTEMIC RESPONSE TO INJURY

Trauma-induced synthesis of plasma glycoproteins in the liver represents a part of systemic response of an animal to injury and hence should be analysed on a broad background of the relevant data supplied by general pathology. Unfortunately this approach exceeds the scope of the present review and the reader interested in the inflammatory process should consult one of the available books, such as that edited by Zweifach, Grant and McCluskey,[287] monographs,[288] or numerous articles concerning this subject.[20,289-291]

When the principal events between the occurrence of trauma and the appearance of newly synthesized AP-reactants in the blood are considered they may be presented in the form of the following chain (p. 103).

It must be remembered that the time scale in this scheme based on experiments with rats is only approximate and for other animals or man it will be different.

Local reaction to injury includes such phenomena as venular dilation, endothelial leakage and oedema, platelet aggregation, fibrin formation, leucocyte accumulation, release of lysosomal enzymes from leucocytes and tissues, formation and release of small molecular weight mediators (histamine, 5-hydroxytryptamine, kinins), mesenchymal cell proliferation, and others. Systemic response includes fever, pain, leucocytosis, increased level of acute-phase reactants, increased ESR, increased function of the pituitary-adrenal system, decreased albumin and serum iron, and others. Glenn et al.[20] suggest that the local reaction has a promotive and contributory effect to the overall process of response to injury and leads to cell death and necrosis, whereas the systemic reaction may exert protective and inhibitory effects resulting in restitution. These authors visualized the entire phenomenon of inflammation as a "wave of leaky membranes": in endothelial cells, plasma cell membranes (leucocytes, mast cells, etc.), intracytoplasmic membranes (lysosomes, etc.).

4.10.2. LYSOSOMES AND LIVER STIMULATION BY HUMORAL FACTORS

It is known that various forms of trauma are associated with an increased permeability of lysosomal membranes and release of acid hydrolases.[292-294] The release of lysosomal enzymes has been referred to by Weissmann[295] as the "final common pathway" in inflammation.

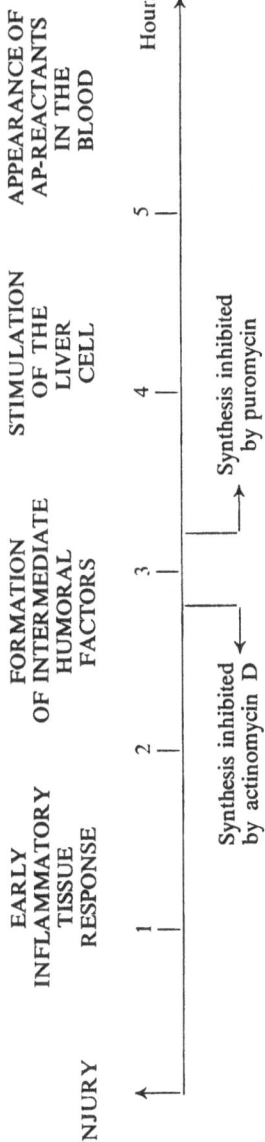

Fig. 4.7. Sequence of events in the AP-protein response.

Moreover, Koj and Allison[86] demonstrated that rabbits treated with various substances labilizing lysosomes (*e.g.* excess of vitamin A) exhibited higher plasma concentrations of fibrinogen, haptoglobin, ceruloplasmin and seromucoid fraction. Hence damage to tissue lysosomes may represent a common "trigger" mechanism linking variety of injuries with increased synthesis of AP-reactants.

Lysosomes may affect acute-phase protein synthesis in any one of several ways. The main possibilities are local effects at the site of injury, effects on cells in the circulation, or direct effects on the liver. Among the local effects can be considered release of lysosomal enzymes leading to the formation in tissues of a factor transported by blood to the liver, where it stimulates the synthesis of AP-reactants. Conceivably lysosomal cathepsins participate in the formation of such a factor; injection of papain can bring about the release of tissue proteases or mimic their effect.[86]

If primary effects of trauma were on lysosomes of blood cells they would probably be due to leucocytes. It is generally accepted that polymorphonuclear leucocytes participate actively in the inflammatory reaction and their hydrolytic enzymes are responsible for secondary tissue injury.[294,296–300] Direct effects of leucocyte lysosomes seem unlikely, since rabbits receiving intravenous injections of preparations obtained from rabbit leucocyte granules showed normal levels of acute-phase reactants (Koj, Regoeczi and Spitznagel, 1968, unpublished observations). On the other hand, Darcy[222] observed moderate increases in serum α_1-AP globulin when intact leucocytes or their subcellular fractions were injected into rats. Eddington *et al.*[301] obtained some positive results after intraperitoneal injections of suitably processed extracts from rabbit polymorphonuclear leucocytes active for pyrogenicity. These observations suggest that leucocytes may exert an indirect effect on the stimulation of the liver by formation (or release) of such chemical mediators as kinins,[302] pyrogens, histamine, or other not yet identified products of proteolysis. Histamine may be either released from mast cells by basic polypeptides present in granulocyte lysosomes[303] or formed at an increased rate by histidine decarboxylase. A significant adaptive increase in the activity of this enzyme has been found shortly after application of various stimuli known to increase the level of AP-reactants, such as local irritation, infection, trauma, administration of endotoxin, catecholamines, *etc.*,[304,305] Lazar and Karady[306] reported that the adaptation to trauma led to a significantly smaller increase of fibrinogen level in rats and at the same time to a decreased activation of histidine decarboxylase. However, experiments in which histamine was added to the perfused rat liver did not provide convincing evidence that this substance is the main mediator responsible for liver stimulation. Although the ratios of radioactivities in fibrinogen: albumin, α_1-AP globulin:albumin (Gordon and Koj, 1968, unpublished observations), and seromucoid:non-seromucoid fraction[307] were significantly higher in the system supplemented with histamine, there was no indication of increased absolute synthesis rates of these acute-phase proteins in the presence of exogenous histamine.

Whether histamine stimulates the synthesis of AP-reactants or not, it should be mentioned here that the necessity of intermediate humoral factors linking the site of peripheral injury and the liver cell has been recognized for a long time. In 1945 Homburger[308] observed that sterile pus from turpentine abscesses contains a thermolabile factor that causes the plasma fibrinogen level to rise when injected intravenously or intramuscularly. Later several authors reported the occurrence of stimulating factors in extracts of connective tissue[162] or in the blood from injured animals.[91,309] Detailed studies indicated, however, that the effect of tissue extracts is probably indirect since it cannot be reproduced during liver perfusion.[162] Also the effect of blood from injured rats appeared to be spurious and the positive results may be explained by some artifacts of tracer experiments.[124,199] Despite these findings, stimulatory factors are probably present in the blood of injured animals, and failure to detect them may be due to their instability and rapid removal from the circulation. Furthermore, blood collected at one time from injured animals may not contain enough of the hypothetical stimulatory factor to lead to a measurable increase in the rate of synthesis in normal liver. This suggestion is indirectly supported by the observations of Moolten and Bucher[310] on the transfer of a humoral agent stimulating the regeneration of rat liver. These authors found that carotid-to-jugular cross circulation between partially hepatectomized and normal rats stimulated incorporation of ^{14}C-thymidine into hepatic DNA in the normal partners when it was maintained for 19 hours at a flow rate 2 ml/min. Cross circulation for 7 hours or less was ineffective. Van Gool and Ladiges[275] reported that production of α_2-AP globulin was stimulated in normal rats after intravenous injection of plasma samples from blood perfusing for 2–3 hours an isolated hind limb.

The difference in the response patterns of particular AP-reactants do not necessarily imply that separate stimulatory factors are released from the site of injury, each responsible for the increased rate of synthesis of each individual protein. Thus differentiation may take place in the liver by one of several possible mechanisms, e.g. variations in the rate of formation or stability of specific mRNA, or hormonal control of protein synthesis.

4.10.3. HORMONAL REGULATION OF AP-PROTEIN SYNTHESIS

Trauma stimulates the pituitary-adrenal system, as well as other endocrine glands, and hormones should be also regarded as humoral factors transported by blood and influencing the synthesis of AP-reactants in the liver. Weimer and Coggshall[48] pointed out that adrenalectomy and administration of cortisol may have multiple effects on the response of serum glycoprotein fractions to tissue injury: (a) regulatory, (b) permissive, (c) homeostatic (anti-inflammatory), (d) catabolic, (e) stimulatory (anabolic), and (f) inhibitory (anti-anabolic). As indicated by the data presented in Table 4.4, based on experiments

TABLE 4.4

Various effects of adrenalectomy on the plasma glycoprotein response to tissue injury in rats* (after Koj[135])

Protein	Effect	Reference
α_2-AP globulin	Total inhibition	31, 48
Haptoglobin	Marked inhibition	163
Seromucoid and fibrinogen†	Slight inhibition	48, 124
Ceruloplasmin	Probably stimulation	183
α_1-AP globulin	Marked stimulation	219, 124

* In other species the response of individual AP-reactants may be different; for example Atencio and co-workers[113] observed that adrenalectomy in rabbits almost totally abolished the fibrinogen increase after injection of endotoxin.

† Adrenalectomy has no effect on the increase of fibrinogen synthesis following partial hepatectomy.[77]

carried out by various authors with rats, the different AP-reactants each require a different level of corticosteroids for maximum synthesis.

The importance of hormones in the regulation of synthesis of acute-phase plasma proteins has been recently demonstrated by John and Miller.[90] They observed that during prolonged perfusions of the isolated normal rat liver, with medium enriched in insulin, cortisol, growth hormone and amino acid mixture, the synthesis of fibrinogen and haptoglobin increased 3-fold between 2 and 6 hours, α_1-AP globulin synthesis increased 3.5-fold between 4 and 8 hours, and α_2-AP globulin synthesis increased 11-fold between 8 and 10 hours. It should also be noted here that injections of ACTH, corticotropin or growth hormone into rats were shown to stimulate the synthesis of fibrinogen *in vivo*.[87,110,113] On the other hand, Werner[54] convincingly demonstrated that manipulation of the nitrogen balance of gastrectomy patients by treatment with an anabolic hormone (methandrosterone) or by infusion of protein hydrolysate does not influence the degree of acute-phase response. Moreover, corticosteroids seem to inhibit the formation of α_1-AP globulin and ceruloplasmin (cf. Table 4.4), so hormones alone cannot be responsible for the trauma-induced synthesis of AP-reactants but rather they modify liver response to hypothetical stimulatory factors of tissue origin.

Several authors drew attention to differences and similarities of effects produced by partial hepatectomy and injury (laparatomy) on the synthesis of AP-reactants. Studies *in vivo*,[77,311] in the perfused liver[9,312] and with the isolated ribosomal fraction[77] demonstrated that partial hepatectomy greatly stimulates not only the synthesis of fibrinogen and seromucoid but also albumin and transferrin. Thus the increased synthesis of cell proteins occurring during regeneration does not prevent a considerable increase in the use of amino acids for formation of AP-reactants. The fibrinogen response after partial hepatectomy was, however, not impaired by adrenalectomy, and the seromucoid fraction showed a biphasic pattern; an early marked increase closely related to the initial stress and a less marked secondary response.[311]

At present the mechanism by which cortisol increases the rates of hepatic synthesis of some acute-phase proteins is obscure, although it has been suggested[313,314] that cortisol may act by enhancing transcription of DNA to produce more functional RNA. The timing and relative magnitude of the changes produced by cortisol may be compared with those during the induction of the hepatic enzymes tryptophan pyrrolase and tyrosine transaminase, both *in vivo*[315] and in the isolated perfused liver.[313,316] The similarity of these two systems is not superficial since Tsukada and co-workers[317] reported an adaptive response of tryptophan pyrrolase and tyrosine transaminase in the rat liver after partial hepatectomy or laparatomy without loading of glucocorticoid or substrate. The rise of tryptophan pyrrolase was inhibited by actinomycin D administered immediately after operation.

4.10.4. EFFECTS OF ACTINOMYCIN D AND PUROMYCIN

The inhibition of increased synthesis of AP-reactants by actinomycin D has been observed many times,[76,221,285,318,319] and it suggests that *de novo* formation of mRNA is involved. Neuhaus *et al.*[76] found that this antibiotic blocked the response of the seromucoid fraction when injected not later than 4 hours after injury and it is known that 2–4 hr are required for the RNA synthesis stimulated by cortisone.[320] In the experiments described by Sarcione[285] actinomycin D was effective in the inhibition of induced synthesis of fibrinogen when administered up to 8 hr after injury, and of α_2-AP globulin up to 16 hr after injection of turpentine.

On the other hand, Balegno and Neuhaus[89] observed that laparatomized rats showed transient increase of albumin synthesis which was critically dependent on the presence of insulin and was not inhibited by actinomycin D. These experiments clearly differentiate stress-responsive glycoproteins, as requiring the synthesis of new RNA molecules, from other proteins such as albumin, which may be formed on more stable templates. However, if increased synthesis of mRNA specific for AP-proteins does occur after trauma, it is not associated with a change of activity of RNA polymerase.[321] Moreover, as emphasized by Chandler and Neuhaus,[321] the results of studies based on using actinomycin D must be treated with caution since this substance may affect various tissues and subcellular components in different ways; it exerts an effect not only on the synthesis of mRNA and ribosomal RNA but also on the stability of polysomes.

Puromycin, which is known to inhibit directly the formation of polypeptide chain when administered to rats within a few hours after laparatomy or turpentine injection, inhibited the synthesis of the seromucoid fraction,[76] α_1-AP globulin[221] and haptoglobin.[319] It reduced also the incorporation of labelled glycine into albumin whereas actinomycin D did not.[89]

4.10.5. STIMULATION OF THE LIVER CELL

Despite a morphological similarity, liver cells may not be synthesizing identical proteins at the same time, as indicated by studies with immunofluorescence techniques.[111,125,147,322,323] In the normal liver fibrinogen was detected in 1–5%, haptoglobin in 2.4% and prothrombin in 10–30% of the hepatocyte population. Stimulation by injury increases the number of cells producing haptoglobin to 11.2%, while stimulation of fibrinogen and prothrombin synthesis showed that all parenchymal cells are capable of producing these proteins. This argues in favour of the hypothesis that changes in the biosynthesis of plasma proteins following trauma depend on recruitment of previously inactive liver cells, although additional stimulation of cells permanently engaged in the formation of a given AP-reactant is also possible. According to Balegno and Neuhaus[89] the stimulation of the liver cell in the consequence of tissue damage includes at least three phenomena:

1. Early increase in cellular permeability, especially to amino acids, maximal at 4–8 hours.[324]

2. Increased synthesis of mRNA, maximal at 8 hours.[321]

3. Increase in the proportion of heavy polysomes, maximal at 18–24 hours, and corresponding increase of amino acid incorporating capacity.[325] A change in the polysomal profile in the rat liver following injury (laparatomy) is shown in Fig. 4.8.

It is well established that liver polyribosomes from partially hepatectomized or injured animals incorporate radioactive amino acids at a much increased rate.[77,321,326,327] A possible explanation is suggested by the experiments of Tsukada et al.[326,327] and of von der Decken.[328] These authors concluded that in normal conditions rates of synthesis of AP-reactants are restricted by inhibitors which are able to limit binding of the amino acyl-tRNA on the ribosome. Instability of such inhibitors may be responsible for the observed increase of relative synthesis rates of fibrinogen and α_1-AP globulin during prolonged perfusion of the rat liver (Gordon and Koj, 1968, unpublished observations).

On the other hand it is possible that injury-stimulated synthesis of AP-reactants is due to longer half-life of mRNA. Shortman[329] and Rahman et al.[330] found in regenerating liver increased amounts of the inhibitor of ribonuclease II which is probably responsible for degradation of mRNA. The decrease of ribonuclease activity was most marked between 8 and 48 hours after hepatectomy. Tsukada[331] observed a similar increase of ribonuclease inhibitor in rat serum during acute experimental peritonitis.

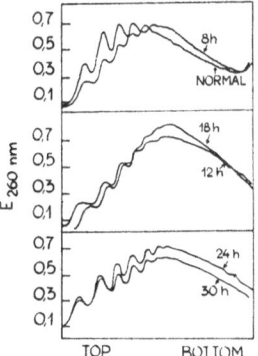

FIG. 4.8. Sucrose density gradient profiles of rat liver polysomes at various times after laparatomy (after Liu and Neuhaus[325]).

Another possible explanation of the mechanism of increased synthesis of AP-reactants arises from the fact that these proteins contain carbohydrates. Some authors postulate that many unrelated diseases and conditions of stress affect the regulatory control of the carbohydration of the polypetide precursors in the liver leading to an accelerated synthesis of glycoprotein molecules.[100] Direct experimental proof for this hypothesis is lacking but Foster[332] observed that injury or adrenalectomy caused changes in the activity of enzymes involved in the metabolism of amino sugars. Moreover, an increase in the enzymatic activity of the liver endoplasmic reticulum is known to occur after trauma, as indicated by enhanced formation of cholesterol.[333]

4.10.6. CONCLUSIONS

Taken together, the currently available evidence suggests that mechanisms regulating AP-protein synthesis in the liver cell involve reversible gene repression, and gene activation or derepression. Neuhaus et al.[76] suggested that formation of AP-reactants may be controlled by a single operon, while those proteins not affected by trauma are under separate genetic control. It should be, however, remembered that AP-reactants include three groups of proteins:

1. "New" proteins, which appear only in pathological states, *e.g.* C-reactive protein.

2. "Foetal" proteins, such as α_2-AP globulin of the rat. In normal adult liver this gene is "switched off" or repressed but becomes operative in consequence of injury.[285]

3. "Normal" constituents of plasma, such as fibrinogen or haptoglobin, synthesized at an increased rate after injury.

Even if the hypothesis of reversible gene activation is correct it cannot explain the highly complicated picture of AP-protein response to trauma. The regulation must undoubtedly involve also the translational level and intracellular organization of protein-synthesizing machinery. Only with this assumption, and taking into account the effect of hormones, the variable synthetic rates of individual glycoproteins may possibly be explained. It is obvious that complete understanding of the mechanism of trauma-induced synthesis of AP-reactants is intimately related to knowledge of general mechanism regulating protein synthesis in the animal organism.

4.11. CLINICAL SIGNIFICANCE OF ACUTE-PHASE REACTANTS

Direct medical application of studies of AP-reactants is limited at present to diagnostic and prognostic assessments, and in much smaller degree to therapy. In most cases changes in plasma level of these proteins are stereotyped, so one cannot distinguish between various types of inflammation or injury. It should be remembered that several commonly used tests of disease activity are related to increased amounts of glycoproteins in plasma: erythrocyte sedimentation rate, serum protein electrophoretic pattern, seromucoid level. The latter is often employed in clinical investigations concerning AP-reactants because of the simplicity of the procedure. However, for reasons previously discussed measurements of individual and well-defined proteins are now recommended.

As pointed out by Crockson *et al.*,[10] in assessing the relative value of the different AP-reactants for practical clinical use as a non-specific indicator of disease the criteria for the ideal protein would be that it is normally absent, so that rapid qualitative screening may be adequate, together with a wide range of quantitative response and a rapid rise and fall, so that secondary responses would be rapidly recorded. C-reactive protein adequately satisfies these criteria since it can be detected in the blood within 24 hours after injury and usually disappears at the onset of convalescence.[10,26] Persistence of CRP always indicates persistence of the disease or its complications, so that it can be used in prognostic assessments.

The rise of haptoglobin and fibrinogen levels following trauma, especially that associated with inflammatory response, are also very steep, and after surgery, myocardial infarction or burns significant differences in comparison with control values are usually observed after 24–48 hours.[10,25,71] The level of plasma fibrinogen is a useful guide in the management of acute myocardial infarction: a raised level is a reliable indication of recent infarction and a return corresponds with clinical recovery.[334] Glenn[335] emphasizes that the plasma fibrinogen response to inflammatory stimuli occurs long before overt clinical evidence of disease. The amount of fibrinogen and related proteins in the blood expressed in "plasma inflammation units" may be used as an objective method for investigating effects of drugs on experimental inflammation and as a diagnostic test.[336,337] It should be remembered, however, that some other factors may reduce the diagnostic value of fibrinogen since its level is decreased by fibrinogenolysis or fibrin deposition. The situation with haptoglobin is similar: if injury or infection are complicated by hemolytic states plasma haptoglobin may be within normal limits, presumably as a result of two opposing effects— injury tending to raise the level and hemolysis tending to lower it.

Several authors[10,11,54,226] reported that after surgery or injection of endotoxin α_1-AT rises rapidly, attaining maximum values after 1–4 days. The α_1-acid glycoprotein reaction is somewhat slower and the highest concentrations are recorded at 4–9 days after operation. The seromucoid fraction behaves similarly to α_1-acid glycoprotein. The changes in ceruloplasmin concentration are small and delayed. The whole picture may be illustrated by Fig. 4.9 based on the data of Werner and Odenthal.[54] In this case α_2-macroglobulin was used as a reference protein.

Individual glycoproteins show also different patterns of return to normal values: haptoglobin and α_1-AT are rather fast in this respect, while increased levels of fibrinogen and orosomucoid may persist in blood for a long time, even in the absence of complications.[23,24,71] The differences in the speed of rise and fall of each protein are related to its rate of turnover, which can be determined with the use of radioisotopes, but such methods are of limited value in routine clinical investigations.

It is known that repeated injuries may significantly alter the response pattern of particular AP-reactants. Weimer and Humelbaugh[19] injected rats with turpentine three times at 16-day intervals and concluded that the fibrinogen response to the final challenge was conspicuously less than that to the primary, whereas α_2-AP globulin values were significantly greater. Serum complement, seromucoid and protein bound hexose concentrations remained significantly elevated in the intervals between the injections of the phlogogenic agent. Darcy[220] reported the enhancement of the α_1-AP globulin response of rat serum to turpentine-induced inflammation following secondary and tertiary

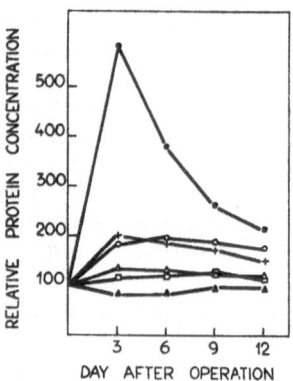

Fig. 4.9. Mean relative serum concentrations of some acute-phase re-actants and α_2-macroglobulin in 12 patients after gastrectomy. ●-CRP, +α_1-AT, o-orosomucoid, △ haptoglobin, □ ceruloplasmin, ▲ α_2-macroglobulin (plotted from the data of Werner and Odenthal[54]).

challenges. On the other hand, second injections of turpentine into rabbits resulted always in a smaller haptoglobin increase.[17] The AP-reactant response is also influenced by the age and body weight of the injured animal[223] and its nutritional conditions.[17]

An important question is how far increased concentrations of AP-reactants are proportional to the severity of injury. Werner and Cohnen[53] studied changes in serum proteins in the immediate post-operative period (after gastrectomy, herniorrhaphy, appendicectomy and thyroidectomy) and found them fairly uniform with apparently no relation between the degree of response and the type and extent of operation. On the other hand, accurate studies on experimental animals indicate that the AP-protein response is proportional to the severity of injury: Mouray[17] found a good correlation between the amount of turpentine injected into a rabbit and maximum level of haptoglobin in the blood 48 hours later (cf. Fig. 4.5b); Darcy[223] reported that the 48 hr response of α_1-AP globulin in the rat is proportional to the logarithm of turpentine dose and to granuloma weight; Wycoff[28] observed also that in rats injected with endotoxin the fibrinogen level measured 24 hr later is proportional to the logarithm of the dose over the range 3–300 μg.

At the same time it is clear, however, that the different injuries cannot be compared on this basis: sublethal and lethal doses of X-rays in rats gave only a slight effect on the concentration of specific glyco-proteins in comparison with turpentine-induced local inflammation.[159]

Macbeth et al.[338] observed that intramuscular injection of formalin into rats produced an intense local inflammatory reaction but without significant seromucoid response. Murray and Connell[151] found that local ulceration after subcutaneous injection of HCl into a rabbit failed to produce marked elevation of haptoglobin. Heim et al.[269] and van Gool and Ladiges[275] observed that only certain types of surgery in rats resulted in the appearance of β_2-AP globulin. These data indicate that systematic studies comparing the response of all main AP-reactants to typical injuries are required.

This situation greatly impaires the usefulness of individual AP-reactants in differential diagnosis of various diseases, although some examples of such applications can already be given. Werner[54] reported that in hepatitis and liver cirrhosis, which are not thought to produce acute-phase reaction, the isolated elevation of α_1-AT and ceruloplasmin is not accompanied by a marked increase of haptoglobin or α_1-acid glycoprotein. An intermediate pattern with elevation of haptoglobin, orosomucoid and ceruloplasmin but not α_1-AT is seen in systemic lupus erythematosus.

To complete the list of medical applications of acute-phase reactants it may be added that they represent a useful model for studying the effects of various anti-inflammatory drugs or hormones, such as indomethacin, aspirin, cortisol, oxyphenbutazone.[20,335,336,338]

At present attempts of using AP-reactants as therapeutical agents are scarce, e.g. in inborn deficiencies of those proteins (afibrinogenemia, α_1-AT deficiency). In most cases full plasma is infused and thus continuous replacement therapy is not feasible.[238] Moreover, AP-reactants show a rapid turnover, so the exogenous protein is quickly eliminated from the body. The progress in this field will depend on the elucidation of the biological functions of acute-phase reactants and employment of purified preparations of these proteins.

4.12. BIOLOGICAL FUNCTIONS OF ACUTE-PHASE REACTANTS

Since protein synthesis requires both energy and information, it may be supposed that trauma-induced formation of AP-reactants represents a useful phenomenon helping the injured organism to survive; otherwise the response would have been eliminated in the process of evolution. This idea is supported by considering some biological functions played by AP-reactants in defence against lethal effects of injury.

As pointed out by Reeve et al.[117] fibrinogen has three main functions: (a) to seal blood vessels and restrain haemorrhage by intra- and extravascular coagulation; (b) to supply a matrix through fibrin formation for spatial organization of fibroblasts and mobilized cells in wound healing and tissue repair; (c) to contribute to the viscosity of the blood. Moreover, fibrinogen and fibrin promote granulocyte emigration[290] and are substrates for numerous proteolytic enzymes (plasmin, leucocyte protease). The products of fibrinogen degradation show anticoagulant activities[339] and may prevent excessive fibrin deposition. Thus increased formation of fibrinogen in the liver would be advantageous not only in vascular injury but also in local inflammation or bacterial infection. On the other hand, increased deposition of fibrin, or its decreased lysis, may play a fundamental role in rheumatoid arthritis[132] or atherosclerosis (cf. Glenn[335]).

The main function of haptoglobin is to bind any haemoglobin that becomes free in the plasma, thus preventing glomerular filtration haemoglobin which leads to both kidney siderosis and irreversible loss of iron from the body. Binding with haptoglobin may also facilitate catabolism of haemoglobin, since Nakajima et al.[340] found that HpHb complex is more susceptible to heme β-methenyl oxygenase. The HpHb complex itself exhibits peroxidase activity which may be useful in inactivation of some agents responsible for tissue damage.

Another AP-reactant—ceruloplasmin—is also an oxidase. Osaki et al.[341] demonstrated ascorbate oxidase activity of ceruloplasmin and suggested that this enzyme maintains a certain level of serotonin, epinephrine and ascorbic acid. Bozhkov et al.[342] reported that quinones obtained after oxidation of serotonin and p-phenylene diamine by ceruloplasmin are effective in restoration of electron transport in the respiratory chain of mitochondria deprived of endogenous ubiquinone. The relationship between ceruloplasmin and the respiratory chain was also pointed out by Broman,[343] who suggested that ceruloplasmin may deliver copper to cytochrome oxidase, and Owen[344] demonstrated that ceruloplasmin is indeed the copper donor for extrahepatic tissues. On this basis Shokeir and Shreffler[345] put forward the hypothesis that some symptoms of Wilson's disease (inborn ceruloplasmin deficiency) are due to abnormally low levels of cytochrome oxidase. Finally, it should be mentioned that ferrous ion is oxidized by ceruloplasmin to the ferric form necessary for optimal binding by apotransferrin. Numerous compounds which reduce trivalent iron (ascorbate, catechol, cysteine, hydroquinone, DOPA) may thus be oxidized by the coupled iron-ceruloplasmin system.[171]

Hokama et al.[264] suggested that C-reactive protein is involved in the formation of haemoproteins, and the level of CRP in the blood may reflect the status of oxidative metabolism, especially that of catalase. Another function of CRP depends probably on stimulating phagocytic activity of leucocytes.[346]

The physiological significance of α_1-acid glycoprotein (orosomucoid) in the regulation of progesterone level in the blood in normal and pathological states and in pregnancy is still obscure, since the hormone is known to be bound also to albumin and cortiscosteroid-binding globulins.[347] Nilsson and Yamashima[348] reported that α_1-acid glycoprotein inhibits transformation of prothrombin to thrombin in a competitive manner and thus may prevent increased clotting in various inflammatory states.

In recent years evidence has accumulated that α_1-antitrypsin is a polyvalent protease inhibitor and among others inhibits leucocyte neutral protease[229,231] which is probably responsible for vascular injury and other tissue damage.[294,298,300] During the inflammatory response, when granulocytes leave the bloodstream, eventually releasing their enzymes at the site of inflammation, α_1-AT may be important as an inhibitor. There is no doubt that the inflammatory response can be greatly reduced by the administration of some protease inhibitors, such

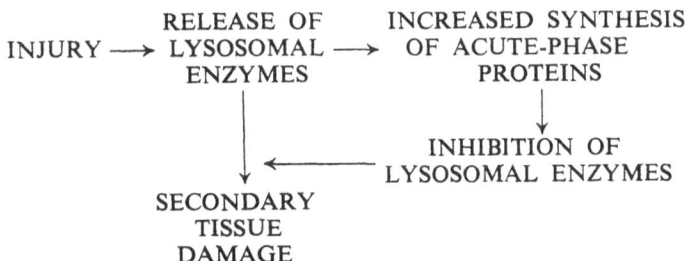

FIG. 4.10. Postulated feed-back mechanism relating lysosomes and acute-phase reactants (after Koj[135]).

as Trasylol.[349-352] As pointed out by Kueppers[228] the strongest argument for a major physiologic role of α_1-AT is chronic obstructive pulmonary disease associated with inborn deficiency states of this protein. The activity of α_1-AT against thrombin and plasmin[235] suggests that α_1-AT participates also in the interplay of formation and lysis of fibrin but its precise role in these reactions is not known.

α_1-AT is not the only one AP-reactant capable of inhibiting lysosomal proteases. Snellman and Sylvén[353] reported that cathepsin B is inhibited non-competitively by haptoglobin during digestion of urea-denatured edestin or haemoglobin in the presence of cystine and EDTA. Koj and Allison (unpublished results) observed that bovine α_1-acid glycoprotein inhibits cathepsin C, and that haptoglobin inhibits proteolysis of haemoglobin by cathepsin D. Darcy[218] postulated, although without experimental evidence, that the biological function of α_1-AP globulin of the rat depends on inhibiting enzymes released from injured cells. It is not known at present whether protein inhibitors of hyaluronidase,[354] alkaline ribonuclease[331] or acid deoxyribonuclease[355] are in any way associated with AP-reactants, but it is possible that in general plasma glycoproteins represent natural inhibitors of variety of lysosomal enzymes. In such a case a feed-back mechanism can be postulated: release of lysosomal enzymes leads to the increased concentration of AP-reactants which inhibit the activity of these enzymes and so help to restore homeostasis (Fig. 4.10).

Several authors postulated that plasma glycoproteins are involved in the healing of wounds and tissue repair by providing material of both carbohydrate and protein moieties.[23,73,356,358] Delaunay and Bazin[359] studied an accumulation of sialoglycoproteins of blood origin in the inflamed site. There is no experimental proof, however, of utilization of AP-reactants for synthetic processes in peripheral tissues, so this hypothesis still remains conjectural.

Possible protective functions of AP-reactants in the systemic response of an animal to injury are summarized in Table 4.5.

TABLE 4.5

Some biological phenomena following injury and associated with them AP-reactants involved in restoring homeostasis
Symbol (?) denotes a disputed function of the protein

Phenomenon	Protein
Bleeding and clot lysis	Fibrinogen α_1-antitrypsin (?) α_1-acid glycoprotein (?)
Haemolysis	Haptoglobin
Leucocytosis and phagocytosis	Fibrinogen C-reactive protein
Enhanced activity of lysosomal hydrolases	α_1-antitrypsin Haptoglobin (?) α_1-acid glycoprotein
Increased requirements for oxidative processes	Ceruloplasmin Haptoglobin C-reactive protein (?)
Wound healing and tissue repair	Fibrinogen Other glycoproteins (?)

An original idea concerning another possible biological function of some plasma sialoglycoproteins of liver origin has been elaborated by Apffel and Peters.[32] The proteins designated by them as "symbodies" include certain AP-reactants particularly rich in carbohydrates (orosomucoid, haptoglobin, α_1-antitrypsin). Histochemical, chemical and immunological evidence suggest that sialoglycoproteins become attached to the surface of tumour cells bringing about the tolerance of the host organism to cellular antigens of the tumour. Apffel and Peters emphasize that cells of inflamed and wounded areas share with cancer cells the capacity of stimulating the synthesis of symbodies in the liver, but they revert to a physiological state after the injury is healed. On the other hand, it is known that normal cells or their constituents damaged in the process of inflammation acquire antigenic properties, and that chronic inflammation is often complicated by immunological reactions. Evidence that persistent rupture of lysosomes can induce the formation of autoantibodies was obtained in experiments with streptolysin S (cf. Weissman[294]). Thus it seems possible that increased synthesis of some acute-phase reactants endowed with immunosuppressive properties may prevent the formation of autoantibodies and the development of various syndromes of immunological autoagression.

REFERENCES

1. H. E. SCHULTZE and J. F. HEREMANS, Molecular Biology of Human Proteins, with special reference to plasma proteins, vol. 1 (Elsevier, Amsterdam-London-New York, 1966)
2. A. H. GORDON and L. N. LOUIS, Biochem. J., 113, 481 (1969)
3. F. KUEPPERS, Humangenetik, 5, 54 (1967)
4. W. G. HEIM, Nature, 217, 1057 (1968)
5. I. KUSHNER and J. A. SOMMERVILLE, Biochim. Biophys. Acta, 207, 105 (1970)
6. R. J. WINZLER (ed., D. Glick), Methods of Biochem. Analysis. (Academic Press, New York 1955), vol. 2, p. 279
7. R. GOT, R. I. CHEFTEL, J. FONT and J. MORETTI, Biochim. Biophys. Acta, 136, 320 (1967)
8. T. DE VONNE LEBRETON, N. GUTMAN and H. MOURAY, Clin. Chim. Acta, 30, 603 (1970)
9. L. E. MUTSCHLER and A. H. GORDON, Biochim. Biophys. Acta, 130, 486 (1966)
10. R. A. CROCKSON, C. J. PAYNE, A. P. RATCLIFF and J. F. SOOTHILL, Clin. Chim. Acta, 14, 435 (1966)
11. M. WERNER and D. ODENTHAL, J. Lab. Clin. Med., 70, 302 (1967)
12. R. J. WINZLER (ed., E. A. Balazs and R. W. Jeanloz), The Amino Sugars, (Academic Press, New York 1965), vol. IIA, p. 337
13. R. J. WINZLER, Clin. Chem., 11, 339 (1965)
14. J. A. OWEN, Adv. Clin. Chem., 9, 1 (1967)
15. R. K. MURRAY and G. E. CONNELL, Nature, 186, 86 (1960)
16. H. E. WEIMER and D. C. BENJAMIN, Am. J. Physiol., 209, 736 (1965)
17. H. MOURAY, Biosynthese de l'haptoglobine chez le lapin, (R. Foulon, Paris, 1966)
18. J. HURLIMANN, G. J. THORBECKE and G. M. HOCHWALD, J. exp. Med., 123, 365 (1966)
19. H. E. WEIMER and C. HUMELBAUGH, Can. J. Physiol. Pharmacol., 45, 241 (1967)
20. E. M. GLENN, B. J. BOWMAN and T. C. KOSLOWSKE, Biochem. Pharmacol. Suppl. "Chemical Biology of Inflammation", (March 1968), p. 27
21. F. E. ASHTON, J. C. JAMIESON and A. D. FRIESEN, Can. J. Biochem., 48, 841 (1970)
22. J. W. DAVIES, C. R. RICKETTS and J. P. BULL, Cli. Sci., 30, 305 (1966)
23. W. DOBRYSZYCKA, R. ZEINEH, E. EBROON and J. C. KUKRAL, Cli. Sci., 36, 231 (1969)
24. R. A. ZEINEH and J. C. KUKRAL, J. Trauma, 10, 493 (1970)
25. M. McCATHIE, J. A. OWEN and A. I. S. MacPHERSON, Scot. Med. J., 11, 83 (1966)
26. D. WIEDERMANN, D. WIEDERMANNOVA and K. CIDL, Prot. Biol. Fluids, 14, 385 (1966)
27. A. KOJ and A. S. McFARLANE, Biochem. J., 108, 137 (1968)
28. H. D. WYCOFF, Proc. Soc. Exp. Biol. Med., 133, 940 (1970)

29. R. A. MACBETH and J. G. BEKESI, *Cancer Res.*, **22**, 1170 (1962)
30. H. E. WEIMER, C. HUMELBAUGH and D. M. ROBERTS, *Can. J. Biochem.*, **45**, 1937 (1967)
31. W. G. HEIM and S. R. ELLENSON, *Nature*, **213**, 1260 (1967)
32. C. A. APFFEL and J. H. PETERS, *Progr. Exp. Tumor Res.*, **12**, 1 (1969)
33. L. L. MILLER, C. G. BLY, M. L. WATSON and W. F. BALE, *J. exp. Med.*, **94**, 431 (1951)
34. L. L. MILLER, H. R. HANAVAN, N. TITTHASIRI and A. CHOWDHURY, *Advan. Chem. Ser.*, **44**, 1740 (1964)
35. S. KRAUSS and E. J. SARCIONE, *Biochim. Biophys. Acta*, **90**, 301 (1964)
36. F. W. PUTNAM (Editor), Plasma Proteins (Academic Press, New York, 1960)
37. V. BOCCI, *Arch. Fisiol.*, **67**, 314 (1970)
38. C. FRANÇOIS, R. D. MARSHALL and A. NEUBERGER, *Biochem. J.*, **83**, 335 (1962)
39. K. B. BJÖRNESJÖ, *Scand. J. Clin. Lab. Invest.*, **7**, 147 (1955)
40. J. A. MOSS, *Biochem. J.*, **61**, 151 (1955)
41. G. BLIX, *Acta Chem. Scand.*, **2**, 467 (1948)
42. T. A. GOOD and S. P. BESSAM, *Analyt. Biochem.*, **9**, 253 (1964)
43. L. SVENNERHOLM, *Biochim. Biophys. Acta*, **24**, 604 (1957)
44. L. WARREN, *J. Biol. Chem.*, **234**, 1971 (1959)
45. A. GOTTSCHALK (Editor), Glycoproteins, their Composition, Structure and Function (Elsevier, Amsterdam-London-New York, 1966)
46. E. A. BALAZS and R. W. JEANLOZ (Editors), The Amino Sugars (Academic Press, New York and London, 1966)
47. H. E. WEIMER, V. COGGSHALL, D. M. ROBERTS and C. HUMELBAUGH, *Prot. Biol. Fluids*, **14**, 403 (1966)
48. H. E. WEIMER and V. COGGSHALL, *Can. J. Physiol. Pharmacol.*, **45**, 767 (1967)
49. E. KÖIW and A. GRÖNWALL, *Scand. J. Clin. Lab. Invest.*, **4**, 244 (1952)
50. G. MANCINI, A. O. CARBONARA and J. F. HEREMANS, *Immunochem.*, **2**, 235 (1965)
51. W. AUGENER, *Prot. Biol. Fluids*, **12**, 363 (1964)
52. K. STÖRIKO, *Blut*, **16**, 200 (1968)
53. M. WERNER and G. COHNEN, *Cli. Sci.*, **36**, 173 (1969)
54. M. WERNER, *Clin. Chim. Acta*, **25**, 299 (1969)
55. W. L. BEEKEN, W. VOLWILER, P. D. GOLDSWORTHY, L. E. GARBY, W. E. REYNOLDS, R. STOGSDILL and R. S. STEMLER, *J. Clin. Invest.*, **41**, 1313 (1962)
56. S. WEISMAN, B. GOLDSMITH, R. J. WINZLER and M. H. LEPPER, *J. Lab. Clin. Med.*, **57**, 7 (1961)
57. F. KUEPPERS and R. J. FALLAT, *Clin. Chim. Acta*, **24**, 401 (1969)
58. T. FREEMAN, *Prot. Biol. Fluids*, **12**, 344 (1964)
59. S. KRAUSS, M. SCHROTT and E. J. SARCIONE, *Am. J. Med. Sci.*, **252**, 184 (1966)
60. H. O'HARA, K. WATANABE and T. WADA, *Clin. Chim. Acta*, **19**, 41 (1968)
61. S. KRAUSS, *Blood*, **33**, 865 (1969)
62. M. KEKKI, P. KOSKELO and E. A. NIKKILÄ, *Nature*, **209**, 1252 (1966)
63. P. KOSKELO, M. KEKKI, E. A. NIKKILÄ and M. VIRKKUNEN, *Scand. J. Clin. Lab. Inv.*, **19**, 259 (1967)
64. A. S. MCFARLANE, D. TODD and S. CROMWELL, *Cli. Sci.*, **26**, 415 (1964)
65. Y. TAKEDA, *J. Clin. Invest.*, **45**, 103 (1966)
66. A. S. MCFARLANE (eds., H. N. Munro and J. B. Allison), Mammalian Protein Metabolism, (Academic Press, New York-London, 1964), vol. I, p. 297
67. V. BOCCI, *Ital. J. Biochem.*, **18**, 346 (1969)
68. J. GROSSMANN, A. A. YALOW and R. E. WESTON, *Metabolism*, **9**, 528 (1960)
69. C. M. E. MATTHEWS, *Phys. Med. Biol.*, **2**, 36 (1957)
70. A. C. ATENCIO, K. JOINER and E. B. REEVE, *Am. J. Physiol.*, **216**, 764 (1969)
71. J. C. KUKRAL, R. ZEINEH, W. DOBRYSZYCKA, J. POLLITT and N. STONE, *Cli. Sci.*, **36**, 221 (1969)

72. G. B. ROBINSON, *Biochem. J.*, **114**, 635 (1969)
73. A. M. CHANDLER and O. W. NEUHAUS, *Am. J. Physiol*, **206**, 169 (1964)
74. H. MOURAY, J. MORETTI and M. F. JAYLE, *C. R. Acad. Sci. Paris*, **258**, 4871 (1964)
75. C. A. ALPER, J. H. PETERS, A. G. BIRTCH and F. H. GARDNER, *J. Clin. Invest.*, **44**, 574 (1965)
76. O. W. NEUHAUS, H. F. BALEGNO and A. M. CHANDLER, *Am. J. Physiol.*, **211**, 151 (1966)
77. C. MAJUMDAR, K. TSUKADA and I. LIEBERMAN, *J. Biol. Chem.*, **242**, 700 (1967)
78. A. S. MCFARLANE, *Biochem. J.*, **89**, 277 (1963)
79. A. S. MCFARLANE, L. IRONS, A. KOJ and E. REGOECZI, *Biochem. J.*, **95**, 536 (1965)
80. R. D. WOCHNER, S. M. WEISSMAN, T. A. WALDMANN, D. HOUSTON and N. I. BERLIN, *J. Clin. Invest.*, **47**, 971 (1968)
81. A. S. TAVILL, A. CRAIGIE and V. M. ROSENOER, *Cli. Sci.*, **34**, 1 (1968)
82. E. A. JONES, A. CRAIGIE, A. S. TAVILL, G. FRANGLEN and V. M. ROSENOER, *Gut*, **9**, 466 (1968)
83. A. CRAIGIE, V. M. ROSENOER, R. A. SMALLWOOD and A. S. TAVILL, (eds., G. Birke, R. Norberg and L. O. Plantin) Physiology and Patophysiology of Plasma Protein Metabolism, (Pergamon Press, Oxford, 1969), p. 61
84. K. N. JEEJEEBHOY, A. M. SAMUEL, B. SINGH, G. D. NADKARNI, H. G. DESAI, A. V. BORKAR and L. S. MANI, *Gastroenterol.*, **56**, 252 (1969)
85. A. KOJ, *Biochim. Biophys. Acta*, **165**, 97 (1968)
86. A. KOJ and A. C. ALLISON, *Folia Biol. (Krakow)*, **17**, 37 (1969)
87. K. N. JEEJEEBHOY, A. BRUCE-ROBERTSON, U. SODTKE and M. FOLEY, *Biochem. J.*, **119**, 243 (1970)
88. H. T. MOURIDSEN, *Cli. Sci.*, **33**, 345 (1967)
89. H. F. BALEGNO and O. W. NEUHAUS, *Life Sci. II.*, **9**, 1039 (1970)
90. D. W. JOHN and L. L. MILLER, *J. Biol. Chem.*, **244**, 6134 (1969)
91. A. H. GORDON and D. A. DARCY, *Brit. J. Exp. Path.*, **48**, 81 (1967)
92. C. A. WILLIAMS, M. C. GANOZA and F. LIPMAN, *Proc. Natl. Acad. Sci. USA*, **53**, 622 (1965)
93. D. C. BENJAMIN and H. E. WEIMER, *Nature*, **209**, 1032 (1966)
94. R. J. WINZLER, J. MOLNAR and G. B. ROBINSON, *J. Biol. Chem.*, **239**, 3157 (1964)
95. G. B. ROBINSON, J. MOLNAR and R. J. WINZLER, *J. Biol. Chem.*, **239**, 1134 (1964)
96. J. MOLNAR, G. B. ROBINSON, and R. J. WINZLER, *J. Biol. Chem.*, **240**, 1882 (1965)
97. E. J. SARCIONE, M. BOHNE and M. LEAHY, *Biochemistry*, **3**, 1973 (1964)
98. G. R. LAWFORD, H. SCHACHTER, *J. Biol. Chem.*, **241**, 5408 (1966)
99. M. SARNECKA-KELLER and J. NOWORYTKO, *Post. Biochem.*, **15**, 3 (1969)
100. R. G. SPIRO, *New Engl. J. Med.*, **281**, 1043 (1969)
101. A. G. MORELL, G. GREGORIADIS, H. SCHEINBERG, J. HICKMAN and G. ASHWELL, *J. Biol. Chem.*, **246**, 1461 (1971)
102. K. JACOBSSON, *Scand. J. Clin. Lab. Invest. Suppl.* **14**, 1 (1955)
103. A. HIRSCH and C. CATTANEO, *Hoppe Seyler's Z. Physiol. Chem.*, **304**, 53 (1956)
104. B. C. ELLIS and A. STRANSKY, *J. Lab. Clin. Med.*, **58**, 477 (1961)
105. A. C. ATENCIO, D. C. BURDICK and E. B. REEVE, *J. Lab. Clin. Med.*, **66**, 137 (1965)
106. E. REGOECZI and B. E. STANNARD, *Biochim. Biophys. Acta*, **181**, 287 (1969)
107. E. REGOECZI, *Cli. Sci.*, **38**, 111 (1970)
108. E. REGOECZI, *Thromb. Diath. Haemorrh.*, **18**, 276 (1967)
109. E. REGOECZI and P. L. WALTON, *Thromb. Diath. Haemorrh.*, **17**, 237 (1967)
110. A. C. ATENCIO and L. LORAND, *Am. J. Physiol.*, **219**, 1161 (1970)
111. W. B. FORMAN and M. I. BARNHART, *J. Am. Med. Ass.*, **187**, 128 (1964)
112. A. J. MARCUS and M. B. ZUCKER, The Physiology of Blood Platelets (Grune and Stratton, New York, 1965)

113. A. C. ATENCIO, P. Y. CHAO, A. Y. CHEN and E. B. REEVE, *Am. J. Physiol.* **216**, 773 (1969)
114. R. A. SMALLWOOD, E. A. JONES, A. CRAIGIE, S. RAIA and V. M. ROSENOER, *Cli. Sci.*, **35**, 35 (1968)
115. A. C. ATENCIO, H. R. BAILEY and E. B. REEVE, *J. Lab. Clin. Med.*, **66**, 1 (1965)
116. E. REGOECZI, G. E. REGOECZI and A. S. MCFARLANE, *Pflügers Archiv.*, **279**, 17 (1964)
117. E. B. REEVE, Y. TAKEDA, A. C. ATENCIO, *Prot. Biol. Fluids*, **14**, 283 (1966)
118. J. H. LEWIS, E. E. FERGUSON and C. SCHOENFELD, *J. Lab. Clin. Med.*, **58**, 247 (1961)
119. J. H. LEWIS, *Proc. Soc. Exp. Biol. Med.*, **114**, 777 (1963)
120. J. GAJEWSKI and B. ALEXANDER, *Circulat. Res.*, **13**, 432 (1963)
121. E. BIERNACKI, *Deutsch. Med. Wschr.*, **23**, 769 (1897)
122. D. P. FOSTER and G. H. WHIPPLE, *Am. J. Physiol.*, **58**, 407 (1932)
123. T. H. HAM and F. C. CURTIS, *Medicine*, **17**, 413 (1938)
124. A. H. GORDON and A. KOJ, *Brit. J. Exp. Path.*, **49**, 436 (1968)
125. M. I. BARNHART, D. C. CRESS, S. M. NOONAN and R. T. WALSH, *Thromb. Diath. Haemorrh. Suppl.* **39**, 143 (1970)
126. L. R. PICKART and L. O. PILGERAM, *Thromb. Diath. Haemorrh.*, **17**, 358 (1967)
127. L. L. MILLER and D. W. JOHN, *Thromb. Diath. Haemorrh. Suppl.*, **39**, 127 (1970)
128. J. E. NILÉHN and P. E. GANROT, *Scand. J. Clin. Lab. Invest.*, **20**, 113 (1967)
129. Y. TAKEDA and A. Y. CHEN, *J. Lab. Clin. Med.*, **70**, 678 (1967)
130. E. REGOECZI, G. E. HENLEY, R. C. HOLLOWAY and A. S. MCFARLANE, *Brit. J. Exp. Path.*, **44**, 397 (1963)
131. S. FISHER, A. P. FLETCHER, N. ALKJAERSIG and S. SHERRY, *J. Lab. Clin. Med.*, **70**, 903 (1967)
132. C. LACK, *Proc. Roy. Soc. Med.*, **59**, 875 (1966)
133. W. GALASINSKI, K. WOROWSKI, S. NIEWIAROWSKI and G. FRANECKI, *Throb. Diath. Haemorrh.*, **18**, 268 (1967)
134. G. MÜLLER-BERGHAUS, *Thromb. Diath. Haemorrh. Suppl.*, **36**, 45 (1969)
135. A. KOJ (eds., R. Porter and J. Knight), The Ciba Foundation Symposium: Energy Metabolism in Trauma, (Churchill, London, 1970) p. 79
136. M. POLONOVSKI and M. F. JAYLE, *C. R. Soc. Biol.*, **129**, 457 (1938)
137. M. F. JAYLE, *Bull. Soc. Chim. Biol.*, **33**, 876 (1951)
138. J. A. OWEN, F. C. BETTER and J. HOBAN, *J. Clin. Path.*, **13**, 163 (1960)
139. M. NYMAN, *Scand. J. Clin. Lab. Invest. Suppl.*, **11**, 39 (1959)
140. M. PANTLITSCHKO and G. WEIPPL, *Clin. Chim. Acta*, **19**, 439 (1968)
141. R. B. ROY, R. W. SHAW, G. E. CONNELL, *J. Lab. Clin. Med.*, **74**, 698 (1969)
142. R. KLUTHE, J. FAUL and H. HEIMPEL, *Nature*, **205**, 93 (1965)
143. G. E. CONNELL adn R. W. SHAW, *Canad. J. Biochem.*, **39**, 1013 (1961)
144. C. LOMBART, J. MORETTI and M. F. JAYLE, *Biochim. Biophys. Acta*, **97**, 262 (1965)
145. R. KLUTHE and H. ISLIKER, *Helv. Physiol. Acta*, **18**, 404 (1960)
146. W. DOBRYSZYCKA, J. MORETTI and M. F. JAYLE, *Bull. Soc. Chim. Biol.*, **45**, 301 (1963)
147. J. H. PETERS and C. A. ALPER, *J. Clin. Invest.*, **45**, 314 (1965)
148. W. D. NOYES and L. GARBY, *Scand. J. Clin. Lab. Invest.*, **20**, 33 (1967)
149. E. C. FRANKLIN, M. ORATZ, M. A. ROTHSCHILD and D. ZUCKER-FRANKLIN. *Proc. Soc. Exp. Biol. Med.*, **105**, 167 (1960)
150. L. GARBY and J. OBARA, *Blut*, **6**, 143 (1960)
151. R. K. MURRAY, G. E. CONNELL and J. H. PERTH, *Fed. Proc.*, **19**, 66 (1960)
152. W. R. KEENE, J. H. JANDL, *Blood*, **26**, 705 (1965)
153. R. ENGLER, J. MORETTI and M. F. JAYLE, *Bull. Soc. Chim. Biol.*, **49**, 263 (1967)
154. E. C. HERMAN, JR, *J. Lab. Clin. Med.*, **57**, 834 (1961)
155. A. C. ALLISON and W. A. REES, *Brit. Med. J.*, **2**, 1137 (1957)
156. M. F. JAYLE and J. MORETTI, *Progress in Hematol.*, **3**, 342 (1962)

157. H. E. SUTTON, *Progress in Med. Chem.*, **7**, 163 (1970)
158. C. LOMBART, M. NEBUT, M. P. OLLIER, M. F. JAYLE and L. HARTMANN, *Rev. Franc. Etud. Clin. Biol.*, **13**, 258 (1968)
159. A. KOJ, *Folia Biol. (Krakow)*, **18**, 275 (1970)
160. R. ENGLER, H. DEGRELLE, P. JANIAUD, M. DOMINGO and M. F. JAYLE, *Clin. Chim. Acta*, **26**, 491 (1969)
161. L. ROBERT, P. MEMBELLONI and D. CROSTI, *Proc. Soc. Exp. Biol. Med.*, **107**, 499 (1961)
162. J. P. BOREL, H. MOURAY, J. MORETTI and M. F. JAYLE, *C. R. Acad. Sci.*, **259**, 3875 (1964)
163. S. KRAUSS, *Proc. Soc. Exp. Biol. Med.*, **112**, 552 (1963)
164. J. A. OWEN, R. SMITH, R. PADANYI and J. MARTIN, *Cli. Sci.*, **26**, 1 (1964)
165. C. G. HOLMBERG and C. B. LAURELL, *Acta Chem. Scand.*, **5**, 476 (1951)
166. O. B. HOUCHIN, *Clin. Chem.*, **4**, 519 (1958)
167. H. A. RAVIN, *J. Lab. Clin. Med.*, **58**, 161 (1961)
168. U. S. SEAL, *Comp. Biochem. Physiol.*, **13**, 143 (1964)
169. F. W. SUNDERMAN and S. NOMOTO, *Clin. Chem.*, **16**, 903 (1970)
170. J. B. BINGLEY and A. T. DICK, *Clin. Chim. Acta*, **25**, 480 (1969)
171. J. A. MCDERMOTT, C. T. HUBER, S. OSAKI and E. FRIEDEN, *Biochim. Biophys. Acta*, **151**, 541 (1968)
172. G. CURZON, *Biochem. J.*, **103**, 289 (1967)
173. R. J. SCHEN, M. RABINOWITZ, *Clin. Chim. Acta*, **13**, 537 (1966)
174. H. F. DEUTSCH, C. B. KASPER and D. A. WALSH, *Archiv. Biochem. Biophys.*, **99**, 132 (1962)
175. L. BROMAN and K. KJELLIN, *Biochim. Biophys. Acta*, **82**, 101 (1964)
176. A. G. MORELL, C. J. VAN DEN HAMER, I. H. SCHEINBERG and G. ASHWELL, *J. Biol. Chem.*, **241**, 3745 (1966)
177. R. P. STOKES, *Clin. Chim. Acta*, **15**, 517 (1967)
178. L. PÉJAUDIER, R. AUDRAN and M. STEINBUCH, *Clin. Chim. Acta*, **30**, 377 (1970)
179. D. W. COX, *J. Lab. Clin. Med.*, **68**, 893 (1966)
180. H. MEIER and A. D. MACPIKE, *Proc. Soc. Exp. Biol. Med.*, **128**, 1185 (1968)
181. I. H. SCHEINBERG and D. GITLIN, *Science*, **116**, 484 (1952)
182. S. H. LAWRENCE, P. J. MELNICK and H. E. WEIMER, *Proc. Soc. Exp. Biol. Med.*, **105**, 572 (1960)
183. G. W. EVANS and R. E. WIEDERANDERS, *Nature*, **215**, 766 (1967)
184. G. W. EVANS and R. E. WIEDERANDERS, *Am. J. Physiol.*, **214**, 1152 (1968)
185. G. W. EVANS, N. F. CORNATZER and W. E. CORNATZER, *Am. J. Physiol.*, **218**, 613 (1970)
186. G. W. EVANS, P. F. MAJORS and W. E. CORNATZER, *Biochim. Biophys. Res. Comm.*, **41**, 1120 (1970)
187. B. STARCHER and C. H. HILL, *Comp. Biochem. Physiol.*, **15**, 429 (1965)
188. G. GREGORIADIS and T. L. SOURKES, *Can. J. Biochem.*, **48**, 160 (1970)
189. A. G. MORELL, R. A. IRVINE, J. STERNLIEB, I. H. SCHEINBERG and G. ASHWELL, *J. Biol. Chem.*, **243**, 155 (1968)
190. C. J. A. VAN DEN HAMER, A. G. MORELL, I. H. SCHEINBERG, J. HICKMAN and G. ASHWELL, *J. Biol. Chem.*, **245**, 4397 (1970)
191. H. MARKOWITZ, C. J. GUBLER, J. P. MAHONEY, G. E. CARTWRIGHT and M. M. WINTROBE, *J. Clin. Invest.*, **34**, 1498 (1955)
192. E. W. RICE, *Clin. Chim. Acta*, **5**, 632 (1960)
193. E. W. RICE, *Clin. Chim. Acta*, **6**, 170 (1961)
194. A. SASS-KORTSAK, *Adv. Clin. Chem.*, **8**, 1 (1965)
195. W. NIEDERMAYER, *Ann. Rheum. Dis.*, **24**, 544 (1965)
196. P. KOSKELO, M. KEKKI, M. VIRKKUNEN, A. LASSUS and T. SOMER, *Acta Rheum. Scand.*, **12**, 261 (1966)
197. B. J. MEYER, A. C. MEYER and E. FRIEDEN, *Am. J. Physiol.*, **194**, 581 (1958)

198. G. W. EVANS, D. R. MYRON and R. E. WIEDERANDERS, *Am. J. Physiol.*, **216**, 340 (1969)
199. W. SCHUMER, J. MOLNAR, J. DOWLING and R. J. WINZLER, *Am. J. Physiol.*, **212**, 184 (1967)
200. B. ROBERT, L. ROBERT and M. F. JAYLE, *Experientia*, **15**, 385 (1959)
201. V. DOLEŽALOVA, Z. BRADA and A. KOČENT, *Clin. Chim. Acta*, **9**, 542 (1964)
202. M. SARNECKA-KELLER, K. WOŹNICZKA and T. CIBA, Proc. XX Confer. of the Diabetologic Section of the Polish Society of Internal Medicine, Krakow 1971
203. E. A. SUGAR, *Clin. Chim. Acta*, **8**, 347 (1963)
204. K. SCHMID, *J. Am. Chem. Soc.*, **72**, 2816 (1950)
205. K. SCHMID, *J. Am. Chem. Soc.*, **75**, 60 (1953)
206. H. E. WEIMER, J. W. MEHL and R. J. WINZLER, *J. Biol. Chem.*, **185**, 561 (1950)
207. R. W. JEANLOZ (ed. A. Gottschalk), Glycoproteins, their Composition, Structure and Function (Elsevier, Amsterdam-London-New York), p. 362
208. K. SCHMID, J. P. BINETTE, S. MAMIYAMA, V. PFISTER and S. TAKAHASHI, *Biochemistry*, **1**, 959 (1962)
209. A. BEZKOROVAINY and R. J. WINZLER, *Biochim. Biophys. Acta*, **49**, 559 (1961)
210. T. KAWASAKI, J. KOYAMA and I. YAMASHINA, *J. Biochem. (Tokyo)*, **60**, 554 (1966)
211. J. MARÇAIS, C. NICOT and J. MORETTI, *Bull. Soc. Chim. Biol.*, **52**, 741 (1970)
212. K. SCHMID, K. TOKITA and H. YOSHIZAKI, *J. Clin. Invest.*, **44**, 1394 (1965)
213. K. TOKITA and K. SCHMID, *Clin. Chim. Acta*, **17**, 39 (1967)
214. K. SCHMID, T. OKUYAMA and H. KAUFMANN, *Biochim. Biophys. Acta*, **154**, 565 (1968)
215. K. SCHMID, R. A. FIELD and H. YOSHIZAKI, *J. Med. Genet.*, **5**, 36 (1968)
216. H. YOSHIZAKI, K. HUNZIKER and K. SCHMID, *Clin. Chim. Acta*, **23**, 147 (1969)
217. R. A. ZEINEH and J. C. KUKRAL, *Clin. Res.*, **17**, 398 (1969)
218. D. A. DARCY, *Brit. J. Exp. Path.*, **45**, 281 (1964)
219. D. A. DARCY, *Brit. J. Exp. Path.*, **46**, 155 (1965)
220. D. A. DARCY, *Brit. J. Exp. Path.*, **47**, 480 (1966)
221. D. A. DARCY, *Brit. J. Exp. Path.*, **48**, 608 (1967)
222. D. A. DARCY, *Brit. J. Exp. Path.*, **49**, 525 (1968)
223. D. A. DARCY, *Brit. J. Exp. Path.*, **51**, 59 (1970)
224. A. G. WEIMER, D. C. BENJAMIN and D. A. DARCY, *Nature*, **208**, 1221 (1965)
225. H. G. SCHWICK, N. HEIMBURGER and H. HAUPT., *Z. ges. inn. Med.*, **21**, 193 (1966)
226. F. KUEPPERS, *Humangenetik*, **6**, 207 (1968)
227. J. BIETH, P. METAIS and J. WARTER, *Clin. Chim. Acta*, **20**, 69 (1968)
228. F. KUEPPERS, *Humangenetik*, **11**, 177 (1971)
229. F. KUEPPERS and A. G. BEARN, *Proc. Soc. Exp. Biol. Med.*, **121**, 1207 (1966)
230. N. HEIMBURGER and H. HAUPT, *Klin. Wschr.*, **44**, 1196 (1966)
231. A. KOJ, J. CHUDZIK, W. PAJDAK and A. DUBIN, *Biochim. Biophys. Acta*, **268**, 199 (1972)
232. S. LINDVALL, O. MAGNUSSON and K. ORTH, *Acta Chem. Scand.*, **23**, 2165 (1969)
233. J. S. BAUMSTARK, *Arch. Bioch. Biophys.*, **118**, 619 (1967)
234. A. Z. EISEN, K. J. BLOCH and T. SAKAI, *J. Lab. Clin. Med.*, **75**, 258 (1970)
235. A. RIMON, Y. SHAMASH and B. SHAPIRO, *J. Biol. Chem.*, **241**, 5102 (1966)
236. F. C. MOLL, S. F. SUNDEN and J. R. BROWN, *J. Biol. Chem.*, **233**, 121 (1959)
237. H. GANS, B. H. TAN, *Clin. Chim. Acta*, **17**, 111 (1967)
238. S. MAKINO and C. E. REED, *J. Lab. Clin. Med.*, **75**, 742 (1970)
239. M. KUNITZ, *J. Gen. Physiol.* **30**, 291 (1946)
240. H. F. BUNDY and J. W. MEHL, *J. Biol. Chem.*, **234**, 1124 (1959)
241. F. C. WU and M. LASKOWSKI, *J. Biol. Chem.*, **235**, 1680 (1960)
242. Y. SHAMASH and A. RIMON, *Biochim. Biophys. Acta*, **121**, 35 (1966)
243. C. B. LAURELL and S. ERIKSSON, *Scand. J. Clin. Lab. Invest.*, **15**, 132 (1963)
244. W. A. BRISCOE, F. KUEPPERS, A. L. DAVIS and A. G. BEARN, *Amer. Rev. Resp. Dis.*, **94**, 529 (1966)

245. P. O. GANROT, C. B. LAURELL and S. ERIKSSON, *Scand. J. Clin. Lab. Invest.*, **19**, 205 (1967)
246. G. M. TURINO, R. M. SENIOR, B. D. GARG, S. KELLER, M. M. LEVI and I. MANDL., *Science*, **165**, 709 (1969)
247. M. K. FAGERHOL and M. BRAEND, *Science*, **149**, 986 (1965)
248. M. K. FAGERHOL and C. B. LAURELL, *Clin. Chim. Acta*, **16**, 199 (1967)
249. N. R. SHULMAN, *J. Exp. Med.*, **95**, 605 (1952)
250. F. C. MOLL, *J. Exp. Med.*, **103**, 363 (1956)
251. C. B. LAURELL, S. KULLANDER and J. THORELL, *Scand. J. Clin. Lab. Invest.*, **21**, 337 (1968)
252. W. TILLET and R. FRANCIS, JR, *J. Exp. Med.*, **52**, 561 (1930)
253. H. ANDERSON and M. MCCARTHY, *Am. J. Med.*, **8**, 445 (1950)
254. L. NILSSON, *Acta Pathol. Microbiol. Scand.*, **73**, 129 (1968)
255. P. HEDLUND, *Acta Med. Scand.*, **169**, Suppl. 361 (1961)
256. G. SCHUMACHER and H. D. SCHLUMBERGER, *Klin. Wschr.*, **40**, 67 (1962)
257. Y. HOKAMA and R. F. RILEY, *Biochim. Biophys. Acta*, **74**, 305 (1963)
258. P. O. GANROT and C. O. KINDMARK, *Biochim. Biophys. Acta*, **194**, 443 (1969)
259. H. D. CROXATTO, E. T. NISHIMURA, K. YAMADA and Y. YOKAMA, *J. Immunol.*, **100**, 563 (1968)
260. E. C. GOTSCHLICH and G. M. EDELMAN, *Proc. Natl. Acad. Sci.*, **54**, 558 (1965)
261. E. M. DAVIS, R. M. BLANKEN and R. J. BEAGLE, *Biochemistry*, **8**, 2706 (1969)
262. R. F. RILEY and M. K. COLEMAN, *Clin. Chim. Acta*, **30**, 483 (1970)
263. R. F. RILEY, M. K. COLEMAN and Y. HOKAMA, *Clin. Chim. Acta*, **11**, 530 (1965)
264. Y. HOKAMA, K. YAMADA, C. SCHELLER and E. T. NISHIMURA, *Cancer Res.*, **30**, 1319 (1970)
265. R. F. RILEY, M. K. COLEMAN and H. D. SNOW, *Experientia*, **25**, 278 (1969)
266. Y. HOKAMA, K. YAMADA, S. N. MOIKEHA and E. T. NISHIMURA, *Cancer Res.*, **29**, 542 (1969)
267. G. H. BEATON, A. E. SELBY, M. J. VEEN and A. M. WRIGHT, *J. Biol. Chem.*, **236**, 2005 (1961)
268. W. G. HEIM, J. M. KERRIGAN and P. H. LANE, *Nature*, **200**, 688 (1963)
269. W. G. HEIM and J. M. KERRIGAN, *Nature*, **199**, 1100 (1963)
270. W. G. HEIM and P. H. LANE, *Nature*, **203**, 1077 (1964)
271. D. J. LAWFORD, *Biochem. Pharmacol.*, **7**, 109 (1961)
272. A. E. BOGDEN, G. A. NEVILLE, W. E. WOODWARD and M. GRAY, *Proc. Amer. Ass. Cancer Res.*, **5**, 6 (1964)
273. M. STANISLAWSKI-BIRENCWAJG, J. URIEL and P. GRABAR, *Cancer Res.*, **27**, 1990 (1967)
274. G. A. BOFFA, J. M. FINE, Y. JACQUOT-ARMAND, F. GAUDIN-HARDING and H. SUBSIELLE, *C. R. Soc. Biol.*, **159**, 1342 (1965)
275. J. VAN GOOL and N. C. LADIGES, *J. Path.*, **97**, 115 (1969)
276. W. G. HEIM, *Nature*, **193**, 491 (1962)
277. D. BRANCENI and J. GONIN, *Rev. Franç. Etudes Clin. Biol.*, **14**, 754 (1969)
278. E. J. SARCIONE, *Cancer Res.*, **27**, 2025 (1967)
279. F. F. MENNINGER, H. J. ESBER and A. E. BOGDEN, *Clin. Chim. Acta*, **27**, 385 (1970)
280. E. J. SARCIONE and M. BOHNE, *Proc. Soc. Exp. Biol. Med.*, **131**, 1454 (1969)
281. W. G. HEIM, *Nature*, **207**, 1403 (1965)
282. D. W. JOHN and L. L. MILLER, *J. Biol. Chem.*, **243**, 268 (1968)
283. E. J. SARCIONE and A. E. BOGDEN, *Science*, **153**, 547 (1966)
284. E. J. SARCIONE, *Biochemistry*, **9**, 3059 (1970)
285. E. J. SARCIONE (eds., M. A. Rothschild and T. Waldmann), Plasma protein metabolism: regulation of synthesis, distribution and degradation (Academic Press, New York, 1970) p. 369
286. W. G. HEIM and S. R. ELLENSON, *Nature*, **208**, 1330 (1965)

287. B. J. ZWEIFACH, L. GRANT and T. MCCLUSKEY (Editors), The Inflammatory Process (Academic Press, New York, 1965)
288. A. BARTELLI and J. C. HOUCK (Editors), Inflammation Biochemistry and Drug Interaction, Proceedings of an International Symposium, (Excerpta Medica Foundation, Amsterdam, 1969)
289. H. SELYE, A. SOMOGYI and P. VEGH, Biochem, Pharmacol. Suppl. (1968), p. 107
290. M. I. BARNHART, Biochem. Pharmacol. Suppl., (1968), p. 205
291. R. W. KELLERMEYER and K. S. WARREN, J. Exp. Med., 131, 21 (1970)
292. A. JANOFF, G. WEISSMANN, B. W. ZWEIFACH and L. THOMAS, J. Exp. Med., 116, 451 (1962)
293. L. BITENSKY, J. CHAYEN and G. J. CUNNINGHAM, Nature, 199, 493 (1963)
294. G. WEISSMANN, Ann. Rev. Med., 18, 97 (1967)
295. G. WEISSMANN, Arthr. Rheum., 9, 834 (1966)
296. J. G. HIRSCH, (eds., B. J. Zweifach, L. Grant and T. McCluskey) The Inflammatory Process (Academic Press, New York, 1965), p. 266
297. S. WASI, R. MURRAY, D. MACMORINE and H. MOVAT, Brit. J. Exp. Path., 47, 411 (1966)
298. A. JANOFF and J. D. ZELIGS, Science, 161, 702 (1968)
299. G. WEISSMANN, I. SPILBERG and K. KRAKAUER (eds., A. Bartelli and J. C. Houck), Inflammation Biochemistry and Drug Interaction (Excerpta Medica Foundation, Amsterdam, 1969), p. 12
300. A. JANOFF, Lab. Invest., 22, 228 (1970)
301. C. L. EDDINGTON, H. F. UPCHURCH and R. F. KAMPSCHMIDT, Proc. Soc. Exp. Biol. Med., 136, 159 (1971)
302. L. M. GREENBAUM, M. C. CARRARA and R. FREER, Fed. Proc., 27, 90 (1968)
303. W. SEEGERS and A. JANOFF, Nature, 213, 144 (1965)
304. R. W. SCHAYER, Am. J. Physiol., 198, 1187 (1960)
305. G. KAHLSON and E. ROSENGREN, Physiol. Rev., 48, 155 (1968)
306. G. LAZAR and S. KARADY, Acta Physiol. Acad. Sci. Hung., 31, 77 (1967)
307. J. W. KEYSER, D. HOOPER, J. R. HARDING, S. THOMAS and J. SHAW, Biochem. J., 118, 26P (1970)
308. F. HOMBURGER, J. Clin. Invest., 24, 43 (1945)
309. E. J. SARCIONE, M. BOHNE and S. KRAUSS, Fed. Proc., 24, 230 (1965)
310. F. L. MOOLTEN and N. L. R. BUCHER, Science, 158, 272 (1967)
311. A. M. CHANDLER and G. A. SNIDER, Proc. Soc. Exp. Biol. Med., 135, 415 (1970)
312. A. H. GORDON (eds., M. A. Rothschild and T. Waldmann), Plasma Protein Metabolism: Regulation of Synthesis, Distribution and Degradation (Academic Press, New York, 1970) p. 351
313. O. BARNABEI and F. SERENI, Biochim. Biophys. Acta, 91, 239 (1964)
314. H. L. STACKHOUSE, C. J. CHETSANGA and C. H. TAN, Biochim. Biophys. Acta, 155, 159 (1968)
315. E. C. LIN and W. E. KNOX, Biochim. Biophys. Acta, 26, 85 (1957)
316. C. B. HAGER and F. T. KENNEY, J. Biol. Chem., 243, 3296 (1968)
317. K. TSUKADA, H. OURA, S. NAKASHIMA and N. HAYASAKI, Biochim. Biophys. Acta, 165, 218 (1968)
318. D. W. JOHN and L. L. MILLER, J. Biol. Chem., 241, 4817 (1966)
319. M. MAUNG, D. G. BAKER, R. K. MURRAY, Can. J. Biochem., 46, 477 (1968)
320. G. WEBER, S. K. SRIVASTAVA and R. L. SINGHAL, J. Biol. Chem., 240, 750 (1965)
321. A. M. CHANDLER and O. W. NEUHAUS, Biochim. Biophys. Acta, 166, 186 (1968)
322. Y. HAMASHIMA, J. G. HARTER and A. H. COONS, J. Cell Biol., 20, 271 (1964)
323. Y. HAMASHIMA, Prot. Biol. Fluids, 14, 245 (1966)
324. Z. SHIHABI, H. F. BALEGNO and O. W. NEUHAUS, Biochim. Biophys. Res. Comm., 38, 692 (1970)
325. A. Y. LIU and O. W. NEUHAUS, Biochim. Biophys. Acta, 166, 195 (1968)

326. K. TSUKADA, T. MORIYAMA, O. DOI and I. LIEBERMAN, *J. Biol. Chem.*, **243**, 1152 (1968)
327. K. TSUKADA, T. MORIYAMA, T. UMEDA and I. LIEBERMAN, *J. Biol. Chem.*, **243**, 1160 (1968)
328. A. VON DER DECKEN, *Biochim. Biophys. Acta*, **166**, 487 (1968)
329. K. SHORTMAN, *Biochim. Biophys. Acta*, **61**, 50 (1962)
330. Y. E. RAHMAN, E. A. CERNY and C. PERAINO, *Biochim. Biophys. Acta*, **178**, 68 (1969)
331. K. TSUKADA, *Biochim. Biophys. Acta*, **186**, 21 (1969)
332. T. S. FOSTER, *Canad. J. Biochem.*, **45**, 1245 (1967)
333. F. DE MATTEIS, *Biochim. Biophys. Acta*, **187**, 422 (1969)
334. R. D. EASTHAM and E. H. MORGAN, *Lancet*, **ii**, 1196 (1963)
335. E. M. GLENN, *Biochem. Pharmacol.*, **18**, 317 (1969)
336. E. M. GLEEN and W. KOOYERS, *Life Sci.*, **5**, 619 (1966)
337. H. BARTFELD, *Arthr. Rheum.*, **3**, 429 (1965)
338.' R. A. MACBETH, N. V. BELUR, G. D. BELL and C. BARRON, *Can. J. Physiol. Pharmacol.* **48**, 123 (1970)
339. D. C. TRIANTAPHYLLOPOUSOS, C. CHEN and E. TRIANTAPHYLLOPOULOS, *Brit. J. Haemat.*, **16**, 589 (1969)
340. H. NAKAJIMA, T. TAKEMURA, O. NAKAJIMA and K. YAMAOKA, *J. Biol. Chem.*, **238**, 3784 (1963)
341. S. OSAKI, J. A. MCDERMOTT and E. FRIEDEN, *J. Biol. Chem.*, **239**, 3570 (1964)
342. B. BOZHKOV, A. WYSOKINSKA-BOROWICZ and J. KRAWCZYNSKI, *Enzym. Biol. Clin.*, **11**, 531 (1970)
343. L. BROMAN, *Acta. Soc. Med. Uppsalien.*, *Suppl.* 7, **69**, 75 (1964)
344. C. A. OWEN, JR., *Am. J. Physiol.*, **209**, 900 (1965)
345. M. H. SHOKEIR and D. C. SHREFFLER, *Proc. Natl. Acad. Sci. USA*, **62**, 867 (1969)
346. P. O. GANROT and C. O. KINDMARK, *Scand. J. Clin. Lab. Invest.*, **24**, 215 (1969)
347. M. GANGULY and U. WESTPHAL, *J. Biol. Chem.*, **243**, 6130 (1968)
348. I. M. NILSSON and I. YAMASHINA, *Nature*, **181**, 711 (1958)
349. B. N. HALPERN, *Proc. Soc. Exp. Biol. Med.*, **115**, 173 (1964)
350. A. BERTELLI, *Biochem. Pharmacol. Suppl.*, (1969), p. 229
351. O. FÖRSTER (eds., A. Bartelli and J. C. Houck), Inflammation Biochemistry and Drug Interaction (Excerpta Medica Foundation, Amsterdam, 1969), p. 53
352. I. SPILBERG and C. K. OSTERLAND, *J. Lab. Clin. Med.*, **76**, 472 (1970)
353. O. SNELLMAN and B. SYLVEN, *Nature*, **216**, 1033 (1967)
354. J. K. NEWMAN, G. S. BERENSON, M. B. MATHEWS, E. GOLDWARE and A. DORFMAN, *J. Biol. Chem.*, **217**, 31 (1955)
355. P. LESCA and C. PAOLETTI, *Proc. Natl. Acad. Sci. USA*, **64**, 913 (1969)
356. M. R. SHETLAR, R. S. BRYAN, J. V. ROSTER, C. L. SHETLAR and M. R. EVERETT, *Proc. Soc. Exp. Biol. (N.Y.)*, **72**, 294 (1949)
357. B. N. WHITE, M. R. SHETLAR and J. A. SCHILLING, *Ann. N.Y. Acad. Sci.*, **94**, 297 (1961)
358. R. ROBERT and B. ROBERT, *Exp. Ann. Bioch. Med.*, **24**, 269 (1963)
359. A. DELAUNAY and S. BAZIN, (eds., A. Bartelli and J. C. Houck), Inflammation Biochemistry and Drug Interaction (Excerpta Medica Foundation, Amsterdam, 1969), p. 94

Addendum to Chapter 4
Acute-Phase Reactants

A. Koj

Since this chapter was submitted numerous publications concerning AP-reactants have appeared. At least some of them should be briefly discussed.

Model systems of plasma protein synthesis and turnover, including some AP-reactants, as well as regulatory mechanisms in protein metabolism, are presented in detail by several authors in relevant books edited by M. A. Rothschild and T. Waldmann,[1] H. Munro[2] and G. Wolstenholme and M. O'Connor.[3]

It appears that the list of AP-reactants should also include the third component of complement (C3). This glycoprotein of molecular weight of 240 000 and containing 2.7% of carbohydrates is present in the serum of healthy persons at a concentration of about 1.2 mg/ml.[4] Alper and co-workers[5] demonstrated that C3 is synthesized in the liver. The level of C3 in plasma increases significantly after cardiac infarction, in pneumonia and in acute phase of inflammatory diseases.[6-9] In contrast to haptoglobin and fibrinogen the increase in C3 concentration in plasma is rather modest, and in certain states such as glomerulonephritis C3 may be even decreased.[10,11] Several authors[12,13] postulate involvement of the complement system in the formation of mediators of inflammation showing prostaglandin-like activity.

The liver of injured animals produces a specific protein which inhibits oedema formation, granulation tissue deposition, delayed hypersensitivity reactions and adjuvant arthritis.[14] It is doubtful, however, whether this anti-inflammatory protein should be listed as AP-reactant since its greatest concentration occurs relatively late after injury. The participation of acute-phase reactants in wound healing has been discussed by Gordon.[15]

Considerable evidence (reviewed by Munro[16,17]) has accumulated within the past few years on the importance of amino acid supply for plasma protein synthesis in the liver. Tryptophan, in particular, appears to be important for polyribosome stability *in vivo*. At the scarcity of essential amino acids dissociation of polysomes and increased breakdown of protein and RNA occur leading to temporary reduced efficiency of protein synthesizing machinery. In relation to these data Neuhaus, Balegno and Milauskas[18] found depressed responses of the proteins and protein-bound hexosamines of some plasma globulin fractions to injury in protein depleted rats. Miller and John[19] observed impaired synthesis of fibrinogen and haptoglobin in the perfused livers obtained from rats fasting for 6 days. It appears, however, that not all acute-phase reactants conform entirely to this pattern. Di Lallo, Haley and Wiliamson[20] observed that feeding a protein-free diet did not influence the increase in the alpha-2-globulin fraction in rats subjected to surgical trauma. Riley and Hokama[21] demonstrated that starvation did not prevent the appearance of C_x reactive protein in rabbits exposed to ionizing radiation. Recently Weimer, Roberts and Comb[22] observed that injection of turpentine to rats fasting for several days produced similar increase of alpha-2-macroglobulin (alpha-2-macrofetoprotein) and of seromucoid fraction as in control, well-fed rats. The authors concluded that the response of these AP-reactants to a phlogogenic stimulus was of such magnitude that it was unaffected by acute starvation. On the other hand, sex and strain differences in the alpha-2-AP globulin response in rats were observed.[23,24]

Various aspects of abnormal metabolism of fibrinogen were thoroughly discussed by Regoeczi.[25–27] Kinetic studies on fibrinogen turnover in rabbits were continued by Reeve and Chen,[28] and in human subjects by Tytgat *et al.*[29,30] Sherman[31] demonstrated multiple pathways of fibrinogen catabolism, either by low grade proteolysis, or limited partial soluble fibrin formation. Bocci and Vitti[32] confirmed circadian fluctuations in the plasma level of fibrinogen using rat as the experimental animal. Considerable age-dependent differences in the metabolism of fibrinogen in rats were also observed.[33]

Turnover of orosomucoid in the dog with local inflammation was studied by Zeineh and co-workers.[34] When using ^{14}C-glucosamine labelled orosomucoid they found shortened half-life of this protein in the animal with a sterile abscess while the EV/IV ratio equal to 2 remained unchanged. Rudman *et al.*[35] observed abnormal orosomucoid in the plasma of patients with neoplastic disease. Gordon and Dykes[36] demonstrated heterogeneity of alpha-1-AP globulin of the rat.

Kampschmidt and co-workers[37,38] continued their studies on induction of alpha-2-AP globulin in rats by injecting partially purified leucocytic extracts. They postulate that activated rabbit leucocytes produce one or more endogenous mediators (LEM) that regulate several biochemical abnormalities associated with acute and chronic disorders. On the same line Pekarek *et al.*[39] carried out experiments suggesting that LEM controls the level of ceruloplasmin in rats.

The effect of hormones on the synthesis of AP-reactants was investigated by several authors, the subject being discussed at length by Miller and John,[19] Jeejeebhoy *et al.*,[40] and by Koj.[41] Seligsohn *et al.*[42] reinvestigated fibrinogen synthesis in adrenalectomized rabbits, and Jeejeebhoy *et al.*[43] confirmed the positive effect of cortisol on the formation of fibrinogen. Reeve and Chen[28] pursued their studies on the effect of ACTH on fibrinogen synthesis in rabbits and reached the conclusion that ACTH peptide fragment α 1–10 does not stimulate fibrinogen secretion in contrast to 1–17 fragment. Alias[44] showed that administration of ACTH to rabbits increases the level of ceruloplasmin while hydrocortisone exerts an opposite effect, probably due to suppression of ACTH production. Barbosa et al.[45] observed that 17 α alkylated anabolic steroids cause in man significant elevations of haptoglobin, orosomucoid, protein-bound sialic acid, plasminogen and beta-glucuronidase, and depression of fibrinogen. Alpha-1-antitrypsin level is significantly increased by oxymethalone.

Additional biological functions have been recently ascribed to some acute-phase reactants: ceruloplasmin was shown to deaminate oxidatively histamine[46] and to be involved in hypoferremia associated with chronic inflammation and endotoxin.[47] Hornung and Fritschi[48] observed stimulation of human lymphocytes in vitro by low doses of C-reactive protein.

REFERENCES

1. M. A. ROTHSCHILD and T. WALDMANN (Editors), Plasma Protein Metabolism, Regulation of Synthesis, Distribution and Degradation (Academic Press, New York and London, 1970)
2. H. N. MUNRO (Editor), Mammalian Protein Metabolism, vol. 4 (Academic Press, New York and London, 1970)
3. G. E. W. WOLSTENHOLME and M. O'CONNOR (Editors), Protein Turnover, Ciba Foundation Symposium 9 (Associated Scientific Publishers, Amsterdam-London-New York, 1973)
4. H. J. MÜLLER-EBERHARD, Advan. Immunol., 8, 1 (1968)
5. C. A. ALPER, A. M. JOHNSON, A. G. BIRTH and F. D. MOORE, Science, 163, 286 (1969)
6. M. HORNUNG and R. C. ARQUEMBOURG, J. Immunol., 94, 307 (1965)
7. C. A. ALPER (eds., M. A. Rothschild and T. Waldmann), Plasma Protein Metabolism, Regulation of Synthesis, Distribution and Degradation (Academic Press, New York and London, 1970) p. 393
8. I. KUSHNER, T. S. EDGINGTON, C. TRIMBLE, H. H. LIEM and U. MÜLLER-EBERHARD, J. Lab. Clin. Med., 80, 18 (1972)
9. F. HARTVEIT, W. BORVE and S. THUNOLD, Acta Path. Microb. Scand., Suppl. 236, 54 (1973)
10. D. K. PETERS (eds., G. E. W. Wolstenholme and M. O'Connor), Protein Turnover, Ciba Foundation Symposium 9 (Associated Scientific Publishers, Amsterdam-London-New York, 1973) p. 273
11. C. A. ALPER and F. S. ROSEN (eds., G. E. W. Wolstenholme and M. O'Connor), Protein Turnover, Ciba Foundation Symposium 9 (Associated Scientific Publishers, Amsterdam-London-New York, 1973), p. 283
12. J. P. GIROUD and D. A. WILLOUGHBY, J. Pathol., 101, 241 (1970)
13. H. J. MÜLLER-EBERHARD, E. H. VALLOTTA, O. GOTZE and T. H. ZIMMERMAN, (eds., I. H. Lepow and P. A. Ward), Inflammation, Mechanisms and Control, (Academic Press, New York and London, 1972), p. 261
14. M. E. J. BILLINGHAM, A. H. GORDON and B. V. ROBINSON, Nature New Biology, 231, 26 (1971)

15. A. H. Gordon (eds., G. E. W. Wolstenholme and M. O'Connor), Protein Turnover, Ciba Foundation Symposium 9 (Associated Scientific Publishers, Amsterdam-London-New York 1973), p. 73
16. H. N. Munro (eds., M. A. Rothschild and T. Waldmann), Plasma Protein Metabolism, Regulation of Synthesis, Distribution and Degradation (Acad. Press, New York and London 1970) p. 157
17. H. N. Munro (ed., H. N. Munro), Mammalian Protein Metabolism vol. 4. (Academic Press, New York and London 1970) p. 299
18. O. W. Neuhaus, H. Balegno and A. T. Milauskas, *Exp. Mol. Pathol.*, **2**, 183 (1963)
19. L. L. Miller and D. W. John (eds., M. A. Rothschild and T. Waldmann), Plasma Protein Metabolism, Regulation of Synthesis Distribution and Degradation (Academic Press, New York and London, 1970), p. 208
20. L. Di Lallo, H. B. Haley and M. B. Wiliamson, *Clin. Chem.*, **9**, 266 (1963)
21. R. F. Riley and Y. Hokama, *Rad. Res.*, **12**, 466 (1960)
22. H. E. Weimer, D. M. Roberts and J. C. Comb, *J. Nutrition*, **102**, 873 (1972)
23. H. E. Weimer and D. M. Roberts, *Comp. Biochem. Physiol.*, **41B**, 713 (1972)
24. H. E. Weimer, D. M. Roberts and J. C. Comb, *Brit. J. Exp. Pathol.*, **53**, 253 (1972)
25. E. Regoeczi (eds., M. A. Rothschild and T. Waldmann), Plasma Protein Metabolism, Regulation of Synthesis, Distribution and Degradation (Academic Press, New York and London, 1970), p. 459
26. E. Regoeczi, *Brit. J. Haemat.*, **20**, 649 (1971)
27. E. Regoeczi and K. L. Wong (eds., G. E. W. Wolstenholme and M. O'Connor), Protein Turnover—Ciba Foundation Symposium 9 (Associated Scientific Publishers, Amsterdam-London-New York 1973), p. 181
28. E. B. Reeve and Y. Chen (eds., G. E. W. Wolstenholme and M. O'Connor), Protein Turnover—Ciba Foundation Symposium 9 (Associated Scientific Publishers, Amsterdam-London-New York 1973), p. 91
29. D. Collen, G. N. Tytgat, H. Claeys and R. Piessens, *Brit. J. Haemat.*, **22**, 681 (1972)
30. G. N. Tytgat, D. Collen, J. Vermylen, *Brit. J. Haemat.*, **22**, 701 (1972)
31. L. A. Sherman, *J. Lab. Clin. Med.*, **79**, 710 (1972)
32. V. Bocci and A. Vitti, *Am. J. Physiol.*, **221**, 719 (1971)
33. I. Nadelhaft and F. Lamy, *J. Lab. Clin. Med.*, **79**, 724 (1972)
34. R. A. Zeineh, B. Barrett, L. Niemirowski and B. J. Fiorella, *Am. J. Physiol.*, **222**, 1326 (1972)
35. D. Rudman, P. E. Treadwell, W. R. Vogler, G. H. Howard and B. Hollins, *Cancer Res.*, **32**, 1951 (1972)
36. A. H. Gordon and P. J. Dykes, *Biochem. J.*, **130**, 95 (1972)
37. C. L. Eddington, H. F. Upchurch and R. F. Kampschmidt, *Proc. Soc. Exp. Biol. Med.*, **139**, 565 (1972)
38. R. F. Kampschmidt, H. F. Upchurch, C. L. Eddington and L. A. Pulliam, *Am. J. Physiol.*, **224**, 530 (1973)
39. R. S. Pekarek, M. C. Powanda and R. W. Wannemacher Jr, *Proc. Soc. Exp. Biol. Med.*, **141**, 1029 (1972)
40. K. N. Jeejeebhoy, A. Bruce-Robertson and H. J. Sodtke (eds., G. E. W. Wolstenholme and M. O'Connor), Protein Turnover, Ciba Foundation Symposium 9 (Associated Scientific Publishers, Amsterdam-London-New York, 1973), p. 217
41. A. Koj (ed., S. Nemeth), Hormones, Metabolism and Stress—Recent Progress and Perspectives, in press
42. U. Seligsohn, N. Alexander and S. I. Rapaport, *Proc. Soc. Exp. Biol. Med.*, **142**, 824 (1973)
43. K. N. Jeejeebhoy, A. Bruce-Robertson, J. Ho and U. Sodtke, *Biochem. J.*, **130**, 533 (1972)

44. A. G. ALIAS, *FEBS Letters*, **18**, 308 (1971)
45. J. BARBOSA, U. S. SEAL and R. P. DOE, *J. Clin. Endocrin. Metab.*, **33**, 388 (1971)
46. J. K. HAMPTON, L. J. RIDER, T. J. GOKA and J. P. PRESLOCK, *Proc. Soc. Exp. Biol. Med.*, **141**, 974 (1972)
47. H. P. ROESER, G. R. LEE and G. E. CARTWRIGHT, *Proc. Soc. Exp. Biol. Med.*, **142**, 1185 (1973)
48. M. HORNUNG and S. FRITSCHI, *Nature New Biology*, **230**, 84 (1971)

Chapter 5

Fibrinogen

E. Regoeczi

Department of Pathology, McMaster University
Hamilton, Ontario, Canada

5.1. INTRODUCTION

In the course of studying plasma proteins it seems almost inevitable that sooner or later research workers should become fascinated by some unique feature of the protein of their individual interest. The resulting change in attitude is an important milestone in anybody's life, for working is now felt as an entertainment rather than a recycling daily necessity. However, when a group of such workers collaborate to piece together a book on plasma proteins, there is the potential danger that the reader will be given the overall impression that plasma is nothing less than a circulating mixture of 'unique' molecules.

On introducing fibrinogen with this in mind, it is probably fair to point out that fibrinogen is neither the largest plasma protein, nor is it the only contractile fibrous element present in the body. It is not even the only protein susceptible to the enzymatic action of thrombin. Nevertheless, it is noteworthy that fibrinogen is very ancient. A coagulable protein is already present in some invertebrates, like protozoa and arthropoda.[1,2] At this stage of the evolution, coagulation is predominantly a function of cells, the coagulocytes (amebocytes). When stimulated with an appropriate agent (*e.g.* endotoxin), coagulocytes aggregate and release their contents. Cinematographic studies on limulus coagulocytes by Grégoire[3] suggest however that in contrast to mammalian platelets, their response is much more elementary-explosive, whereby coagulable material from these cells spreads out in a wide area and gelates the haemolymph. Today, this ancient intracellular fibrinogen provides the most sensitive substrate for detecting endotoxin.[4]

During further evolution fibrinogen reappeared in vertebrates, skipping the group of animals between arthropoda and vertebrates on the phylogenetic tree. However, coagulation and cellular responses to injuries of the vascular system (commonly known as platelet aggregation) now emerge more clearly as two separate principles: fibrinogen becomes an extracellular protein synthesized by the parenchymal cells in the liver together with numerous new proteins, the clotting factors, which greatly facilitate its controlled activation. Doolittle[5] temporarily regards this development as an example of parallel evolution, but this does not preclude the possibility that the phylogenetic principle of cellular coagulation is probably still present in mammals in an inactive (rudimentary?) form; it is known that platelets contain a small amount of intracellular fibrinogen,[6] and isotopic studies suggest that this does not originate from the intravascular fibrinogen pool.[7]

By purposefully adventuring into the problems of fibrinogen and evolution, it is to be hoped that one might become aware of the manifold aspects of this protein and the necessity of selecting only a few of them for discussion. The chemistry and physicochemistry of fibrinogen, as well as its normal and abnormal transition to fibrin, have been well documented in recent monographs. Effort is therefore concentrated here first of all on questions relating to its physiological metabolism. This is of course the field of radioactive techniques, where both evaluation of the work of others and active participation depends on understanding the advantages and shortcomings of the methods used. Since many of the relevant techniques are widely scattered in the literature (and not infrequently in a much too condensed form which hampers easy reproducibility), emphasis is given to technical details. To comply with editorial policy, only the central references are quoted.

5.2. SOME PROPERTIES OF THE FIBRINOGEN MOLECULE

5.2.1. SHAPE

Despite extensive efforts by numerous investigators with the electron microscope, the shape of the fibrinogen molecule is still controversial. One of the reasons for this is thought to be that at the low concentrations which are needed for electron microscopic studies the molecule is probably unstable. Dealing with fibrin strands is technically simpler, but then deductions on the structure of the basic unit become dependent on the correct interpretation of a complex polymeric system.

The fibrinogen model which has so far met with the widest approval was proposed in 1959 by Hall and Slayter.[8] In shadow-cast preparations of bovine fibrinogen, the molecules appeared as a linear structure of 3 nodules held together by a very thin thread. In the dried state, the overall length of the array was 475 ± 25 Å, the diameter of the terminal nodules 65 Å, that of the central nodule 50 Å. The connecting pieces could not be measured accurately, but they could be assessed as being between 8 and 15 Å thick. Two fibrinogen molecules in end-to-end position as drawn to the above scale in Fig. 5.1 demonstrate that the postulated structure could well explain the axial periodicity of an adjacent fibrin strand. The longitudinal arrangement of the nodules would also fit with the results of various hydrodynamic measurements which indicate that fibrinogen is an asymmetrical molecule.

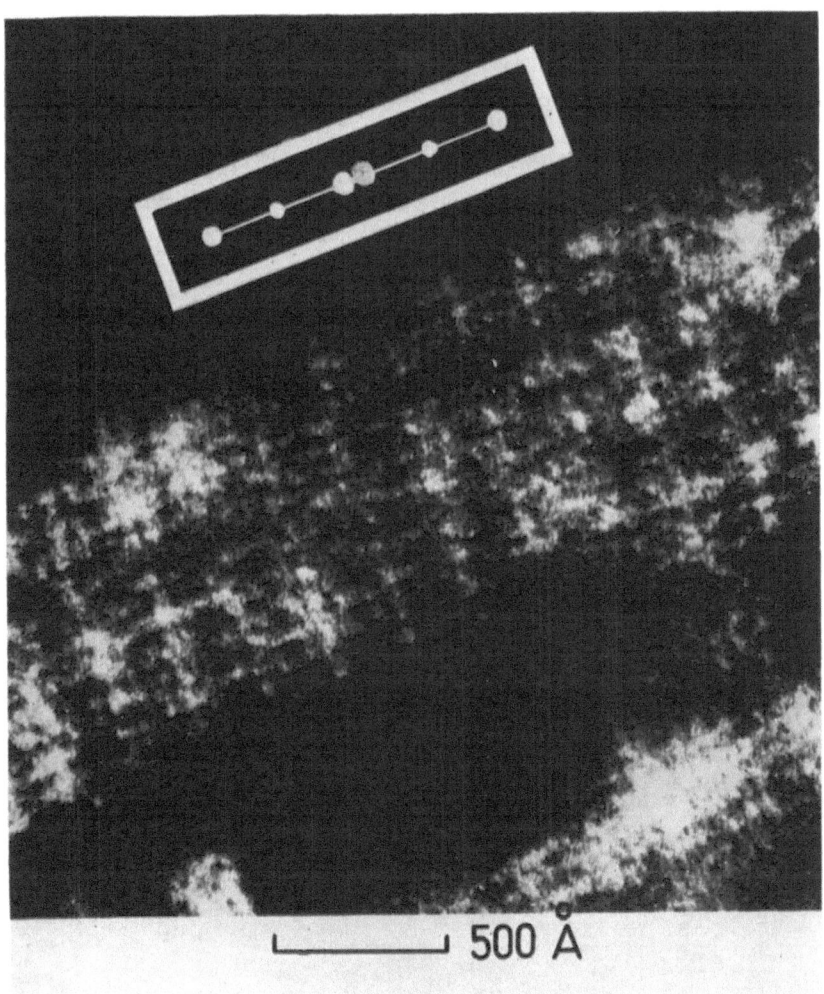

FIG. 5.1. Purified human fibrinogen clotted with bovine thrombin (negative staining, 500 000 ×). The electron micrograph was kindly prepared by the late Dr. R. C. Valentine, N.I.M.R., London. The drawing in the top section shows two fibrinogen molecules scaled to the model by Hall & Slayter[8] as they would appear at the present magnification.

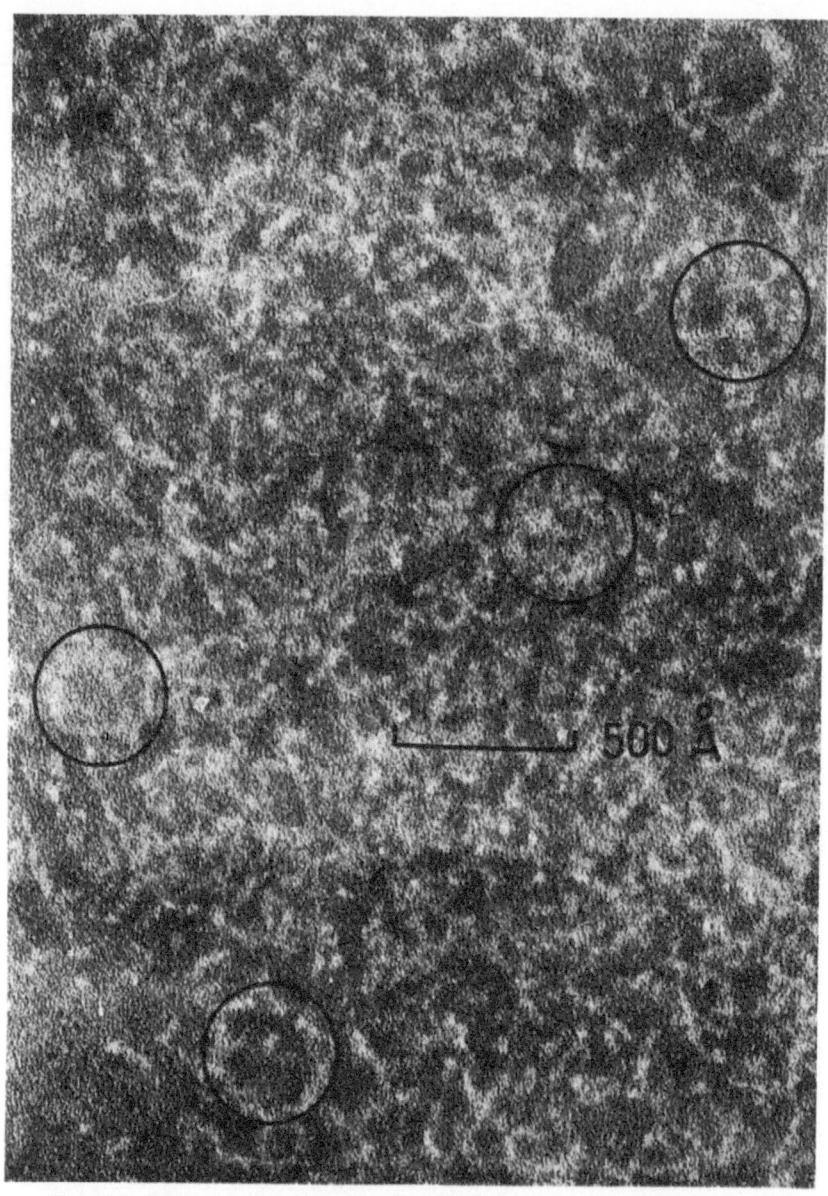

FIG. 5.2. Electron micrograph of rabbit fibrinogen, obtained by the kind collaboration of Dr. R. R. Dourmashkin, N.I.M.R., London. Negative staining, 497 000×. For explanations see text.

Taking all this into consideration it seems rather a pity that subsequent investigators have not been able to confirm the existence of the trinodular structure.[9,10] The electron micrograph shown in Fig. 5.2 should make this point more easily understood. It was taken from a rabbit fibrinogen preparation containing 97.6% thrombin-clottable protein. Before dilution, the fibrinogen solution was chilled in ice water and was kept near the freezing point until drying on the grid was complete. The diluent contained calcium.

The picture in Fig. 5.2 does not allow definite conclusions regarding the size and shape of the fibrinogen molecule. Nonetheless, it shows clearly that the basic layout of the protein material is more comparable with the model proposed by Köppel[10] than with a trinodular arrangement. Using a special printing technique aimed to improve the contrast between protein and background, this author found that fibrinogen has the shape of a pentagon dodecahedron (*i.e.* a corpuscle surrounded by 12 pentagonal planes). This form is thought to be established in such a way that the polypeptide chains (3 pairs) are wound along the edges (70–75 Å each), whereas the space within the particle as well as the planes themselves are 'empty'. When rotated around one of its axes, the projection of a pentagon dodecahedron is manifold, and consequently its diameter can adopt values between 170 and 260 Å.

The overall impression from Fig. 5.2 is that the protein material was present primarily in the form of fine threads or filaments. These were neither straight nor curved but were arranged in numerous angles. Many of the angled filaments followed a polygonal structure, so that the whole picture resembles an irregular honeycomb. Several pentagons are clearly visible, some of which are marked by an arrow. In certain areas integration of the polygonal elements into a higher-grain or bead-like structure can also be observed (see circles).

By not having applied special techniques to harden the photographs, many details in Fig. 5.2 regarding line continuities have been lost against the background at the price of a clear molecular image. But even so, the filamentous appearance of the basic material forming numerous angles and polygons seems to be well reconcilable with the idea that the polypeptide chains of fibrinogen are arranged along the edges of a hollow geometrical form—perhaps of the type favoured by Köppel. Certain stereochemical illustrations published by Blombäck[11] lend fibrinogen the appearance of a Y-shaped molecule, and it is not difficult to realize that the structure visible in Fig. 5.2 could have also been established by units of this kind.

Neither in this, nor in other similar micrographs could a trinodular arrangement of particles be observed, regardless whether the preparations were fixed with formaldehyde. Admittedly, it may be argued that despite the precautions taken at processing the native protein, the material has disintegrated on contact with the carbon film—if not earlier. In accordance with the laws of probability, a few molecules however should have escaped distortion even in the case of severe denaturation. Noncovalently bonded subunits would make the molecule particularly vulnerable, but since the molecular weight of fibrinogen does not change in 5M guanidine-HCl (an agent which completely unfolds it), the existence of subunits of this kind is thought to be unlikely.[12]

5.2.2. PHYSICOCHEMICAL ASPECTS

Only a few points which are relevant to the handling of the protein will be mentioned here.

Fibrinogen (m.w. 340 000) is chemically a dimer inasmuch as it is composed of a twin set of 3 polypeptide chains, called α (A), β (B) and γ, of approximately equal molecular weights (50 000–65 000).[13,14] According to Blombäck et al.,[15] the N-terminal ends of the chains are held together by a 'disulphide knot', the molecule in this region containing 3 inter- and 2 intrachain disulphide bonds. If fibrinogen is subjected to partial proteolysis by plasmin, the first chain to undergo alterations is the α (A), and from this Gaffney[16] concludes that the α (A) chain is located on the outer surface of the molecule. Fibrinogen also contains a few percent carbohydrate in the forms of hexose (mannose and galactose), D-glucosamine and sialic acid. However, it contains no free SH-groups.

Fibrinogen can be electrophoresed relatively easily both in starch and acrylamide, although its mobility is slow. Fibrinogen is also suitable for gel filtration, but it is important to use a correct gel grade; Sephadex-type gels, including G-200, are only useful to separate fibrinogen from smaller proteins (IgG, caeruloplasmin, or smaller). Separation of fibrinogen co-polymers on one hand and the large fragments (X, Y) of fibrinogen on the other from the native molecule can be attempted on Sepharose 6B columns, though even here, molecular species may considerably overlap and give rise to shouldering or the formation of complex peaks.[17] The primary application of Sepharose 6B and similar columns is therefore less in the preparative and more in the analytical field, particularly in combination with one or more radioactive column-markers. The same applies also to the separation of native fibrinogen from plasma proteins of higher molecular weight such as α_2-macroglobulin.

For most practical purposes it is convenient and sufficiently accurate to determine the concentration of purified fibrinogen solutions by their absorption in ultraviolet light. Extinction coefficients ($E_{0.1}$) of human, rabbit, dog, bovine, and goat fibrinogens are all close to 1.5–1.6.[18] To minimise the error one should keep in mind that (1) the position and the intensity of the absorption peak is influenced by the pH of the solvent, and (2) the extinction coefficient of various plasma proteins differs considerably. Contamination with other proteins can therefore affect the results disproportionately.

When fibrinogen is converted to fibrin, the fibrinopeptides are released. The biochemistry of these peptides is relevant to working with iodine-labelled fibrinogen since in any species in which one of these peptides (B or B-analog) contains a tyrosyl residue the phenolic hydroxyl group of which is unesterified some of the label is substituted into this residue.[19] As only a limited number of the tyrosyl residues in the fibrinogen molecule are available for iodination, the radioactivity carried by fibrinopeptide B represents a significant fraction of the total activity in preparation. In canine fibrinogen, for example, this fraction amounts to 11–15%.[19,20]

5.3. THE CHOICE OF LABEL FOR STUDIES *IN VIVO*

Basic considerations in relation to selecting an optimal label for fibrinogen *in vivo* are essentially the same as those for other plasma proteins. Accordingly, tracer techniques can be identified as belonging to one of the following two main groups: (1) a marked, natural constituent of the protein is incorporated biosynthetically, or (2) an artificial marker is associated with the protein *in vitro*. For practical purposes it is important to distinguish in group (1) two alternatives, (1/a): the labelled precursor is injected into the subject in whom the measurements are to be performed, and (1/b): the precursor is given to a donor and the resulting labelled protein is passively transferred into the recipient under investigation. Both of these techniques have been applied to fibrinogen.

Procedures of the type (1/a) involving ^{35}S-methionine or cystine as a precursor, were characteristic for the nineteen fifties, i.e. the epoch of pioneer studies on plasma proteins *in vivo*. However, as improved marking techniques were forthcoming, labelled amino acids have largely been abandoned because of the theoretical difficulties due to reutilization of the precursor.[21] What is known today about fibrinogen metabolism in a quantitative sense is the result of studies with iodine-labelled preparations. But in spite of the well known fact of amino acid reutilization, studies using labelled precursors (^{75}Se-methionine, ^{14}C-glutamic acid) are still occasionally performed, probably reflecting a not unrealistic concern among some investigators about the possible adverse effects of isolating fibrinogen from plasma.

While reutilization is considerable if the label spreads before and during the formation of biosynthetically labelled fibrinogen, recycling from the preformed protein is practically negligible. Passive transfer of carbon-labelled fibrinogen, although excluded from clinical use on ethical grounds, may be looked upon therefore as the most critical experiment designed to test the behaviour of iodinated fibrinogen. The existing proof that iodination *per se* does not alter the metabolic properties of fibrinogen originates from experiments with simultaneously injected ^{14}C- and ^{131}I-fibrinogens.[22]

5.4. THE TURNOVER OF FIBRINOGEN

5.4.1. METHODS FOR MEASURING THE TURNOVER OF FIBRINOGEN

(a) *General considerations*

Nowadays many methods are available to evaluate data obtained *in vivo* with tracer proteins; they have been the subject of excellent reviews by Bianchi *et al.*[23] and Donato.[24] Almost every method is based on a different set of assumptions which may appeal to one mind but not to another. Nevertheless, a best choice should in any instance take into consideration the metabolic characteristics of a particular protein. In the case of fibrinogen these are: (1) high metabolic rate, (2) a small extravascular compartment, and (3) first order diffusion and catabolic rates.

Since labelled terminal hydrolysis products of proteins (iodotyrosines and radioiodide) are eliminated from the body at finite rates, rapid metabolism is likely to create problems if studies are based on excreted radioactivities (cf. Fig. 5.3). An advantage arising from the predominantly intravascular distribution of fibrinogen is that the error due to inevitable assumptions concerning the size of the extavascular pool is minimized. With fibrinogen there is no need, therefore, to subject the data to complicated multicompartmental analysis. The first order kinetics of fibrinogen metabolism provide a basis for studies in 'steadily non-steady states', which is a distinct advantage considering that fibrinogen is an acute-phase protein with a labile synthesis rate.

When examined in the light of the above general considerations some of the most widely known techniques can be evaluated as follows:

(b) *U/P methods*

These methods relate 'catabolized' (= urinary) radioactivity to intravascular (protein-bound) activity over identical intervals.[25,26] In a recent modification, an attempt has been made to allow for the activity already liberated by catabolism but not yet excreted from the body.[27]

It is to their credit that these methods do not accept any radioactivity which has disappeared from the plasma as catabolized until this fact is evidenced by the appearance of the terminal breakdown products. Hence, the danger of mistaking sequestration of labelled proteins at extravascular sites for catabolism (edema) is with U/P methods correspondingly small.

Yet U/P methods are not the best choice for studies with fibrinogen. Theoretically, the relation of urinary radioactivity to the plasma curve over the *same* period would only be correct if catabolized activity were excreted instantly. However, this is not the case, so that urinary and plasma activities, in a chronological sense, are more or less out of phase.

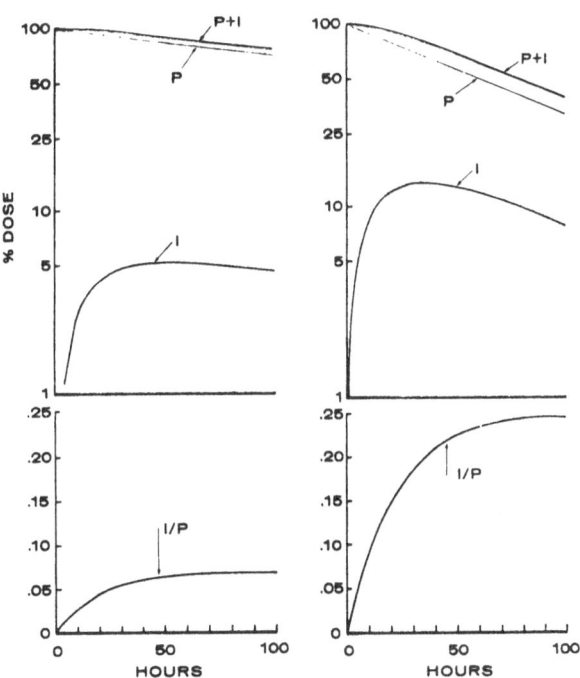

FIG. 5.3. Computer output for total body protein-bound (P) and iodide (I) activities of a slowly ($t_{1/2} = 8^d$, left panel) and a rapidly ($t_{1/2} = 2.5^d$, right panel) catabolized protein, the total body half-life of radioiodide being in each case 12 hours. Such curves would be representative of albumin and fibrinogen in an average rabbit in which no special measures were taken to speed up excretion of iodide. For further explanations see text.

The resulting error is negligible if these methods are applied, as originally intended, for slowly catabolized proteins, because the change in the plasma protein slope is minute compared with the clearance rate of the catabolized label. However, the situation is quite different when the protein slope is also fast. This is illustrated in Fig. 5.3 by comparing theoretical total body protein-bound and iodide activities for a slowly and a rapidly catabolized protein (albumin and fibrinogen), the clearance rate for radioiodide in both cases being the same. It will be seen that iodide not only contributes to a noticeably larger extent to the total body radiation in the case of the protein having a short half-life, but also that the ratio of radioiodide to labelled protein (I/P) increases continuously. In other words, production of catabolized label and its elimination from the body fail to attain equilibrium.

The above theoretical problem is usually further complicated in practice by the difficulty of collecting urine quantitatively. The latter is not merely a hardship with laboratory animals, but perhaps even more so in prolonged studies on humans. Unfortunately, experimenters often do not appreciate what they expect their patients to perform until they undertake a study on themselves. No doubt, without certain special arrangements it is nearly impossible to avoid losing urine during de-faecation. Additional problems with collecting urine frequently arise in elderly persons of both sexes.

(c) *Plasma slope methods*

Methods based on assaying intravascular protein-bound activities are at present probably the best available option to evaluate experiments with rapidly catabolized proteins, including fibrinogen. The mammillary compartmental model by Matthews,[28] the multiple pipeline model by Reeve and Bailey[29] and the plasma slope integral method by Nosslin[30] appear to be equally useful and to give identical results. In any case, *clottable* protein-bound activity values rather than the curve of plasma protein-bound activity should be used for calculations, although the two slopes, provided that the labelled preparation was reasonably homogeneous, are usually close.[31]

Equations used in the plasma slope methods are listed in the appendix (section 5.7.2), and their solution is also shown by the stepwise evaluation of a clinical case.

In recent years, the wider application of these methods has been greatly facilitated by the fact that desk computers became available as an intermediate between electric calculators on one hand and computing centres on the other. The Wang 720 desk computer, for example, has 249 memory stores and executes programs up to 1984 steps at one time. The scheme of analysing plasma slopes (cf. section 5.7.2) is programmed once and is recorded on a magnetic tape. To operate the program only the variables (*i.e.* the radioactivity value and the time of the samples) have to be keyed in, whereas the optimal function fitting the data and the sequence of equations are solved automatically within seconds.

Before the era of desk computers curves were often evaluated using simplified formulae. Thus the fractional catabolic rate (FCR) was obtained as the product of the rate constant of the terminal slope and the pool ration: $FCR = \ln 2/t_{1/2} \cdot (R + 1/R)$, where R is i.v./e.v. fibrinogen. R was estimated by extrapolating the terminal slope to zero time (cf. Fig. 5.4).

To see how far older data can be regarded reliable in the light of contemporary evaluation techniques, the data from ten randomly selected human experiments were evaluated both ways. The results are summarized in Table 5.1 and they show systematic differences, the Sterling technique giving consistently higher values. It would appear that while this difference regarding fractional catabolic rates is not too serious and is probably acceptable for most types of work, the disagreement concerning the size of the extravascular fibrinogen pool is

Comparison of the fractional catabolic rate (FCR) and the distribution of fibrinogen between intra- and extravascular spaces as obtained by exponential analysis (A) and the sterling technique (B) in a random group of normal (n) and abnormal (a) clinical cases

Case No.	Type	A FCR (%/day)	e.v./i.v.	B FCR (%/day)	e.v./i.v.	Δ FCR (%)	Δ Ratio (%)
1	n	24.4	0.24	25.2	0.32	+3.2	+33.3
2	n	26.4	0.37	27.4	0.50	+3.8	+35.1
3	n	22.9	0.15	25.4	0.37	+10.9	+146.6
4	n	19.3	0.33	20.3	0.46	+5.1	+39.3
5	n	20.1	0.29	21.5	0.46	+6.9	+58.6
6	n	20.8	0.26	21.5	0.34	+3.3	+30.7
7	a	35.7	0.26	38.8	0.47	+8.6	+80.7
8	a	39.1	0.29	43.3	0.56	+10.7	+93.1
9	a	40.4	0.10	41.8	0.16	+3.4	+60.0
10	a	60.5	0.29	63.9	0.43	+5.6	+48.2
Mean		30.9	0.26	32.9	0.41	+6.1	+62.5

more fundamental: exponential analysis suggests that the mean extravascular fibrinogen pool in the group equalled 20.2% of the total body pool, in contrast to 28.6% by extrapolation. The two methods of calculating extravascular fibrinogen are illustrated and explained in Fig. 5.4. The magnitude of the difference between A and B data in Table 5.1 does not seem to depend on the rate of catabolism.

5.4.2. NORMAL TURNOVER RATES IN MAN

A synopsis of the parameters which may be regarded as characteristic for the turnover of this protein in healthy human beings is given in Table 5.2. It is based on 33 studies, including observations from Amris and Amris,[32] Baker, Rubenberg, Dacie and Brain,[33] Davies, Ricketts and Bull,[34] Hickman,[35,36] Takeda[37] and the present writer. To put catabolic and diffusion rates in their correct perspective, a detailed analysis of the possible errors is drawn up in the subsequent section (5.4.3).

It will be noticed that, except for the absolute catabolic rate, only fractional rates were used to describe distribution and metabolism of fibrinogen in Table 5.2. Since F_c and $F_{i,e}$ are of a first order kinetic process,[38] this should not be too much of a surprise.

A rough idea about the synthetic activity of the hepatocytes can be obtained by calculating the number of fibrinogen molecules which are produced by them a day. The synthetic rate of 34 mg/kg/day (cf. Table 5.2) amounts to a total of 2.38 g fibrinogen in a 70-kg man

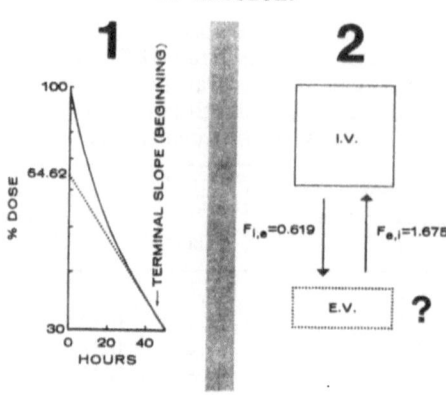

FIG. 5.4. Calculating extravascular fibrinogen by (1) extrapolation and (2) exponential analysis. Ad (1): This method assumes that the size of the e.v. pool is proportional to the initial drop of the plasma activity curve. In the presented case (cf. also Fig. 5.8), 64.6 % of the total body fibrinogen (= 100 %) is intra- and $100 - 64.6 = 35.4 \%$ is extravascular. The pool ratio (e.v./i.v.) is therefore 0.54. Ad (2): The e.v. pool is calculated from the ratio of the transcapillary rate ($F_{i,e}$) and the return flow ($F_{e,i}$), each expressed as a fraction of the pool from which the transfer takes place. Hence the difference in their numerical value, though they represent identical quantities of protein. Since I.v. pool × 0.619 = e.v. pool × 1.675, the e.v. pool = I.v. pool × 0.919/1.675. By this method the pool ratio for the same case is 0.37.

(= 4.22×10^{18} molecules). Since the number of parenchymal cells in a 1500-g liver is about 1.8×10^{11} (120 million cells/g tissue[39]), the contribution of one cell must be in the order of 23.5 million molecules a day, or 271 per second. One fibrinogen molecule consists of nearly 3000 amino acids, therefore the number of amino acids assembled into fibrinogen must be around 785 000/cell/second.

This calculation is based on two assumptions, namely that the synthesis rate is steady throughout the day and that each hepatocyte takes an even share. If these assumptions are false (and they probably are), then the true synthetic activity of the parenchymal cells engaged into

TABLE 5.2
Normal rates of fibrinogen metabolism in man

Number	F_c $\bar{x}$	S.D.	$F_{i,e}$ $\bar{x}$	S.D.	$F_{e,i}$ $\bar{x}$	S.D.	e.v./i.v. $\bar{x}$	S.D.	ASR* $\bar{x}$	S.D.
33	0.23	0.03	0.45	0.25	1.93	1.25	0.23	0.09	34	11

$\bar{x}$ = mean, S.D. = standard deviation (±).

* ASR = absolute synthesis rate in mg/kg/day as inferred from the pool size and the catabolic rate

making fibrinogen is even higher. Obviously, science has a long way to go: in human-made assembly lines it takes many hours to link together amino acids in the right order to obtain polypeptide chains of very modest lengths.

5.4.3. THE RELIABILITY OF FIBRINOGEN TURNOVER MEASUREMENTS

As explained in the appendix (section 5.7.2), the plasma curve of labelled fibrinogen can be described as the sum of two exponential terms with the constants C_1, C_2 and the exponents b_1, b_2. Thus the success of studies *in vivo* depends on the accuracy with which these values are determined.

A computer program elaborating these values should always include a subroutine to check the reliability of the slopes. Useful measures for this purpose are the standard error of the slope, expressed as percentage of the regression coefficient, and the degree of determination. The latter is the square of the correlation coefficient and it can have values between 0 and 1 (or 0 and 100 if expressed as a percentage). The attractive feature about it is that it gives an immediate impression of the proportion of changes in radioactivity which is referable to changes in time.

To illustrate the magnitude and distribution of error encountered in human experiments, some data are presented in Table 5.3. They are from the same group of randomly selected studies which have already

TABLE 5.3

Analysis of the accuracy in determining plasma slopes of labelled fibrinogen in humans

No	1st Exponential a	b	2nd Exponential a	b	$C_1 + C_2^*$
1	3.6	99.3	9.6	97.2	0.987
2	4.8	99.2	20.0	92.5	0.939
3	2.1	99.5	32.1	76.3	0.999
4	2.7	99.6	8.1	99.3	0.977
5	6.4	99.1	23.5	94.7	0.979
6	2.2	99.5	17.0	97.1	0.989
7	3.9	99.3	27.8	86.5	0.984
8	3.7	99.0	28.0	76.0	0.952
9	2.6	99.7	21.7	95.4	0.946
10	3.7	99.0	20.9	88.3	0.999
Mean	3.5	99.3	20.8	90.3	0.975
S.D.	1.3	0.2	7.7	8.4	0.021

a = standard error of the slope, expressed as % of the regression coeff.
b = degree of determination (%).
* = both constants obtained by regression analysis.

been introduced in Table 5.1. These data show that the first exponential term—representing catabolism—can be determined at a satisfactory level of accuracy, so that both C_1 and b_1 may be regarded as reliable values. Unfortunately, the situation is considerably less favourable with regard to the 2nd exponential (diffusion), as the averaged standard error of 20% seems to be rather high even for a biological experiment. That the experimental error mainly affects the second exponential term was also found by Atencio, Joiner and Reeve[40] in studies on rabbits.

The last set of values in Table 5.3 is a control for the applicability of the widely accepted double-exponential model for fibrinogen. According to this model, labelled fibrinogen disappears from the plasma either as a result of catabolism or of diffusion into a single extravascular pool ("single" in a kinetic rather than an anatomical sense). C_1 is the constant for the fraction undergoing catabolism, and C_2 for the one that diffuses. Logically, if there are no other pathways for fibrinogen to leave the plasma, then $C_1 + C_2$ must equal 1. For this reason it is an accepted custom to calculate C_2 as $1 - C_1$. In contradistinction to this custom, the C_2 values in the sums: $C_1 + C_2$ in Table 8.3 are those which are consistent with the best fit of the points defining the 2nd exponential. It is noteworthy that the average of the sums: $C_1 + C_2$ is by 2.5% short of the theoretical value of 1, and that random values exceeding 1 are absent. It seems likely therefore that despite the difficulties in obtaining accurate measurements of the second exponential, this discrepancy is probably genuine.

If so, the addition of a small and fast third exponential term to the formula now used to describe the plasma fibrinogen curve may become warranted. For the time being, this question is more of theoretical than practical interest since, as shown by the data in Table 5.4, it makes relatively little difference in practice whether $C_1 + C_2$ is exactly one or is somewhat less.

On the basis of the above considerations it is justifiable to say that catabolism of fibrinogen can be measured quite reliably. This circumstance which, no doubt, is mainly due to the predominantly intravascular distribution of body fibrinogen, is also reflected in the low standard deviation of catabolic rates among healthy subjects (cf.

TABLE 5.4

Comparison of the mean catabolic rate (F_c), capillary transfer rate ($F_{i,e}$), return flow ($F_{e,i}$) and pool ratio (e.v./i.v.) of fibrinogen in ten human studies using two different constants for the second exponential term

Constants	F_c	$F_{i,e}$	$F_{e,i}$	e.v./i.v.
$C_1 + (1 - C_1)$	0.309	0.320	1.223	0.26
$C_1 + C_2^*$	0.312	0.288	1.214	0.23

* see last column in Table 5.3

Table 5.2). The comparatively less satisfactory situation concerning the determination of the diffusion rates can be explained to a large extent by the short half-life of the second exponential ($\simeq 13 \pm 5$ hour). This, together with the normally small constant C_2, means that the second exponential term rapidly approaches zero, whereby changes in the plasma curve become solely a function of the catabolic process. Calculations show that, on the average, the contribution of diffusion to changes in plasma radioactivities in humans falls below 1% at 60 hours following the injection of labelled fibrinogen. However, for reliable measurements, the contribution of this term should be at least 5%, which will usually be reached at 31 hours. As a result of the limited time available to withdraw samples before the second exponential term expires, their number is in most cases a modest 3–5. This fact affects the error function unfavourably because the number of points underlying the confinement of a slope bears a significant weight in the calculations.

The short time is also a major obstacle towards the clarification of the precise nature of the small, but nevertheless regularly demonstrable discrepancy between the two-exponential theoretical model and the results of studies *in vivo*. It seems that we are dealing with an additional component with a C-value around 0.02–0.03 and a b-value in the order of 5–6 day^{-1}, but this will have to be confirmed in special studies.

The physiological meaning of this hypothetical component is even less clear: it may represent a small amount of denatured, viz. partially proteolyzed but still clottable fibrinogen (X-fragment); if this were so, its regular occurrence would almost inevitably imply that fibrinogen cannot be removed from the body without causing a small proportion of the molecules to fragment. On the other hand, the presence of a small but rapidly exchanging extravascular fibrinogen compartment or adsorption of some of the fibrinogen on intravascular surfaces (platelets, endothelium, etc.) appears to be equally feasible possibilities.

5.5. MAINTENANCE OF THE NORMAL METABOLIC BALANCE

The experimental evidence available today suggests that the body fibrinogen pool is maintained by a remarkable biological system characterized by regulated synthesis, primarily unregulated catabolism and no apparent coordination between production and elimination.[38,40,41] The details of how these conclusions have been arrived at over the last few years will not be described here. Instead, consideration will be given to less well understood aspects of the model and areas of diverging opinions.

5.5.1. CATABOLISM

The pronounced dependence of the absolute catabolic rate on the plasma concentration of fibrinogen has led to the working hypothesis that this protein is being removed for catabolism by a blind, droplet-ingesting process (pinocytosis). The life-span of fibrinogen, unlike that of erythrocytes,[42] does not seem to be "preset", as appears from studies with labelled fractions which were prepared from acutely hyperfibrino-genaemic animals consisting mainly of molecules of a relatively narrow range of age. Furthermore it appears that sudden changes in pool size are followed by equally rapid changes in the absolute catabolic rate, whereby it becomes unlikely that fibrinogen catabolism is the mani-festation of utilization in a biological process. The description as a random destructive process of unknown destination is probably more realistic.

The fact that this catabolic pattern is superimposed by a similar diffusion kinetic lends support to the hypothesis that catabolism and diffusion of fibrinogen are likely to have a common pathway. In visualizing this, it is assumed that the fraction of plasma fibrinogen which actually crosses the capillary endothelial barrier is the sum represented by both the catabolic and diffusion rates ($F_c + F_{i,e}$); during or shortly after passage F_c undergoes degradation, while $F_{i,e}$ survives and recirculates. The catabolic process may take place either in the endothelium itself,[43,44] or perhaps in the pericapillary histio-cytes. The latter possibility emerged indirectly from the observation that experimentally induced hypercatabolism of fibrinogen was accom-panied by hyperendocytosis of iodinated polyvinylpyrrolidone.[45] Histiocytes take up a large fraction of the polymer following intra-venous injection.[46] Needless to say, this would not exclude the participa-tion of other cells or tissues.

A somewhat different picture, which is seemingly in conflict with the catabolic mechanism as outlined above, was proposed recently by McFarlane and Koj.[27] These authors studied the catabolism of ioai-nated fibrinogen (and albumin), in the early hours after injecting the labelled protein into rabbits by relating catabolized radioactivities to the mean protein-bound activity in the circulation over the same period of time. The radioactivity liberated by catabolism was either obtained directly as the sum of the activity excreted and of that believed to be in the body iodide pool, or indirectly, that is, by simultaneously injecting an independent marker for inorganic iodine and applying the kinetics of its distribution and elimination to the non-precipitable fraction of the labelled protein in the plasma.

With these techniques it was found that the fractional catabolic rate was very low to start with and gradually rose to normal values over the first 24–36 hours of injection. To explain this finding, the authors were inclined to adopt the view that there may exist an intracellular catabolic pool of significant protein content which is independent of the plasma and extravascular pools, and consequently, the presently held theory, that the specific activity of labelled proteins at the catabolic site is close to that in the plasma, would no longer be tenable.

The existence of a sizeable pool of fibrinogen awaiting catabolism in the intracellular compartment would necessarily invalidate most of the kinetic models in current use to quantify turnover. For this reason it is of a considerable practical importance to examine whether there is an alternative explanation for the phenomenon of 'rising catabolic rates'.

A hitherto unexploited source of additional information pertinent to this question is presented by total body radioactivity measurements. Obviously, if the full catabolic rate is not reached until about the first experimental day, this must be reflected in a delayed onset of the single-exponential decay of the total body radiation. Moreover, it should be possible to assess the overall delay (that is as if catabolism had suddenly turned from zero to normal speed after a lag-phase) by calculating the time at which the regression line of the total body activity extrapolates to 100% (Fig. 5.5).

The relevant values from 9 recently recorded total body curves[38] averaged at 5.7 (S.D. ± 3.2) hours, implying that the full involvement of the injected dose into catabolism could not have been delayed more than that. In fact, it is predictable that the true delay time must be even shorter, considering that labelled breakdown products in the body significantly contribute to the total body gamma radiation[47] and displace its slope in the same way as delayed catabolism would. When an allowance of 4–6% is made for the contribution of these products to the total body radiation,[48] the regression line of the resulting theoretical

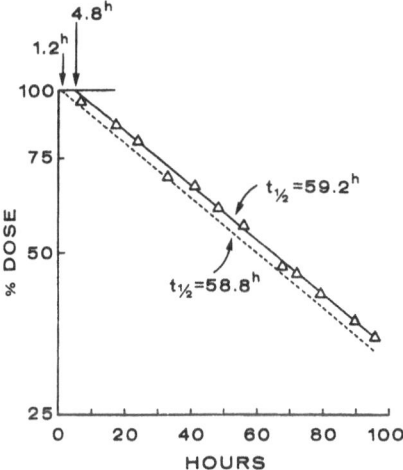

FIG. 5.5. Total body activities in a rabbit following the injection of
^{131}I-fibrinogen. The dotted line was obtained by allowing 5 % for the contribution of labelled breakdown products to the body γ-radiation. The arrows mark the times at which the curves intercept the full dose level.

total body protein-bound activity curve intercepts the 100% value within 1.5 hours of the injection time.

Thus total body radioactivity measurements do not favour the assumption that the apparently low initial catabolic rates of fibrinogen are due to a slowly equilibrating intracellular pool of this protein. Nor does the fact that sudden injection into rabbits, which were preinjected with labelled fibrinogen, of exogenous fibrinogen sufficient to nearly halve the intravascular specific activity, fails to shift the total body slope:[38] in terms of the hypothesis under discussion, this treatment should have had the opposite effects of those following injection of labelled protein.

Since fibrinogen exhibits this behaviour only if early catabolic rates are expressed in terms of liberated radioactivity, it seems likely that the phenomenon of rising catabolic rates is an artefact arising from the peculiar pathway of iodide in the body, much of which is poorly understood. Thus, plasma and total body slopes of a single intravenous dose of iodide are different, in animals with blockaded thyroid more radioactivity leaving the plasma than the body. No doubt, this is due, at least partly, to the fact that some 35% more iodide is being secreted into the upper gastronintestinal tract than excreted with the urine during the same time,[49] and that the radioactivity trapped in the contents of the gut remains unavailable for elimination until reabsorption from the colon has taken place. A convexity frequently seen in the plasma iodide curves 2–6 hours following the injection of radioiodide (cf. Fig. 5.6) could be due to the first wave of radioactivity returning via the intestinal pipeline.

The situation is probably even more complicated when radioiodide is not injected but is produced *in vivo* by the catabolism of labelled proteins because of the uncertainty about the extravascular distribution volume of the breakdown products: (1) Iodide space estimations are based on radioiodide injected into the central (plasma) compartment. However, an unknown fraction of the radioactivity liberated during protein catabolism is released directly into the extravascular space. Since transfer rates for iodide activity are not the same in both directions, the ratio of intra- to extravascular free radioactivity is likely to be different in protein turnover studies compared with the situation after intravenous injections of iodide, (2) The intracellular distribution volume and specific activity of labelled terminal breakdown products of proteins cannot be assessed by injecting labelled iodide because most cells are impermeable to this anion. This is a probable explanation for the observation that the quantity of nonprecipitable radioactivity in catabolically active organs may considerably exceed on a weight basis the corresponding value in the plasma. Unpublished experiments showed, for example, that the spleen of rabbits injected with *Trypanosoma brucei* contained 24 hours after receiving labelled fibrinogen nonprecipitable radioactivity equivalent to 2–3 ml plasma per g of tissue.

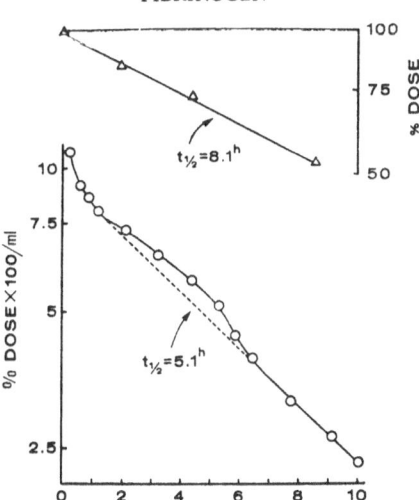

FIG. 5.6. Total body (△) and plasma (○) radioactivities in a 3 kg rabbit following the intravenous administration of Na^{131}I. The drinking water contained 3.3×10^{-5} M NaI before and during the experiment.

5.5.2. DIRECT OR INDIRECT FIBRINOGEN CATABOLISM?

The problem to be discussed here briefly is whether the physiological pathway of fibrinogen catabolism includes a priming step, that is some kind of alteration of the molecule making it liable to removal from the circulation and subsequent hydrolysis.

The idea is not altogether new: theorizing along these lines has started well before the development of the present molecular approach to plasma protein catabolism and has culminated in the concept that the fluidity of blood is being maintained by a subtle dynamic balance between (continuous) intravascular coagulation and fibrinolysis. What this concept implies is that the difference between the ways of handling fibrinogen under normal and abnormal conditions is quantitative rather than qualitative. An excellent review of the theory of continuous formation and lysis of fibrin in the body was recently presented by Astrup.[50]

Compared with the weight of this issue as a biological principle, the support originally offered for it appears surprisingly meagre. It was postulated that the relatively short biological half-life of fibrinogen reflects 'consumption' in coagulation. It is true that some 15 years ago the rapid turnover of fibrinogen was indeed in contrast to the slow catabolic rates of albumin and gamma globulin, the only other proteins available for metabolic comparison at that time. However, from later studies it is now clear that the speed of the turnover of many plasma proteins comes close to or even surpasses that of fibrinogen. This holds, for example, for prealbumin, caeruloplasmin, haptoglobin and various classes of lipoproteins—all of which are outside the haemostatic chain. With considerably more metabolic information at hand today, one is inclined to reverse the emphasis by asking 'why is the life time of albumin and IgG so unusually long'?

Hjort and Hasselback[51] and Hjort,[52] who thoroughly surveyed the field of clinical and experimental haemostasis in seeking evidence for the theory under discussion, came to negative conclusions. Reviewing the isotopic aspects of the problem a few years ago Regoeczi[43] shared their view. No data have emerged since then which would require reconsideration. On the contrary, Takeda's[53] observation that normal catabolism of prothrombin does not involve preliminary conversion to thrombin lends further support to the view that fibrinogen is catabolized directly.

Thus while continuous formation and lysis of fibrin appears to be improbable at present as a physiological pathway for the catabolism of fibrinogen, small fractions of it are likely to be utilized in the form of fibrin for normal repair processes. As can be inferred from the example of severe haemophilics, our life is probably always associated with minor blunt traumas to the vascular bed arising from various forms of physical activity. However these normally remain unnoticed—obviously because repair with platelets and fibrin is possible before haematomas develop.

5.5.3. SYNTHESIS

As already pointed out, fibrinogen catabolism is a steady process and its rate is probably determined by another fundamental biological process, namely the endocytic activity of certain cells. The latter is grossly insensitive to the body's requirements on fibrinogen: this is the explanation for some striking observations, as for example that afibrinogenaemic subjects are unable to economise on infused fibrinogen by reducing catabolism, or that the useless hyperfibrinogenaemia accompanying rheumatoid arthritis is not diminished by compensatory hypercatabolism.

In view of the insensitivity of the catabolic mechanism it seems unlikely that the supply of fibrinogen in the body is solely controlled by degradation. Indeed, the pronounced responsiveness of the synthesis rate to a variety of stimuli makes synthesis appear a far more important regulator of the fibrinogen pool.

The question of regulation is complex: one may ask as how the basic synthesis rate is modified by stimuli such as tissue injury on one hand, and what maintains the fibrinogen output of the liver under normal conditions on the other.

At present there is no satisfactory answer to either of these problems; moreover, it is equally obscure whether experimentally induced changes reflect an augmentation of the physiological synthesis promoter or the appearance of factors which are normally absent. The fibrinogen-synthesising activity of a non-stimulated liver runs at about 12–20% of the full capacity, and it is tempting to speculate that the majority of the available synthetic sites are normally repressed (cf. Jacob and Monod[54]). The physiological stimulus could be envisaged in this system as a substance capable of maintaining 12–20% of the synthesis sites unrepressed. Repressor synthesis itself could be operational under a feedback mechanism, enhanced repressor inactivation (= excess stimulation) leading to increased repressor production.

The idea of a repressor-controlled synthetic mechanism for fibrinogen is, for the time being, hypothetical. Nevertheless, certain observations, for example the relative refractory phase following preceding stimulation,[55] are perhaps more easy to comprehend this way than by the assumption that the stimulus acts primarily as an inducer.

If the physiological substance maintaining fibrinogen synthesis is to be a hormone, then ACTH,[55] thyroxine,[56] or growth hormone[57] could compete for this role. Another hitherto unexplored possibility arises from scanty observations in the literature on links between lipid metabolism and fibrinogen synthesis. Pilgeram and Pickart[58] studied the incorporation of ^{14}C-glycine into fibrinogen in human liver biopsy samples and found that addition of free fatty acids (FFA) to the incubating medium significantly increased the rate of uptake. Long, saturated chains were more effective than short, unsaturated ones. According to Thorp, Cotton and Oliver,[59] the FFA/albumin ratio and fibrinogen concentration in plasma change in the same direction under a variety of conditions. By the virtue of competing for the binding sites of FFA on albumin, p-chlorophenoxybutyrate inhibits the release of FFA from adipose tissue and its transport to the liver[60]; administration of the precursor of this anion in vivo results in decreased fibrinogen concentration. Decreased fibrinogen synthesis can also be inferred from the data obtained by Hickman[35] with iodine-labelled fibrinogen following the administration of phenformin; although this drug increases plasma levels of FFA, it does so by reducing FFA oxidation.[61]

Much attention has been focused over the last decade or so on the possible role that fibrinogen may play by contributing to the regulation of its own synthesis rate. That results are conflicting is understandable considering the wide application of unphysiological approaches which often border on injury to blood constituents. To qualify as a reliable contribution, in the writer's opinion a test preparation of fibrinogen or its derivatives should fulfill the following criteria: (1) The protein should be prepared under sterile conditions. The use of pyrogen-free reagents is essential because fibrinogen synthesis is extremely sensitive to fragments of bacterial wall (endotoxin) and these are not removed by ordinary sterilizing filters. (2) It should be devoid of active enzymes, since the platelet-release reaction can be induced with traces of thrombin, (3) It should be free of oligopolymers of fibrin "cryoprofibrin" which can initiate platelet aggregation (S. Niewiarowski, E. Regoeczi and J. F. Mustard, to be published) and possibly also other cellular reactions, (4) The protein should be homologous, heterologous fibrinogen behaving unphysiologically (cf. section 5.6).

The stringent precautions outlined above clearly show the task involved in this field of investigations, and at present there is no published work that would comply with all of them. Nevertheless, results by several independent groups of investigators suggest that the synthesis rate of fibrinogen is not affected either by raised or by diminished fibrinogen concentrations.[40,58,62,63]

Increased synthesis of fibrinogen following pathological stimulation is accompanied by a similar response of certain other plasma proteins (haptoglobin, caeruloplasmin, α_1-antitrypsin, etc.), usually referred to as acute-phase reactants. The theoretically intriguing question is whether the synthesis of this group of proteins is regulated by an operon. Unfortunately, the number of studies involving the whole range of the acute-phase reactants is not yet sufficient to piece a final picture together. Liver perfusion studies imply however that the response is probably sequential, synthesis of fibrinogen changing first, that of α_1-globulin last.[64]

Pathological stimulation of fibrinogen synthesis is likely to be initiated by a cellular factor of unknown composition and localization,[65] so that 'tissue injury' can be defined as a cellular release reaction leading to an acute-phase plasma protein response. A list of astonishingly heterogeneous and apparently unrelated agents described in the literature to elicit the response was given elsewhere.[45] Two preliminary conclusions can be drawn from available information: (1) 'tissue injury' is by no means equivalent to trauma in a surgical sense, and (2) the cellular constituent responsible for the reaction is not confined to one particular organ but it is probably widely spread over the body. Ubiquitous tissues are the formed elements of blood and the vascular tree, and these are most likely to be affected by the simplest kind of atraumatic stimulation, such as haemodilution.[66,67] Attempts to identify the stimulating agent at a subcellular level have so far been unsuccessful. Lysosomes are an attractive possibility, but in unpublished experiments, Koj and Regoeczi found no conclusive evidence for their involvement by injecting disrupted lysosomes from peritoneal leucocytes to rabbits intravenously.

Clinical hyperfibrinogenaemia is often associated with elevated body temperature, this posing the question whether hyperthermia alone could stimulate fibrinogen synthesis. Ham and Curtis[68] investigated this possibility over 30 years ago by studying plasma fibrinogen concentrations in humans before and after physically induced hyperthermia. They found that elevated temperature *per se* did not give rise to increased synthesis of fibrinogen. The same conclusion is supported by unpublished experiments by the present writer using etiocholanolone. When

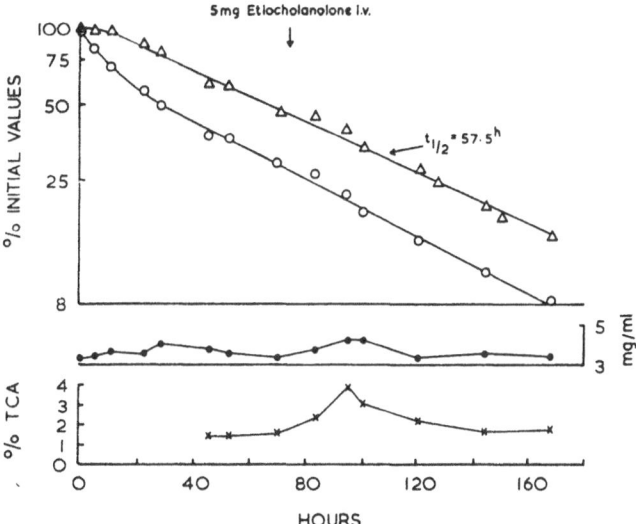

FIG. 5.7. Lack of influence of elevated body temperature on fibrinogen metabolism in a rabbit, the curves denoting total body activity ($\triangle$), clottable intravascular activity ($\bigcirc$), fibrinogen concentration ($\bullet$) and plasma non-protein activity ($\times$). Following the injection of etiocholanolone, the rectal temperature (not shown but continuously recorded with a thermocouple) rose 1.5–2°C for a period of 9 hours. The rise in TCA-values was probably due to reduced renal clearance of iodide as a result of the fever. The other curves show no significant changes.

dissolved in dimethylsulphoxide and injected intravenously, this steroid sometimes provokes a marked pyrogenic response in rabbits. However, there are no concomitant changes in the turnover of fibrinogen (Fig 5.7).

5.6. HETEROLOGOUS FIBRINOGEN

In view of the current efforts in medicine to surmount immuno-logical barriers, this survey of fibrinogen metabolism would not be complete without a brief outline of the heterologous systems.

In species which are phylogenetically apart, fibrinogen is a good anti-gen: Following a single intravenous injection of not more than a few milligrams of human fibrinogen into rabbits, antibodies can be detected in the circulation within 4–5 days. Coinciding with the onset of the immune response there is a sharp break in the slope of intravascular heterologous fibrinogen activity, after which the foreign protein dis-appears with a half-life of between 6 and 8 hours. This process is accompanied by a steep rise in non-precipitable plasma radioactivities indicating that the protein which disappeared has indeed been hydrolyzed.

Metabolic observations relevant to the behaviour of heterologous (human) fibrinogen *before* the onset of the immune response are summarized in Table 5.5. By comparison with the corresponding values for

TABLE 5.5

Plasma disappearance rate and distribution of human fibrinogen (3 mg/kg) in rabbits before the onset of the immune response

Expt.	F_c (%/day)	$F_{i,e}$ (%/day)	$F_{e,i}$ (%/day)	e.v./i.v.
1	59.6	162.7	420.5	0.38
2	56.6	148.7	382.8	0.39
3	54.1	158.5	537.1	0.29
4	49.8	132.9	463.7	0.28
5	63.0	116.4	423.9	0.27
Mean	56.6	143.8	445.6	0.32
Mean for homologous fibrinogen*	39.0	45.0	187.0	0.24

* from Regoeczi[38]

homologous fibrinogen it is shown there that the heterologous protein disappeared from the plasma at 1.45 times the normal rate before any immunological clearance mechanism came into effect. Unfortunately, total body radioactivity curves are not available from these experiments, so that it is impossible to be certain whether the fibrinogen which disappeared in excess underwent immediate catabolism or persisted for some time in cells recognizing it as a foreign protein. The body distribution of heterologous fibrinogen differed only slightly from that of the homologous protein, whereas its diffusion rates appeared to be markedly higher. The latter finding is difficult to understand without assuming some form of selective or active transport for heterologous fibrinogen across the capillary barrier. Further studies are clearly required.

From a practical point of view it would be important to know if the enhanced disappearance of heterologous fibrinogen—and, as a matter of interest, of any heterologous protein—from the plasma can be taken as an advance sign for an antibody response to follow. A few studies performed by the writer on rhesus monkeys do not seem to support this idea: human ^{131}I-fibrinogen (2 mg/kg) was eliminated 1.3 times faster from the circulation in this species than simultaneously injected rhesus ^{125}I-fibrinogen, despite the fact that no immune response could be detected.

5.7. APPENDIX

5.7.1. PREPARATION OF LABELLED FIBRINOGEN

The procedure involves 3 steps: (1) preparing the protein for labelling, (2) the labelling process, and (3) storage. These will be discussed in the above order.

(a) *Fractionation of plasma to obtain fibrinogen*

There are numerous ways of isolating fibrinogen from plasma, and there is good reason to believe that metabolically satisfactory preparations can be obtained by more than one method provided that no denaturing agents are used.

For experiments on humans, the donor plasma must originate from a healthy subject. Great care should be taken to exclude carriers of hepatitis, the virus being easily transferred with fibrinogen.[69] Blood from donors with low-grade intravascular coagulation or pathologically enhanced fibrinolytic activity may contain large molecular weight fragments of fibrinogen, and the complete separation of these from the intact protein is extremely tedious.

Since autologous and homologous fibrinogens are catabolized at identical rates,[70] the isolation of the patient's own fibrinogen for labelling is normally of no special advantage, except from the hygienic aspect.

The plasma to be fractionated should preferably be fresh, and in no case more than 1–2 days old. In the standard ACD-solution, widely used to preserve blood for transfusion, fibrinogen is decomposed rapidly.[71] Fibrinogen from such plasma may contain some rapidly catabolized material.[70]

So far, the preponderance of metabolic work on fibrinogen has been carried out with protein fractions which were obtained by various techniques involving fractionation with ammonium sulphate. Whichever method is adopted, it is advisable to first eliminate the low-solubility fibrinogen fractions ('contractinogen'[72]), as they not only interfere with storage but are also catabolized at rates which are different from those of uncomplexed fibrinogen. The details of such a technique which has been developed from a previously published method[19] are given below.

(a) Blood is collected into a 2% (w/v) solution of oxalates (40% potassium oxalate and 60% ammonium oxalate: Heller and Paul[73]) using 1 volume of the anticoagulant for 4 volumes of blood.

(b) The plasma is separated by centrifugation and its volume is measured.

(c) Prothrombin is removed by adsorption to $BaSO_4$ or $Al(OH)_3$-gel. Of the former 5 g is added to each 100 ml of centrifuged plasma and stirred for 7–8 min; then an equal quantity of $BaSO_4$ is added and stirring is continued for another 7–8 mins. $Al(OH)_3$ is used as a suspension: 20 g of a 10–20% gel is made up to 100 ml with water and 3–6 ml of this is added to 100 ml of plasma. Adsorption is complete in 5–10 mins.

Deprothrombinization is best done in a beaker using a thick rod. Both $BaSO_4$ and $Al(OH)_3$ are heavy, thus tending to sediment despite keeping the solution in motion. This can be avoided by inspecting the beaker from beneath from time to time.

(d) Plasma is cleared of $BaSO_4$ *viz.* $Al(OH)_3$ by high-speed centrifugation and its volume is re-measured.

(e) Ammonium sulphate is added to 0.18 saturation while gently stirring and the mixture is left to stand at room temperature for 15–20 mins.

It is convenient to use for this purpose, and also for the subsequent fractionation steps, a 2.05 M solution of $(NH_4)_2SO_4$ which corresponds to half-saturation at 25°C. The pH of a freshly prepared solution of ammonium sulphate is usually around 5. Neutralization can be achieved with a few drops of ammonia. It should be remembered that the degree of saturation of $(NH_4)_2SO_4$ solutions is dependent on temperature. Fluctuations in the laboratory temperature not exceeding $\pm3°C$ make little difference; at more extreme temperatures the correct amount of $(NH_4)_2SO_4$ to be used can be calculated from tables, such as the International Critical Tables.

The low-solubility fibrinogen fractions precipitating at 0.18 saturation are rough and fibrous. Depending on the source and handling of the blood, their proportion may vary from nil to over half of the total fibrinogen present. Sometimes the low-solubility fractions do not separate from the plasma on standing at 0.18 saturation. Instead, the plasma turns opaque. However, if such a preparation is subjected to a brief, vigorous shaking at the end of the precipitation time, the low-solubility fractions suddenly appear as fibrous aggregates.

(f) Low-solubility fractions, if any, are removed either by winding onto a glass rod or by centrifugation. The volume of the fractionation mixture is remeasured and more $(NH_4)_2SO_4$ is added to raise the degree of saturation to 0.24. The 'high-solubility' fibrinogen now precipitating has the appearance of fine snowflakes.

(g) One hour is allowed for precipitation, after which the mixture is *gently* centrifuged. Compact pellets are undesirable as they are unsuitable for washing.

(h) The supernatant is discarded, and if there are several pellets (because of a large initial fractionation volume) they are pooled in one or two 100-ml transparent polycarbonate centrifuge tubes for washing with ammonium sulphate of 0.24 saturation. For an efficient washing the pellet is gently dispersed in the washing fluid with a glass rod of medium thickness. The optimal amount of washing fluid depends on the size of the preparation; for a pellet from 200–300 ml of plasma 30–50 ml aliquots of washing fluid should suffice. As soon as the pellet has dispersed, the suspension is sedimented by gentle centrifugation. The supernatant is decanted and the washing process is repeated 2–3 times.

At this stage it may happen that part of the protein precipitate floats off rather than sediments on centrifugation. The floating material is the same as the sedimenting one, except that it has air bubbles adsorbed onto it. Formation of the floating protein can be minimized by avoiding too vigorous dispersion of the pellet.

(i) The last centrifugation is carried out at a higher speed so that the resulting pellet is compact. After decanting the supernatant, the tube is drained for 1–2 min. by positioning it upside down on a pad of filter paper. Subsequently the pellet is dissolved either in 0.9% NaCl containing 0.005 M trisodium citrate or in 0.15 M phosphate buffer pH 7.8. The volume of the solvent is about 1/10th of the starting volume. The protein solution is transferred into a Visking dialysis tubing ($\frac{8}{32}''$ inflated diameter) for dialysis against the same solvent. With 2–3 changes at half-hourly intervals and by applying a 10–20 fold buffer excess in the outer compartment, the protein solution becomes virtually free of ammonium sulphate within 2 hours.

(j) The volume of the dialysate is adjusted to 20–30% of the original fractionation volume with one of the above solvents, and the fractionation procedure is repeated as from the saturation at 0.24. The progress made during the subsequent washings is conveniently controlled in the following way: 2 ml undisturbed washing fluid is added after each centrifugation to a test tube containing an equal volume of 20% (w/v) trichloroacetic acid; the turbidity of this mixture parallels the protein content of the washings. Ideally, the washing fluid should remain nearly or entirely water-clear when added to TCA.

(k) The last washing is completed by a hard centrifugation and the tube is drained as described above. The pellet is dissolved in a minimum of solvent (cf. above) and is dialyzed against the same overnight at 2–4°C.

(l) Anything remaining undissolved in the dismantled dialysate is removed by high-speed centrifugation.

(m) For estimating protein concentration in the preparation, 0.05 to 0.1 ml of the dialysate is added to 3.95 or 3.90 ml of the dialyzing fluid, and absorption is measured at 280 nm in a spectrophotometer using the dialyzing fluid as a blank. The concentration of protein is calculated from the specific absorption coefficient of fibrinogen (cf. 5.2.2).

With a little experience, the above fractionation scheme can yield preparations which contain 90–98% clottable protein and exhibit satisfactory metabolic behaviour. The principal contaminants are plasminogen, factor XIII, and occasionally some alpha$_2$-M, but no IgG. As previously pointed out,[43] ammonium sulphate cannot be expected to result in absolutely pure fibrinogen preparations, and its use is clearly a compromise between biochemical purity and the necessity to avoid biological denaturation.

One of the main advantages of salt fractionation is speed. If plasma is received for example 9 a.m., fractionation can reach the stage of the final dialysis by 6 p.m. In case an interruption is unavoidable the protein should not be left in ammonium sulphate solution, and a pellet must not get dry under any circumstances. Good stages for a pause are the deprothrombinized plasma and either of the dialyzing steps.

The perceptive reader will have noticed that in the whole process of preparing fibrinogen the use of citrate is kept to a minimum or avoided altogether. This is so intentionally, because citrate ion is a slow activator of prothrombin[74] as well as of plasminogen.[75] Citrate is likely to be wholly or partly responsible for the relatively rapid decomposition of certain clotting factors, particularly factors I, II, and VIII., in stored blood anticoagulated with ACD.[71] Solutions of purified fibrinogen should be regarded as being particularly vulnerable to even minute amounts of proteolytic enzymes because its natural 'protectors': plasma antithrombins and antiplasmins, have been removed in the course of the fractionation.

Iodination

Since iodination techniques are being dealt with in detail in another chapter of this book, the present discussion will be restricted to a few points arising with iodinating fibrinogen.

Fibrinogen can be and has been labelled using chloramine-T or electrolysis, although most workers, as apparent from the literature of the last decade or so, incline to iodine monochloride. No convincing evidence has so far been produced in favour of any of the above methods for labelling fibrinogen. Years of personal experience with one particular method is probably more important than the differences between methods. This may need emphasizing in the case of chloramine-T because its possible oxidizing side-effects have repeatedly been criticised. Fibrinogen iodinated using chloramine-T was used by Hickman[36] in some of his studies, and the body distribution and catabolic rate of this material was not significantly different from that of fibrinogen labelled with ICl.

Those who favour iodine monochloride for its simplicity and speed, and this includes the writer, probably find the low levels at which substitution of fibrinogen is permissible (≤ 0.5 atoms of I/mole of protein) to be a practical problem. Indeed, at this ratio not more than 0.1867 μg of I may be associated with each mg of the protein, and this amount of iodine is contained in 0.444 μl of the standard ICl solution ($= 0.42$

mg I/ml). Since the iodinating power of ICl weakens on dilution, the better alternative is using concentrated fibrinogen solutions (30–50 mg/ml). The optimal amount of fibrinogen to be iodinated at one time is around 10 mg in a volume of 0.3 ml or less.

While the above maximal permissible substitution level seems to be well established for rabbit fibrinogen,[22,36] Blombäck, Carlson, Franzén and Zetterqvist[77] used fibrinogen iodinated with up to 2.7 I/mole in studies on humans and found no obvious signs of overiodination. At present there is no plausible explanation for this difference, which is probably related to species.

Finally, it should be pointed out that substitution of several iodine atoms into one fibrinogen molecule *per se* does not interfere with the thrombin-fibrinogen reaction or the polymerization of fibrin monomers.[17] Therefore if fibrinogen is only labelled for *in vitro* use, there is no reason why the level of substitution should not be raised to 2–3, or even higher.

Storage

Irrespective of whether fibrinogen is an intrinsically unstable molecule or only appears to be so because of trace-contamination with proteolytic (pro-)enzymes, the storage properties of this protein in solution at 2–4°C are disappointing.

Thus storage should be in the frozen state at lower temperatures. Storability depends to some extent on the species and it decreases in the order: rabbit > rhesus monkey > man > dog. Human fibrinogen solutions usually keep all right at temperatures between −15 and −22°C for 2–3 weeks. Longer storage requires deep freezing at −60 to −70°C. Deep frozen fibrinogen solution remains metabolically intact for at least 6 months, possibly much longer. To avoid damage due to extreme pH and salt effects in the subfreezing zones, freezing and thawing should be done rapidly.

It is safer to divide a preparation into 'n' aliquots at the beginning, each being sufficient for one iodination, and store them individually, rather than freeze and thaw a bulk of protein 'n' times for subsampling; though experimental evidence exists that fibrinogen can repeatedly be subjected to freezing-thawing without affecting its metabolic behaviour, this may not hold for excessive repetitions.

5.7.2. ANALYSIS OF THE PLASMA CURVE OF FIBRINOGEN

In this last section of the current chapter routine calculations which are necessary to quantify fibrinogen metabolism are summarized from a purely practical point of view. Equations always rest on assumptions, and a system of assumptions creates a kinetic model; for the ones relevant to the following presentation the reader should consult

Matthews,[28] Atencio, Bailey and Reeve,[76] Atencio, Joiner and Reeve,[40] Nosslin[30] and Donato.[24]

The plasma curve of labelled fibrinogen

It is generally agreed that the plasma curve of fibrinogen is satisfactorily described by a double-exponential function of the type

$$f(x) = C_1 e^{-b_1 t} + C_2 e^{-b_2 t} \qquad (5.1)$$

C_1 and C_2 are constants, and theoretically $C_1 + C_2 = 1$. The first term reflects replacement, the second one diffusion. The purpose of measuring the turnover is to estimate the constants C_1, C_2 and the exponents b_1, b_2. Since b_2 is much larger than b_1, the second term rapidly approaches zero. Experiments have to be continued beyond this point for some time so that a reliable estimate of the first term is obtained.

C_1, C_2, b_1 and b_2 are calculated by fitting the set of experimental values to equation 5.1. Some years ago this was done graphically, but nowadays it is much faster and more accurate to have it accomplished by a computer. A computer-fitted curve for a set of values from a clinical case of accelerated fibrinogen turnover is plotted on a semilog scale in Fig. 5.8. The linear plot of the same function in Fig. 5.9 is to illustrate the basic features of double-exponential curves, the mathematical solutions being given below.

Following the injection of a dose 'D' into a plasma volume 'V' at time t_0, the initial intravascular distribution of the dose is D/V and this will be referred to as $D_{(0)}$ further on. In practice, $D_{(0)}$ is the radio-activity in a plasma sample obtained 10 min after injecting the labelled fibrinogen (the blood being taken on the side opposite to the one dosed).

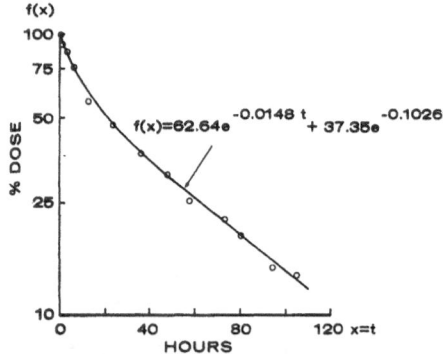

FIG. 5.8. Optimal fit of equation (5.1) by computer to a set of experimental points from a clinical case of increased fibrinogen catabolism and therapeutic hypofibrinogenaemia following the administration of Arvin.[62] For explanations see text.

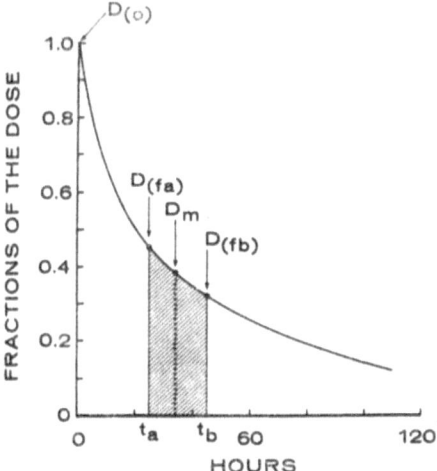

FIG. 5.9. Linear plot of equation 5.1 for the values from Fig. 5.8, showing some characteristic points of double-exponential curves, the equations of which are given in the text.

It is a convenient custom to express radioactivities in subsequent samples either as a fraction or a percentage of $D_{(0)}$.

The remaining dose in the circulation at $t = a$ is $D_{(fa)}$, that is the value of equation (5.1) at $\Delta t = t_a - t_0$:

$$D_{(fa)} = C_1 e^{-b_1 \Delta t} + C_2 e^{-b_2 \Delta t}, \qquad (5.2)$$

whereas the activity lost from the plasma up to this time is

$$D_{(0)} - D_{(fa)} = C_1(1 - e^{-b_1 \Delta t}) + C_2(1 - e^{-b_2 \Delta t}). \qquad (5.3)$$

The area under the curve between points $D_{(fa)}$ and $D_{(fb)}$ is

$$\int_{D_{(fa)}}^{D_{(fb)}} C_1 e^{-b_1(t_b - t_a)} + C_2 e^{-b_2(t_b - t_a)} \, dx$$

$$= \left[\frac{C_1}{b_1}(1 - e^{-b_1 t_b}) + \frac{C_2}{b_2}(1 - e^{-b_2 t_b}) \right]$$

$$- \left[\frac{C_1}{b_1}(1 - e^{-b_1 t_a}) + \frac{C_2}{b_2}(1 - e^{-b_2 t_a}) \right] \qquad (5.4)$$

The mean of the integral given in equation (5.4) is:

$$D_m = \frac{\left[\frac{C_1}{b_1}(1 - e^{-b_1 t_b}) + \frac{C_2}{b_2}(1 - e^{-b_2 t_b}) \right] - \left[\frac{C^1}{b_1}(1 - e^{-b_1 t_a}) + \frac{C_2}{b_2}(1 - e^{-b_2 t_a}) \right]}{t_a - t_b} \qquad (5.5)$$

The meaning of the value calculated by equation (5.5) is hypothetical: it shows that the plasma activity would have been between times a and b if the same amount of radioactivity, which actually was decreasing exponentially, had been maintained at a constant level. $D_{\hat{m}}$ is no longer an area value but a point on the slope. Its application is primarily in the U/P methods to find the mean specific activity of the plasma curve for urine collecting periods.

The area under the whole curve is

$$\int_{t_0}^{+\infty} C_1 e^{-b_1 t} + C^2 e^{-b_2 t}\, \mathrm{d}x = \frac{C_1}{b_1} + \frac{C_2}{b_2} \qquad (5.6)$$

Calculating the fractions undergoing catabolism or transfer from and to the plasma compartment

There are at least 3 different types of processes in which a fibrinogen molecule may be taking part: it may be on its way to extravascular sites, it may be returning from there or it may become degraded. We will write F for a fraction behaving in either of these ways and specify them with subscript c for catabolism, *i.e.* for the movement from intra- to extravascular sites and e, i for the movement in the opposite direction. Thus we obtain:

$$F_c = \frac{1}{C_1/b_1 + C_2/b_2} \qquad (5.7)$$

$$F_{i,e} = C_1 b_1 + C_2 b_2 - F_c \qquad (5.8)$$

$$F_{e,i} = b_1 + b_2 - (F_c + F_{i,e}) \qquad (5.9)$$

F is always a fraction of the protein in the compartment under question, that is, F_c and $F_{i,e}$ mean fractions of the intravascular and $F_{e,i}$ of the extravascular fibrinogen. Since b values are ln 2/half-life, the dimension of the F values with respect to time depends whether the intervals between samples were expressed in minutes, hours or days. Conversion from an hour to a day scale is done by division by 24, and it is immaterial whether the b values are converted or the final answer.

It will be noticed that F_c is the reciprocal of the integral given by equation (5.6). It is important that integration be from 0 to ∞, and not only up to the end of the experiment, otherwise catabolic rates turn out erroneously high. $F_{i,e}$ and $F_{e,i}$ can also be calculated using somewhat different formulas,[28] namely

$$F_{i\,e} = \frac{C_1 C_2 (b_2 - b_1)^2}{C_1 b_2 + C_2 b_1} \qquad (5.10)$$

and

$$F_{e,i} = C_1 b_2 + C_2 b_1 \qquad (5.11)$$

These equations give the same answers as equations (5.8) and (5.9) respectively. However it is easier to solve equations (5.7)–(5.9) in that order than equations (5.10) and (5.11).

Total body and extravascular fibrinogen

Total body fibrinogen (TBF) is given by the formula:

$$\text{TBF} = \frac{C_1/b_1^2 + C_2/b_2^2}{(C_1/b_1 + C_2/b_2)^2} \tag{5.12}$$

from which the extravascular fibrinogen (EVF) is obtained by subtracting the intravascular fibrinogen. The latter is taken either 100 or 1, depending on whether the plasma activity curve was plotted as a percentage or as a fraction of the dose.

$$\text{EVF} = \text{TBF} - 100 \text{ (or 1)} \tag{5.13}$$

Extravascular fibrinogen can also be derived directly from the ratio of the transfer rates:

$$\text{EVF} = \frac{F_{i,e}}{F_{e,i}} \tag{5.14}$$

Total body fibrinogen is then

$$\text{TBF} = \frac{F_{i,e}}{F_{e,i}} + 1 \tag{5.15}$$

Absolute rates

The task now remaining is to find the actual amount of fibrinogen represented by each of the above fractional rates. In a steady state this is no problem, provided that the plasma fibrinogen concentration and the intravascular distribution volume of the dose are known. The fibrinogen concentration should be measured in each sample withdrawn so that a mean value is obtained which is representative of the whole experiment.

Frequently however, serially measured fibrinogen concentrations exhibit systemic fluctuations to a degree at which a conventional steady state cannot be assumed any longer. Since absolute catabolic and transfer rates are directly proportional to plasma fibrinogen concentration, mean values for the latter can be used in such cases. Arithmetical means are permissible only if the samples are equidistant (hardly ever the case), otherwise the mean of the integrated fibrinogen concentration curve should be used. To obtain the latter, a linear plot of fibrinogen concentrations versus time is prepared on any suitable scale and the area under this curve is estimated either by planimetry or by cutting out and weighing (cf. Fig. 5.10). The value for the total

E. REGOECZI

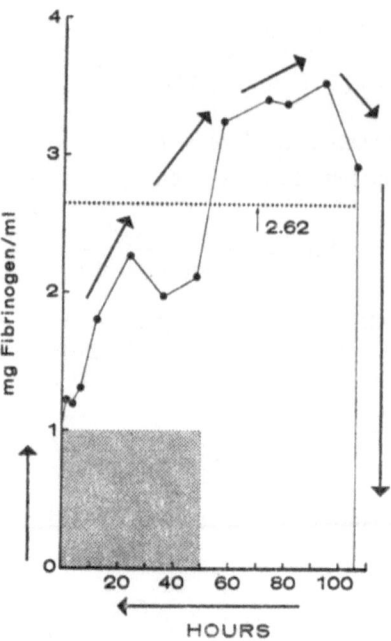

FIG. 5.10. Gravimetric estimation of the mean of the integrated plasma fibrinogen concentration. The irregular curve, which was obtained by plotting the serially estimated fibrinogen concentrations on millimetre paper, was copied onto transparent paper, cut out along the curve and the sides indicated by the arrows, and weighed on an analytical balance (= 1195.5 mg). A square (shadowed), corresponding to the area of 1 mg × 50 hrs, was drawn on another piece of the same paper, cut out and weighed (= 215.5 mg); 1/50th of this (= 4.31 mg) is the area-index. The mean of the integral, obtained using equation 5.16, is shown by the dotted line. This curve is from the same patient whose radioactive data are illustrated in Fig. 5.8.

area (mm² or mg) is then divided by the product of the area-index and the length of the observation period:

$$\text{Mean } \phi \text{ conc. (mg/ml)} = \frac{\text{total area (mm}^2 \text{ or mg)}}{\text{area index} \times t} \qquad (5.16)$$

The area-index is the value (in mm² or mg) of a 1 mg fibrinogen × 1 hr strip under the curve. Using transparent paper of sufficient homogeneity and making cuts with a scalpel, the reproducibility of the gravimetric method is ±1.5%.

The biological half-life of native fibrinogen

Because of diffusion into the extravascular space and recirculation from there after a mean delay time, a labelled protein tends to some

extent to prolong its own half-life in the plasma compartment. Therefore, the isotopic and the biological half-lives (T) of a protein are not
exactly the same.

According to Nosslin,[30] this artifact can be eliminated using the
following formula:

$$T = \frac{C_1/b_1^2 + C_2/b_2^2}{C_1/b_1 + C_2/b_2} \qquad (5.17)$$

Summary of the plasma curve analysis

As a demonstration of the use of the method, an evaluation of the
case shown in Figs. 5.8, 5.9, and 5.10 is given below.

In Fig. 5.8 we have $C_1 = 62.64$, $C_2 = 37.35$, $b_1 = 0.0148$ and $b_2 = 0.1026$. For subsequent calculations we transform C values from
percentages into fraction ($= \div 100$) and express b values in days
($= \div 24$): $C_1 = 0.6264$, $C_2 = 0.3734$, $b_1 = 0.3555$ and $b_2 = 2.4624$.

Using these values the catabolic rate by equation (5.7) is 0.522
($= 52.2\%$ of the i.v. pool day^{-1}), the capillary transfer rate by equation
(5.8) $= 0.619$ ($= 61.9\%$ of the i.v. pool day^{-1}), and the return flow by
equation (5.9) $= 1.675$ ($= 167.5\%$ of the e.v. pool day^{-1}).

The extravascular fibrinogen by equation (5.14) is 0.369, and the
total body fibrinogen by equation (5.15) $= 1.369$. The extravascular
fibrinogen pool, expressed as $\%$ of total body fibrinogen, is 26.9%
($= 0.369 \times 100/1.369$), the corresponding value for the i.v. pool being
$100 - 26.9 = 73.1\%$.

The mean of the integrated fibrinogen concentration curve (cf. Fig.
5.10) using equation (5.16) is $1195.5/4.31 \times 106 = 2.62$ mg/ml. At
the prevailing i.v. distribution volume for the dose ($= 2210$ ml), this
corresponds to an i.v. pool of 5.79 g fibrinogen, an absolute catabolic
rate of 3.02 g day^{-1} and an absolute capillary transfer rate of 3.58 g.
The extravascular fibrinogen pool is 2.13 g ($= 0.619 \times 5.79/1.675$, or
$5.79 \times 26.9/73.1$) and the total body pool $5.79 + 2.13 = 7.92$ g. These
absolute values are only valid for the observation period as a whole, the
early values being much lower and the last ones much higher because
of the regeneration of the pool.

The biological half-life by equation (5.17) is 43.6 hours as contrasted
by 46.7 hours for the half-life of the final exponential.

REFERENCES

1. A. E. NEEDHAM, (ed., R. G. MacFarlane), The Haemostatic Mechanism in Man
   and Other Animals (Academic Press, London, 1970) p. 19
2. J. LEVIN and F. B. BANG, *Bull. Johns Hopkins Hosp.* **115**, 337 (1964)
3. C. GRÉGOIRE, (ed., R. G. MacFarlane), The Haemostatic Mechanism in Man and
   Other Animals (Academic Press, London, 1970) p. 45
4. J. LEVIN, P. A. TOMASULO and R. S. OSER, *J. Lab. clin. Med.*, **75**, 903 (1970)
5. R. F. DOOLITTLE, *Thrombos. Diathes. haemorrh. (Stuttg.) Suppl.* **39**, 25 (1970)

6. S. LOPACIUK and N. O. SOLUM, *Thrombos. Diathes. haemorrh.* (*Stuttg.*) **21**, 409 (1969)
7. P. A. CASTALDI and J. CAEN, *J. clin. Path.*, **18**, 579 (1965)
8. C. E. HALL and H. S. SLAYTER, *J. biophys. biochem. Cytol.*, **5**, 11 (1959)
9. D. KAY and B. J. CUDDINGAN, *Brit. J. Haemat.*, **13**, 341 (1967)
10. G. KÖPPEL, *Zschr. f. Zellforschung*, **77**, 443 (1967)
11. B. BLOMBÄCK, (ed., R. G. MacFarlane), The Haemostatic Mechanism in Man and Other Animals (Academic Press, London, 1970) p. 167
12. E. MIHÁLYI, *Thrombos. Diathes. haemorrh.* (*Stuttg.*) *Suppl.* **39**, 43 (1970)
13. A. HENSCHEN, *Arkiv Kemi*, **22**, 1 (1963)
14. E. MIHÁLYI, (ed., K. Laki), Fibrinogen (Dekker, New York, 1968) p. 61
15. B. BLOMBÄCK, M. BLOMBÄCK, A. HENSCHEN, B. HESSEL, S. IWANAGA and K. R. WOODS, *Nature* (*Lond.*), **218**, 130 (1968)
16. P. J. GAFFNEY JR., personal communication (1970)
17. E. REGOECZI, *Brit. J. Haemat.*, **20**, 649 (1971)
18. R. J. DELLENBACK and S. CHIEN, *Proc. Soc. exp. Biol. Med.*, (*N.Y.*), **134**, 353 (1970)
19. E. REGOECZI and P. L. WALTON, *Thrombos. Diathes. haemorrh.* (*Stuttg.*), **17**, 237 (1967)
20. B. BLOMBÄCK and E. ZETTERQVIST, *Acta chem. scand.*, **23**, 1137 (1969)
21. A. S. MCFARLANE, *Bull. Swiss Acad. Sci.*, **21**, 173 (1965)
22. A. S. MCFARLANE, *J. clin. Invest.*, **42**, 346 (1963)
23. R. BIANCHI, L. DONATO, P. MANCINI, G. MARIANI, A. PILO and F. VITEK, (eds., M. A. Rothschild and T. Waldmann), Plasma Protein Metabolism (Academic Press, New York, 1970) p. 25
24. L. DONATO, (eds., L. Donato, G. Milhaud and J. Sirchis), Labelled Proteins in Tracer Studies (Euratom, Brussels, 1966) p. 375
25. R. M. CAMPBELL, D. P. CUTHBERTSON, C. M. MATTHEWS and A. S. MCFARLANE, *Int. J. appl. Radiat.*, **1**, 66 (1956)
26. S. A. BERSON and R. S. YALOW, *Fed. Proc.*, **13**, 135 (1957)
27. A. S. MCFARLANE and A. KOJ, *J. clin. Invest.*, **49**, 1903 (1970)
28. C. M. E. MATTHEWS, *Physics Med. Biol.*, **2**, 36 (1957)
29. E. B. REEVE and H. R. BAILEY, *J. Lab. clin. Med.*, **60**, 923 (1962)
30. B. NOSSLIN, (ed., S. B. Andersen) Metabolism of Human Gamma Globulin (Blackwell, Oxford, 1964).
31. E. REGOECZI, *Thrombos. Diathes. haemorrh.* (*Stuttg.*), **18**, 276 (1967)
32. A. AMRIS and C. J. AMRIS, *Thrombos. Diathes. haemorrh.* (*Stuttg.*), **11**, 405 (1964)
33. L. R. I. BAKER, M. L. RUBENBERG, J. V. DACIE and M. C. BRAIN, *Brit. J. Haemat.*, **14**, 617 (1968)
34. J. W. L. DAVIES, C. R. RICKETTS and J. P. BULL, *Clin. Sci.*, **30**, 305 (1966)
35. J. A. HICKMAN, *J. clin. Path.*, **23**, 797 (1970)
36. J. A. HICKMAN, *Clin. Sci.*, (in press).
37. Y. TAKEDA, *J. clin. Invest.* **45**, 103 (1966)
38. E. REGOECZI, *Clin. Sci.*, **38**, 111 (1970)
39. R. DAOUST, (ed., R. W. Brauer), Liver Function (Amer. Inst. Biol. Sci., Washington, D.C., Publication no. 4, 1958) p. 3
40. A. C. ATENCIO, K. JOINER and E. B. REEVE, *Amer. J. Physiol.*, **216**, 764 (1969)
41. E. B. REEVE, Y. TAKEDA and A. C. ATENCIO, *Protides Biol. Fluids Proc. Coll. Bruges*, **14**, 283 (1966)
42. M. POLLYCOVE and R. MORTIMER, *Clin. Res. Proc.*, **4**, 51 (1956)
43. E. REGOECZI, (eds., L. Donato, G. Milhaud and J. Sirchis), Labelled Proteins in Tracer Studies (Euratom, Brussels, 1966) p. 85
44. V. BOCCI and G. P. PESSINA, *Archivio di Scienze biologiche*, **54**, 45 (1970)
45. E. REGOECZI, (eds., M. A. Rothschild and T. Waldmann), Plasma Protein Metabolism (Academic Press, New York, 1970) p. 459
46. N. JANCSÓ, Speicherung (Akadémiai kiadó, Budapest, 1955) p. 149

47. G. Terres, W. L. Hughes and W. Wolins, *Amer. J. Physiol.*, **198**, 1355 (1960)
48. E. B. Reeve and J. E. Roberts, *J. gen. Physiol.*, **43**, 445 (1959)
49. M. T. Hays and D. H. Solomon, *J. clin. Invest.*, **44**, 117 (1965)
50. T. Astrup, *Thrombos. Diathes. haemorrh. Suppl.*, **20**, 71 (1966)
51. P. F. Hjort and R. Hasselback, *Thrombos. Diathes. haemorrh. (Stuttg.)*, **6**, 580 (1961)
52. P. F. Hjort, *Thrombos. Diathes. haemorrh. (Stuttg.)*, *Suppl.* **20**, 15 (1966)
53. Y. Takeda, *J. Lab. clin. Med.*, **75**, 355 (1970)
54. F. Jacob and J. Monod, (ed., M. Locke), Cytodifferentiation and Macromolecular Synthesis, The 21st Growth Symposium (Academic Press, New York, 1963)
55. A. C. Atencio, P.-Y. Chao, A. Chen and E. B. Reeve, *Amer. J. Physiol.*, **216**, 773 (1969)
56. J. C. Hoak, W. R. Wilson, E. D. Warner, E. O. Theilen, G. L. Fry and F. L. Benoit, *J. clin. Invest.*, **48**, 768 (1969)
57. K. N. Jeejeebhoy, A. Bruce-Robertson, U. Sodtke and M. Foley, *Biochem. J.*, **119**, 243 (1970)
58. L. O. Pilgeram and L. R. Pickart, *J. Atherosclerosis Res.*, **8**, 155 (1968)
59. J. M. Thorp, R. C. Cotton and M. F. Oliver, *Progr. biochem. Pharmacol.*, **4**, 609 (1967)
60. A. M. Barrett, *Brit. J. Pharmacol.*, **26**, 363 (1966)
61. S. Muntoni, M. Duce and G. U. Corsini, *Life Sciences*, **9**, 241 (1970)
62. W. R. Bell and E. Regoeczi, *J. clin. Invest.*, **49**, 1872 (1970)
63. M. I. Barnhart, D. C. Cress, S. M. Noonan and R. T. Walsh, *Thrombos. Diathes. haemorrh. (Stuttg.)*, *Suppl.* **39**, 143 (1970)
64. A. H. Gordon and A. Koj, *Brit. J. exp. Path.*, **49**, 436 (1968)
65. A. Koj, (eds., R. Porter and J. Knight), Energy Metabolism in Trauma. A Ciba Foundation Symposium (Churchill, London, 1970) p. 79
66. D. P. Foster and G. H. Whipple, *Amer. J. Physiol.*, **58**, 393 (1921)
67. F. C. Monkhouse and S. Milojevic, *Amer. J. Physiol.*, **199**, 1165 (1960)
68. T. H. Ham and F. C. Curtis, *Medicine (Baltimore)*, **17**, 413 (1938)
69. N. R. Boeve, L. C. Winterscheid and K. A. Merendino, *Ann. Surgery*, **170**, 833 (1969)
70. E. Regoeczi and B. A. Stannard, *Biochem. biophys. Acta (Amst.)*, **181**, 287 (1969)
71. R. Janda, *Folia histochemica et cytochemica*, **4**, 315 (1966)
72. I. R. Morrison, *Amer. J. Med. Sci.*, **211**, 325 (1946)
73. V. G. Heller and H. Paul, *J. Lab. clin. Med.*, **19**, 777 (1934)
74. W. H. Seegers, *Proc. Soc. exp. Biol. Med.*, *(N.Y.)*, **72**, 677 (1949)
75. T. Astrup and I. Sterndorff, *Thrombos. Diathes. haemorrh. (Stuttg.)*, **4**, 462 (1960)
76. A. C. Atencio, H. R. Bailey and E. B. Reeve, *J. Lab. clin. Med.*, **66**, 1 (1965)
77. B. Blombäck, L. A. Carlson, S. Franzén and E. Zetterqvist, *Acta med. scand.*, **179**, 557 (1966)

Chapter 6

The Biochemistry of Blood Clotting Factors

D. E. G. Austen
and
C. R. Rizza

Oxford Haemophilia Centre,
Churchill Hospital,
Headington,
Oxford

6.1. INTRODUCTION

When blood is drawn from a vein and placed in a glass tube it clots within 4–5 minutes. The change from the fluid to the solid state is very dramatic and has understandably attracted the attention and stimulated the curiosity of generations of physiologists, biochemists and physicians, not to mention the layman. It is now well established that the ability of the blood to clot firmly and rapidly requires the presence of certain plasma factors as well as platelets. At present there are thought to be eleven plasma factors including calcium ions involved in the production of a firm stable clot. By International agreement those various factors have been assigned a Roman Numeral (Table 6.1). In the case of nine of the factors, deficiency is associated not only with poor clotting in the test tube but also with defective haemostasis in the person concerned. Much work has been undertaken to discover the sequence and kinetics of the reactions between the various factors and a major step forward was achieved as a result of the study of the clotting action of Russell's Viper Venom. In this work it was shown that the coagulant activity of Russell's Viper Venom was due to its being able to split factor X and to change it from an inactive to a highly active form. Subsequently it was shown that factor X was similarly activated by a sequence of reactions involving plasma clotting factors with or without tissue extract. Those observations led to the cascade hypothesis[1,2] in which it was proposed that the normal blood clotting reaction is based on a biochemical chain mechanism involving two reaction pathways. In the "extrinsic" pathway tissue juice in the presence of factor VII is responsible for activating factor X while, in the "intrinsic" system, this activation is achieved through a succession of reactions involving factors XII, XI, IX and VIII.[3,4] Once factor X is activated

TABLE 6.1

Factor	Alternative Names
Factor I	Fibrinogen
Factor II	Prothrombin
Activated Factor II	Thrombin
Factor V	Ac-globulin, Labile Factor, Proaccelerin
Factor VII	Proconvertin, Autoprothrombin I Stable Factor, Serum Prothrombin Conversion Accelerator (SPCA)
Factor VIII	Antihaemophilic Globulin (AHG). Platelet Cofactor I. Thromboplastinogen A
Factor IX	Christmas factor, Antihaemophilic Factor B, Platelet Cofactor II. Thromboplastinogen B. Autoprothrombin II
Factor X	Stuart-Power Factor. Autoprothrombin III
Activated Factor X	Autoprothrombin C, Coagulation Product I, Prephase Accelerator (PPA)
Factor XI	Plasma Thromboplastin Antecedent (PTA), Antihaemophilic Factor C,
Factor XII	Hageman Factor, Surface Factor,
Factor XIII	Fibrin Stabilising Factor (FSF), Fibrinase, Laki-Lorand Factor

there is a common path in which the activated factor converts pro-thrombin (factor II) to thombin in the presence of calcium, phospholipid and factor V. In its turn thrombin acts upon fibrinogen (factor I) to form the fibrin clot. This can be represented diagrammatically as follows:—

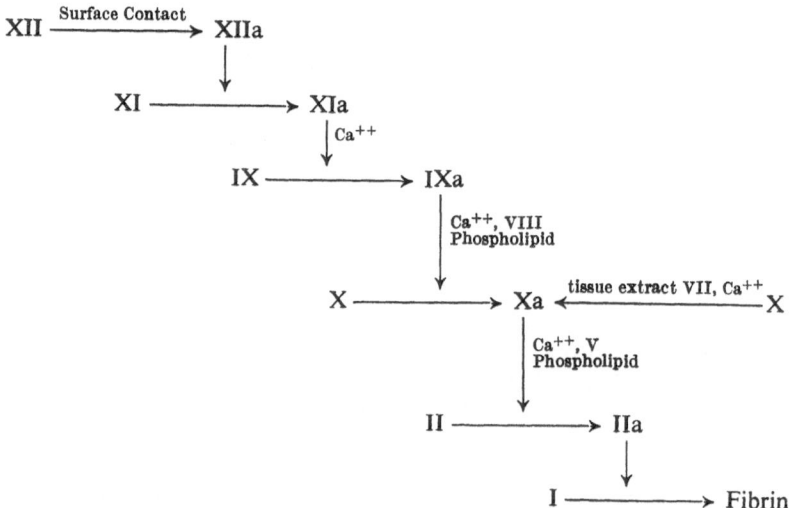

TABLE 6.2
Some properties of blood clotting factors

Factor	Site of synthesis	In vivo $\frac{1}{2}$-life	Adsorption by Al(OH)$_3$, BaSo$_4$	Precipitated by (NH$_4$)$_2$SO$_4$ (% saturation)	Present in Cohn's fraction
I	Liver	3–6 days	Not adsorbed	25	I
II	Liver	2–3 days	Adsorbed	50	III and IV
V	?Liver	15–24 hours	Not adsorbed	50	III
VII	Liver	4–6 hours	Adsorbed	50	III and IV
VIII	?throughout reticulo-endothelial system	12–18 hours	Not adsorbed	33	I
IX	Liver	18–30 hours	Adsorbed	50	III and IV
X	Liver	48–60 hours	Adsorbed	50	III
XI	?	60 hours	Not adsorbed	33	III and IV
XII	?	48–72 hours	Not adsorbed	50	III and IV
XIII	?	?3 days	Not adsorbed	33	I

It will be seen that many of the factors operate by becoming activated and then activating the subsequent factor in the reaction chain. Some, however, appear to act more as co-factors.

In the following sections, the biochemistry of these clotting factors will be considered together with the effect of factor deficiency *in vivo* and replacement therapy. Some of the more basic properties of these factors are recorded in Table 6.2.

6.2. FACTOR XII

Factor XII is a protein present in normal human plasma[5] and in the plasma of most mammals and amphibians. Absence of this factor from human blood prolongs the whole blood clotting time in glass tubes but does not, in general, give rise to a bleeding disorder and there is no need for protein concentrates to treat factor XII deficiency. The site of factor XII synthesis is not known although plasma levels of the factor are reduced by liver damage which would suggest that synthesis occurs mainly in the liver. In the clotting process, factor XII is activated by contact with a foreign surface and the active factor, in its turn, activates factor XI. This latter reaction does not require calcium ions and is believed to be enzymic. In addition to its role in blood coagulation factor XII can participate in several physiological and pathological reactions; it increases vascular permeability; dilates blood vessels and contracts smooth muscle and is involved in the production of kinins; it induces leucocyte migration through blood vessel walls and can activate the first component of "complement".

In vitro, factor XII can be activated by a wide variety of surfaces including glass, bentonite, asbestos, collagen, skin, elastin and diatomaceous earths. It is also activated by ellagic acid, silicic acid and long chain fatty acids. Surfaces which do not result in significant activation are the inert non-polar surfaces such as oils, waxes, silicones, some plastics and vascular endothelium. At first it was considered that the surfaces which could cause this activation were those which were easily wettable by aqueous solutions. However, the modern view is that an active surface is one which possesses electronegative sites and is suitable for molecule unfolding. It is suggested that factor XII is a folded molecule which straightens on adsorption to reveal its active sites.[6]

Many different values have been quoted for the molecular weight[7,8] of factor XII ranging from 20 000 to 140 000. It is interesting that workers reporting the low molecular weights found the factor not to be inhibited by DFP whereas some of those recording high values showed that there was considerable inhibition. This has led to the suggestion that factor XII has an inactive monomer of 20 000 molecular weight and that these compound into larger molecules when adsorbed onto surfaces.

Factor XII has an isoelectric point between pH 7.8 and 8.0. The molecule contains carbohydrate but there is disagreement as to the amount. Values between 2% and 14% are recorded. Amino-acid studies show high amounts of serine and glutamic acid but terminal amino-acid studies are not yet conclusive. There has been some difference of opinion as to whether activated factor XII has esterolytic properties. Recent experiments have suggested that the esterolytic actions which have been observed are not due to active factor XII but due to some other enzyme which is itself activated by the active factor XII.[3]

Concentrates of factor XII can be prepared using a combination of adsorption, chromatographic and precipitation techniques. Preparation usually starts by adsorbing plasma with aluminium hydroxide to remove factors II, VII, IX and X and, in addition, factor XI may be removed by adsorption on a restricted quantity of celite. A limited degree of concentration is obtained using precipitation techniques either with ammonium sulphate or with zinc acetate in the presence of ethanol but the main purification step is one of chromatography. This can be carried out using a wide variety of materials since factor XII is readily adsorbed and eluted from many surfaces. Several good fractionation methods have been described in the literature, for example, those of Speer et al.[7] and Schoenmakers et al.[10] With all methods of purifying factor XII it is essential to use plastic or siliconised equipment but even with these precautions the factor will always become activated to some degree.

6.3. FACTOR XI

In the sequence of clotting reactions Factor XI is activated by active factor XII and in its turn the active factor XI activates factor IX.

The site of synthesis of factor XI in the body is not known. Deficiency of this factor can give rise to a serious bleeding disorder but the severity of bleeding is not well correlated with the extent of the factor deficiency as measured by assays based on clotting tests. This is difficult to explain. Factor XI deficiency is usually hereditary and an unusually large number of those suffering from the condition are of Jewish origin. Treatment of bleeding in factor XI deficiency is extremely easy since haemostasis can be achieved simply by giving transfusions of fresh frozen plasma in a dose of 15–20 ml/kg. body weight.

Among the clotting factors, factor IX is one of intermediate stability and is partially inactivated by heating to 56°C for thirty minutes.[11] The clotting process consumes factor XI but there is sufficient quantity in plasma for large amounts of it to remain in serum after clotting. It migrates between the β and γ globulins on electrophoresis. It is readily adsorbed onto celite and can be easily eluted again, although such steps always result in some activation of the factor. Adsorption from plasma onto celite is used in the preparation of so called "Contact Product" which is a mixture of activated factors XI and XII and is used as a reagent in certain clotting assays.[12] In Cohn fractionation factor XI is found in fractions III and IV-I. With ammonium sulphate precipitation it is precipitated at 30–50% saturation.

Activated factor XI has an esterolytic action on TAME and on benzoyl-1-arginine ethyl ester.[13] The reaction in which activated factor XI is formed by the action of activated factor XII does not require calcium. The subsequent reaction in which activated factor XI activates factor IX does require calcium and is inhibited by DFP.

Concentrates of factor XI can be prepared, but if the starting material is normal plasma then factor XII can never be completely removed. For this reason nearly all of the information on factor XI has been obtained from studies of samples contaminated with factor XII. Factor XI, free from factor XII, can only be prepared by fractionating plasma from a person with congenital factor XII deficiency. Preparation of factor XI concentrates starts by adsorbing plasma or serum with aluminium hydroxide or barium sulphate to remove factor II, VII, IX and X. The supernatant material is heated to 56° for thirty minutes to remove fibrinogen and to reduce factor V and VIII levels. The product is then adsorbed on carboxymethyl cellulose in sodium acetate solution at 5.2 pH and eluted in phosphate-buffered saline. The eluate can be treated with ammonium sulphate (30–50% saturation range) to precipitate the factor which is then dialysed and rechromato-graphed on carboxymethyl cellulose.[14]

6.4. FACTOR IX

Factor IX is present in relatively constant amount in normal plasma. Unlike factor VIII its level is unaffected by exercise and fear, and most of the normal population have a factor IX level between 80% and 120% of the average. Absence of this factor causes a bleeding diathesis very similar to haemophilia from the point of view of heredity and clinical manifestations. The condition is called Haemophilia B or Christmas disease. As in haemophilia the severity of the haemorrhagic mani-festations correlate fairly well with the degree of factor deficiency (Table 6.3). Treatment of haemorrhages in those patients consists of

TABLE 6.3

Relationship between plasma level of factor VIII or IX and haemorrhagic manifestations

Factor VIII level (% average normal)	Bleeding symptoms
50–100	None
25–50	Excessive bleeding occasionally after major surgery or serious accident
5–25	Excessive bleeding after minor injuries and surgery; e.g. dental extraction and tonsillectomy
1–5	Severe bleeding after minor surgery; occasional haemarthroses
0	Severe haemophilia with spontaneous haemorrhage into muscles and joints with possible crippling

giving intravenous transfusions of plasma or plasma protein concentrates rich in factor IX. There is now good evidence to suggest that at least a small proportion of Christmas disease patients have an inactive variant of factor IX in their plasma.[15] This variant is inactive in the clotting process but it will still react with factor IX antibodies and it can be detected by means of this reaction.

Factor IX is one of the more stable coagulation factors and in its normal plasma environment or in the form of concentrated therapeutic materials, it will maintain its activity for many hours at room temperature. If such materials are frozen at $-20°C$ the factor IX activity is almost indefinitely stable. In the circulation factor IX has a half-life of 18–24 hours. In general it is found in close association with factors II, VII and X; their physical properties are similar and they are difficult to separate from each other completely. Another feature which they have in common, is that their plasma levels are depressed in liver disease and in Vitamin K deficiency making it highly likely that the liver is the main site of their synthesis. Certain oral anticoagulant drugs such as the coumarins also depress their levels.

In the cascade chain of clotting reactions factor IX is activated by activated factor XI in the presence of calcium ions in what is considered to be an enzymic, possibly an esterase reaction and this reaction is inhibited by DFP.[16] Activated factor IX in its turn activates factor X in the presence of factor VIII, phospholipid and calcium ions. When factor IX is activated it loses a small peptide fragment and develops a different electrophoretic mobility. The activated material is relatively stable at pH 7 to 9. Clotting activity of factor IX does not appear to depend upon serine or thiol groups but activity is lost when the factor is chemically reduced by reagents such as dithiothreitol and this would suggest the presence of disulphide bonds.

Factor IX has a molecular weight of approximately 50 000[17] as measured by sedimentation and diffusion experiments and amino-acid analysis. Gel filtration methods give higher values of up to 200 000. The factor contains about 20% carbohydrate in the molecule and the N and C terminal acids are believed to be proline and tyrosine respectively. Its electrophoretic mobility is consistent with its being a globulin; in plasma or serum it migrates between the α_1 and α_2 bands while in purified form it migrates with the β globulins. Factor IX survives the clotting process and is present in normal serum. It is readily adsorbed from solution by such materials as aluminium hydroxide, barium sulphate or calcium phosphate and can be easily eluted again without any apparent change in its properties. It is precipitated from plasma by 50% saturated ammonium sulphate and in Cohn fractionation it is recovered in fractions III and IV. Using Kekwick and MacKay's ether fractionation method it is found in the G_2 fraction.

Concentrates of factor IX can be prepared using a combination of adsorption, precipitation and chromatographic procedures. Factors II, VII, IX and X are adsorbed from plasma onto barium sulphate or calcium phosphate and, after elution, these factors are further concentrated by precipitation with ammonium sulphate, ether or ethanol. Using DEAE-cellulose chromatography factor VII can be separated from the other three factors but factor IX concentrates even after chromatography will always contain significant amounts of factors II and X. Products without factor II have been prepared by starting from serum but the product then contains activated factors which may not be desirable in preparations used for experimental or therapeutic purposes. Concentrates for therapeutic use in Christmas disease or liver disease have been prepared until recently by precipitation and adsorption techniques alone[18] but recent advances have incorporated large scale chromatographic separations[19] and have resulted in therapeutic concentrates with a potency of 50 units/ml or more. (1 unit/ml of factor IX activity is defined as the amount present in 1 ml of fresh citrated normal plasma.) With such products the average dose size is small enough to be injected intravenously by syringe.

6.5. FACTOR VIII

Factor VIII is protein present in the blood of normal humans and reduced in the blood of haemophiliacs and patients suffering from von Willebrand's disease. It is essential for normal clotting and haemostasis.

The amount of factor VIII in normal human plasma is approximately 1.2×10^{-4} mg/ml but there is wide variation of the normal human level, ranging from 50% to 150% of the average figure. Exercise, pregnancy or fear will increase the amount of factor VIII *in vivo*, and certain clinical conditions, notably cancer and thyrotoxicosis will also elevate it. The mechanism by which activity is increased is not clearly understood. It is not known whether this represents additional factor VIII which is released from storage, activation of an inactive precursor or new synthesis of factor VIII. The site of factor VIII synthesis is not known and attempts to demonstrate synthesis of factor VIII by means of tissue slices *in vitro* have in general failed. Unlike factor II, VII, and IX and X this factor is not reduced in the blood of patients with liver disease until the terminal stages of this illness. Plasma concentrations vary considerably between different species. The blood of cows and pigs contains 10–15 times the amount of factor VIII found in human blood and blood from these animals is an important source of factor VIII concentrates used for the treatment of haemophilia.

Factor VIII deficiency if severe enough will result in a haemorrhagic state and is found classically in haemophilia where the deficiency is hereditary. Severe classical haemophilia manifests itself as a sex-linked recessive bleeding disorder which affects males and is transmitted by apparently normal females. Spontaneous haemorrhages into muscles and joints are the hall-mark of the condition and the patients also bleed persistently after even the most minor of injuries. The severity of the bleeding symptoms varies with the degree of the factor deficiency state (Table 6.3) and treatment consists of giving intravenous transfusions of fresh normal plasma or plasma protein concentrates rich in factor VIII activity (Table 6.4). On rare occasions, the spontaneous formation of factor VIII antibodies in previously normal people can reduce the blood factor VIII level and give rise to a haemophilia-like condition.[20] Factor VIII is also often reduced in von Willebrand's disease but the heredity of this condition is unlike that of haemophilia being autosomal dominant. Also the bleeding manifestations are different from haemophilia

TABLE 6.4
Therapeutic material used in the management of
bleeding in haemophilia

Material	Approximate factor VIII content (Units*/ml)
Whole Blood	0.3
Fresh Frozen Plasma	0.6
Cryoprecipitate	3–4
Lyophilised human AHG concentrate	3–4
Lyophilised Animal AHG	10–15

* 1 unit of factor VIII is the amount of factor VIII present in 1 ml. of fresh citrated normal plasma.

in that haemarthroses are very uncommon but bleeding from the mucous membrane of the gastro-intestinal tract and genito-urinary tract is common.

Recently it has been shown that antibodies can be raised in rabbits to a material which is closely associated with factor VIII and which has been termed factor-VIII-related antigen. Using these antibodies the antigen can be quantitatively measured using the Laurell electrophoretic technique. It has been shown that haemophiliacs possess normal amounts of this antigen while von Willebrand's patients have reduced amounts roughly in proportion to their level of factor VIII. The relationship between factor-VIII-related antigen and factor VIII activity is not known at present.

In the clotting cascade factor VIII is involved in the reaction whereby activated factor IX activates factor X. Phospholipid and calcium ions are required in this reaction. It was originally considered that this activation of factor X might involve an activated form of factor VIII but current opinion favours a mechanism in which factor VIII acts more as a cofactor to factor IX.

The molecular weight of factor VIII has been variously reported. Values calculated from sedimentation and diffusion studies suggest a molecular weight between 180 000 and 196 000.[21] However, gel filtration experiments suggest figures in excess of 200 000 and recent estimates have put the molecular weight as high as 2 000 000. This apparent discrepancy is paralleled by similar variation in sedimentation coefficients; some workers report values between 6.0 and 6.7 S while others record 12 to 19 S. The most reasonable explanation seems to be that factor VIII can easily form molecular aggregates and that gel filtration experiments reflect molecular weights corresponding to such

aggregates. The electrophoretic mobility of factor VIII is that of a globulin but it has been shown, by different workers, to migrate with the
α or β or γ globulins depending on the conditions of the experiment.
It must be concluded that its mobility alters with the conditions of the
experiment and in particular with the nature of the supporting medium.
This is not inconsistent with the hypothesis that factor VIII is a large
and complex molecule.

The possibility of factor VIII being a lipoprotein has been raised but
at present the evidence for this is not very strong. Several workers have
reported the presence of a carbohydrate constituent in the molecule
and the precipitation properties of factor VIII with lectins such as
concanavillin are consistent with its being a glycoprotein.[22]

Factor VIII activity appears to be destroyed by thrombin, plasmin
and many other proteolytic enzymes and activity is not present in
serum. Variations of pH, both acid and alkaline, also destroy it and
maximum stability is achieved in the range of pH 6.2–7.2. It is also
inactivated by oxidation and reduction reactions indicating that for
maximum stability, factor VIII should be maintained within a restricted range of redox potential. In plasma there are buffering mechanisms which give some protection against changes in both pH and redox
potential but in purified factor concentrates added protection may have
to be provided in the form of buffers and stabilising additives. Thiol
groups have been shown to be present at the sites involved in clotting
and oxidation or chemical substitution of these groups results in a loss
of factor VIII activity.[23] Factor VIII is inactivated also by haemato-
porphyrin when irradiated by light but this is probably an example of
photo-oxidation.

One of the most prominent characteristics of factor VIII is its lability;
this has already been mentioned when considering its sensitivity to
oxidation. The lability of factor VIII is of considerable importance
since it is this feature which seriously hinders purification and detailed
chemical investigation. Stability appears to decrease as purity increases
and in particular the removal of fibrinogen seems to markedly reduce
stability. Factor VIII in human plasma is stable for several months at
−20°C but when stored at 4°C loses considerable activity over only a
few days. Numerous attempts have been made to find additives which
will increase storage stability and some success has been reported with
the use of materials such as glycerol, albumin, fructose and glycine
methyl ester. Antioxidant additives might be thought to be useful too
but so far no advantage for them has been demonstrated. The simple
act of freezing and thawing also alters factor VIII activity and various
degrees of loss of activity and sometimes increases have been observed.
It is usually found that there is a 20% loss of activity following freezing
and thawing.

Factor VIII can be spun down in the ultracentrifuge and some workers
using this technique have been able to obtain several fractions each

with factor VIII activity. On the basis of this work they suggest the existence of molecular variants of factor VIII. However, their observations did not exclude the possibility that these apparent variants were the same molecule in different geometric and spatial configurations. In addition, as has already been suggested, molecular aggregation of factor VIII may take place and it is not known how this would influence those results and their interpretation.

Factor VIII is slightly adsorbed by barium sulphate, aluminium hydroxide and bentonite but this requires a large excess of adsorbent. ECTEOLA cellulose and DEAE cellulose adsorb the factor easily and have been used as the basis of separation procedures. However, the factor is inactivated by these procedures and a relatively small proportion of activity is recovered after elution. Gel filtration can be used to separate factor VIII from smaller molecular species. Using gel filtration techniques the factor VIII appears to be completely excluded from all of the gels available today. Factor VIII can be separated from fibrinogen in this way and about 50% of the factor VIII activity can be recovered. As mentioned before the fibrinogen-free product is much reduced in stability even when stored at $-40°C$.

The purification of factor VIII is of immense importance for effective treatment of haemophilia and considerable effort has been devoted to it.[24,25] Methods which have been employed include isoelectric precipitation, and precipitation by alcohol, ether, β-alanine and polyethylene glycol. In addition, therapeutic materials are prepared by cryoprecipitation, whereby frozen plasma is allowed to thaw at 4°C and melted plasma is poured off to leave factor VIII activity in the cryoproteins.[26] This cryoprecipitation technique has found widespread application in the concentration of factor VIII because it is simple and can be easily adapted to the routine of blood transfusion centres. More highly purified materials can be obtained for research purposes,[25,27] usually by a combination of the above techniques plus gel filtration or ion exchange chromatography.

6.6. FACTOR VII

Factor VII is present in the blood of normal humans and is necessary for normal clotting in the presence of tissue thromboplastin. It is closely associated with factor II, IX and X and like them is thought to be made in the liver. Deficiency of factor VII, congenital or acquired, results in a bleeding disorder.[28] The congenital form is very rare but the acquired form is commonly seen in association with deficiency of factor II, IX and X in liver disease and in patients taking coumarin type anticoagulant drugs. Treatment of congenital factor VII deficiency consists of giving intravenous transfusions of plasma or plasma protein concentrate rich in the factor. Effective treatment may be difficult in some situations because of the extremely short half-life of factor VII in the circulation, approximately 3–4 hours.

Factor VII activity can be destroyed by heating to 56°C. It is unstable below pH 3 and above pH 9. On electrophoresis it migrates between the α and β globulins. It is readily adsorbed onto the surfaces of inorganic precipitates such as aluminium hydroxide, barium sulphate, and calcium phosphate from which it may be easily eluted. In Cohn fractionation it separates with fraction III.

Factor VII survives the clotting process and the molecular weight of factor VII separated from serum is found to be significantly different from that isolated from plasma. The serum factor has a molecular weight of 48 000 when measured by gel filtration, the plasma factor has a molecular weight of 55 000–63 000. Molecular weights measured by sedimentation techniques are somewhat less, the serum factor being 34 000.[29,30] Calculations based on this value together with amino-acid determinations indicate that the serum factor contains 221 amino-acid residues. N terminal and C terminal acids appear to be the same and consist of glycine and serine.[31] It is interesting that similar terminal acids have been reported for unactivated factor X. Other points of similarity between factors VII and X are seen in the degradation products after trypsin degradation and in the molecular weight of plasma factor VII compared to unactivated factor X. Factor VII is reported to have an high carbohydrate content of 51%. Most of this is said to be in the form of hexoses with 1.2% hexosamines.[32]

In the series of clotting reactions factor VII is required in the activation of factor X by tissue factor. This tissue factor or tissue thromboplastin as it has come to be called is present in a variety of tissues but for use in most laboratory assay techniques the preferred material is made by macerating human brain in saline.

In the activation of factor X by tissue factor, factor VII is required probably in the role of a co-factor and calcium ions are also necessary. The existence of an activated form of factor VII has not been demonstrated. Many workers consider the tissue factor reaction to be enzymic with factor X as substrate, although it has not been directly proved. There is an analogy with the action of Russell's Viper Venom on factor X and it is interesting that factor X which has been completely activated by tissue factor and factor VII cannot be further activated by Russell's Viper Venom.

Factor VII can be separated from plasma or serum. By using plasma as starting material a product is obtained which is free of activated factors and this is preferred in the preparation of therapeutic concentrates. Initially the mixture of factors II, VII, IX and X is separated from plasma as described in the purification of factor IX. This product is then chromatographed on an ion exchange medium such as DEAE cellulose when factor VII can be separated from the other three factors. Useful methods of concentrating factor VII are given by Prydz[33] and Tischkoff et al.[30] while Bidwell et al.[24] give a detailed review of the preparation of the concentrate containing factors II, VII, IX and X.

6.7. FACTOR X

Factor X is found in normal plasma and serum and is essential for normal clotting and haemostasis.[34] Deficiency of the factor may be congenital or acquired and if severe enough may result in a haemorrhagic disorder. In some instances of the congenital deficiency state the bleeding may be as severe as that found in haemophilia. Acquired factor X deficiency is most commonly seen in liver disease and in patients taking coumarin type drugs in which situation it is part of a combined deficiency of factors II, VII, IX and X.

Treatment of bleeding due to congenital factor X deficiency requires transfusions of plasma or protein concentrates rich in factor X. The half-life of factor X in the circulation is 50–60 hours and because of this transfusion need not be given as frequently as in factor VIII or factor VII deficiency. Treatment of acquired deficiency of factor X consists of correcting the underlying liver disease or withdrawing anticoagulant drugs as the case may be.

Factor X is a relatively stable clotting factor usually found associated with factors II, VII and IX and resembles these factors in many physicochemical aspects.[35] It migrates electrophoretically with the α_1 globulins and has an isoelectric point of pH 4.75. It is adsorbed by inorganic precipitates such as aluminium hydroxide, calcium phosphate and barium sulphate and it is easily eluted from them. Molecular weight determinations have given values between 52 000 and 87 000, most support being given for the higher end of the range. The N terminal acids have been found to be glycine along with serine or alanine or possibly all three. C terminal amino-acids are reported to be glycine and serine. These results suggest that factor X has two chains in its molecule.

In the cascade hypothesis of clotting reactions, factor X occupies a unique position since it may be activated by both the so-called intrinsic and extrinsic systems. In the intrinsic system this activation is achieved by the active form of factor IX in the presence of factor VIII, phospholipid and calcium. In the extrinsic mechanism factor X is activated by a combination of tissue extract, factor VII, phospholipid and calcium. A constituent of Russell's Viper Venom can also activate this factor in the presence of a phospholipid. In addition trypsin can cause activation and factor X can apparently become active spontaneously in strong citrate solutions. Some workers hold the view that factor X exists as a complex with prothrombin factor VII and factor IX, the combination being termed "prothrombin complex". This point of view is discussed in greater detail in the section concerned with factor II (prothrombin).

The activated form of factor X is usually referred to as factor Xa. It has a reported molecular weight between 25 000 and 38 000, which is about half of the molecular weight of inactive factor X. This suggests that factor X could be a dimer of the active factor Xa. The N terminal acids are alanine and glycine along with another residue which some workers report as leucine or isoleucine and others as serine.

Active factor X has clearly defined esterase activity as can be shown by its action on TAME. In the clotting process it activates factor II in the presence of factor V and calcium ions to yield thrombin. Its esterase and its coagulant activity can be inhibited by soya bean trypsin inhibitor, by thrombin and by DFP. The latter inactivation suggests that there is a serine group at the active site.[36] Oxidising or reducing agents and carbonyl-group reagents can destroy the activity of factor Xa but thiol blocking reagents have no effect.[37] A natural anti-Xa activity exists in plasma and presumably regulates the level of the active factor in the circulating blood. This anti-factor can be clearly demonstrated in vitro and it is possible to measure its level by clotting assay techniques.

High potency concentrates of factor X have been prepared by a combination of precipitation and chromatographic techniques. Preparation is started by treating oxalated plasma with barium sulphate which adsorbs factors II, VII, IX and X. This mixture of factors is eluted off the adsorbent using sodium citrate solution. Further purification can be obtained by isoelectric precipitation or by precipitation with ammonium sulphate or with acetone. However, in order to separate factor X adequately from factors II, VII and IX either preparative electrophoresis, or chromatography on an ion exchange medium such as DEAE cellulose, must always be included in the isolation procedure. Several effective methods of preparation are given in the literature, for example, the methods of Esnouf and Williams[35] and Jackson et al.[38]

6.8. FACTOR V

Factor V is one of the relatively labile plasma clotting factors. Estimates of its concentration in human plasma are in the range 1 to 7×10^{-2} mg/ml. The factor is essential for normal clotting as well as haemostasis and deficiency is associated with a severe haemorrhagic disorder. Congenital absence of factor V is extremely rare and affects males and females equally. Fortunately treatment of haemorrhage in such cases is relatively easy and consists of giving transfusions of fresh plasma.

The site of synthesis of factor V is unknown. The depression of this factor seen in patients or animals with liver damage, suggests that the liver plays some part in its synthesis. However, experiments to demonstrate synthesis of factor V in vitro using liver slices have given equivocal results.

The molecular weight of factor V has been variously reported, the values obtained depending on the method of measurement.[39] Most values fall between 98 000 and 300 000 but in one experimental study two active species have been suggested, one of 70 000 molecular weight and one of 350 000. As is often found with other clotting factors, molecular weights determined by gel filtration techniques are usually higher than those obtained by other methods. There is evidence to suggest the existence of a considerable non-protein entity in factor V which could amount to 25% of the molecule. Determinations of amino-acid sequence made so far show considerable disagreement.

Factor V has an isoelectric point of pH 5.2–5.3. It is not removed from plasma by Seitz filtration and it is only slightly adsorbed on to the inorganic precipitates used to remove factors II, VII, IX and X, from plasma. It is markedly adsorbed by soya bean lecithin.

The electrophoretic mobility of factor V appears to vary with the experimental conditions especially with the nature of the support medium. With paper electrophoresis the factor travels between the α and β globulins but with continuous free film electrophoresis it is found in the albumin fraction.

Human factor V is very labile, much more so than the bovine material. It has been suggested that this lability involves a process of physical inactivation. Calcium ions are important for stabilisation of this factor and a common method of selectively removing factor V from human plasma is to store it in the presence of oxalate ions. Similarly chelating agents such as EDTA will inactivate factor V by removing calcium. Human factor V is destroyed by thrombin and accordingly does not survive the clotting process in human plasma.[40] Bovine factor V on the other hand is resistant to thrombin and is found in large amounts in bovine serum.

Factor V has been shown to contain thiol groups and thiol blocking reagents or oxidising agents will destroy its activity.[41] However, storage properties of the factor are not improved by the addition of reducing agents, suggesting that oxidation of thiol groups does not significantly contribute to the labile nature of factor V. Stabilising additives have been sought and the best one found so far is 50% glycerol either in water or buffer. Sucrose and albumin have some stabilising influence and dialysis against water or buffer also seems to make factor V more stable.

In the original "cascade hypothesis", the existence of an activated factor V was considered. However, current opinion favours a mechanism in which factor V and calcium are involved in a reaction between factor II (prothrombin) and activated factor X. In this role it appears to act more as a co-factor.

Factor V may be concentrated from plasma and the methods of Esnouf and Jobin[39] and Barton and Hanahan[42] are examples of effective procedures. In the case of the bovine product preparation may also start from serum since bovine factor V survives the clotting process. The first stage of purification is to remove factors II, VII, IX and X, usually by adsorption of oxalated plasma using barium sulphate. Most methods for preparing high purity material include ion exchange chromatography of the adsorbed plasma usually on TEAE-cellulose and some procedures additionally include purification by isoelectric precipitation or ammonium sulphate precipitation.

6.9. PROTHROMBIN (FACTOR II)

Prothrombin or Factor II is one of four clotting factors in human plasma requiring vitamin K for their synthesis. Those four factors, factors II, VII, IX, and X behave in a similar way during health and disease. At birth they are present in reduced amounts in plasma and there may be a further fall during the first week of life which can result in haemorrhagic disease of the newborn. In patients with liver disease or malabsorption of vitamin K the concentration of factors may be depressed. In Western Countries, the commonest cause of reduced levels of those factors is the ingestion of Coumarin type anticoagulant drugs.

Isolated congenital deficiency of prothrombin is extremely rare and there are only 2 or 3 authenticated cases in the world literature. Patients have a moderately severe bleeding disorder characterised by bleeding after surgery, and accidents. Treatment consists of giving transfusions of plasma or plasma protein concentrates rich in prothrombin.

Prothrombin is the inactive precursor of thrombin. It is activated by active factor X in the presence of factor V, phospholipid and calcium ions. Some workers consider that a distinct intermediate product is formed and that this in its turn is converted to thrombin. This intermediate has been termed prethrombin and will be discussed in a separate section. *In vitro*, prothrombin can also be activated to thrombin by sodium citrate together with factor X. It has a molecular weight approximately twice that of thrombin indicating that activation probably involves a molecular cleavage. This is supported by the fact that proteolytic enzymes such as trypsin will also convert prothrombin to thrombin. It is not universally agreed that thrombin is the sole product of prothrombin activation. Some workers consider that factors II, VII, IX and X are co-existent in the form of a "prothrombin complex" and that these are released as separate entities upon activation.[43] However, the molecular weight found for the "prothrombin complex" is much less than the combined molecular weights of factors II, VII, IX and X and this would seem to exclude the possibility of these factors being combined chemically in the complex. The "prothrombin complex" displays clotting properties of factors VII, IX and X but this is most likely due to these latter factors being present as impurities. These impurities would be detected by clotting tests which are extremely sensitive compared with chemical methods of analysis and which detect factor levels that are minute in terms of molar proportions. Factors II, VII, IX and X all have similar physico-chemical properties, and this makes complete separation particularly difficult. However, after chromatography on DEAE Sephadex, prothrombin may be obtained which does not have the clotting properties of these other factors.

Prothrombin is believed to be an α globulin with a molecular weight of approximately 69 000 for the bovine and human materials.[44] Higher molecular weights have been reported for equine prothrombin. Determinations of molecular weight on the purified factor are not markedly lower than those for the prothrombin complex suggesting that the other members of the complex, factors VII, IX and X, are present in only small amounts. By the electron microscope it appears globular with a height of 105 Å although earlier calculations had favoured an ellipsoid 119–134 Å long by 34–35 Å wide. Its N-terminal aminoacid is alanine while the C-terminal acids are believed to be tyrosine and glycine. It contains disulphide linkages estimated at 8 per molecule and a carbohydrate entity which probably represents 10–12% of the molecular weight.

A variety of methods is available for the preparation of high potency factor II concentrates but the principles involved are similar. Factors II, VII, IX and X are adsorbed from plasma onto a material such as barium sulphate, barium citrate or magnesium hydroxide. The mixed factors are released from the adsorbant by elution, or in the case of barium citrate by the action of EDTA or an ion exchange resin, or in the case of magnesium hydroxide by dissolution of the material using carbon dioxide. The released factors can then be further purified by isoelectric precipitation or by precipitation with alcohol or ammonium sulphate. Good examples of such techniques are given by Seegers[45] and Magnusson.[46,47] The product at this stage has been referred to as the "prothrombin complex" and this has been discussed above. To separate the factors further ion exchange chromatography is necessary and this is usually carried out on DEAE cellulose.[48] In this way factor II can be prepared free from factors VII, IX and X as detected by clotting assay techniques.

The degree of purification which is obtained can vary considerably with the animal species. Bovine factor II is easier to purify than the human material and in addition it is possible to remove factor VII and much of the factor X from bovine plasma merely by Seitz filtration. With equine plasma very pure preparations are possible and after the ion exchange chromatography stage the product may be crystallised from a chloroform/methanol mixture or from n-butanol.

6.10. PRETHROMBIN

Prethrombin is the name given to an intermediate product in the activation of prothrombin to thrombin.[49] It is not designated as a particular clotting factor and no bleeding states have so far been associated with its absence. It is usually prepared by the activation of prothrombin by thrombin and purified by chromatographic and pre-cipitation steps. It appears to be homogeneous in the analytical ultra-centrifuge and on electrophoresis. By gel filtration its molecular weight is 43 000–61 000. It has two N-terminal acids, lysine and threonine, the latter also occurring as an N-terminal aid in thrombin. Its isoelectric point of pH 5.5 is close to that of thrombin. Prethrombin has no pro-teolytic activity and will not clot fibrinogen. It cannot be converted to thrombin unless activated factor X is present.[50] Thus from several points of view prethrombin occupies a position intermediate between prothrombin and thrombin. It appears to be a discrete chemical entity but there is no evidence as yet to suggest that it deserves separate biological consideration.

6.11. THROMBIN (ACTIVATED FACTOR II)

Thrombin is a proteolytic enzyme which, in the normal clotting process, acts upon fibrinogen to produce the fibrin clot. It is the pro-duct of prothrombin activation brought about by the action of acti-vated factor X in the presence of factor V, phospholipid and calcium ions. In vitro, 25% sodium citrate solution in the presence of factor X can also activate prothrombin. In its clotting action thrombin splits arginyl-glycyl bonds at the N terminal end of the fibrinogen molecule. It exhibits esterase activity[46,47,51] against the methyl esters of tosyl-L-arginine (TAME), benzoylarginine and tosyl-L-lysine as well as against a series of acyl-arginyl and peptidyl-arginyl amides. DFP destroys both its clotting and esterase activity demonstrating that serine constitutes part of the active centre. Thrombin activity is also inhibited by 1-chloro-3-tosylamido-7-amino-2-heptanone (TLCK) indicating that histidine also occurs in the active centre.

Most investigations have placed the molecular weight of thrombin between 30 000 and 34 000.[52,53] It is interesting that this is about half the value for prothrombin and speculations have been made that pro-thrombin may be a dimer of thrombin or that it may consist of thrombin combined with another clotting factor. Thrombin is believed to be a 2-chain molecule the chains being linked by disulphide bridging. The N-terminal aminoacids are isoleucine and threonine and there are a total of 2-3 disulphide links per molecule. A carbohydrate entity has been detected amounting to 9.7% of the total molecular weight and containing hexoses, hexosamines and sialic acids.

Chromatography on IRC-50 resin results in a modified thrombin with a sedimentation coefficient of 3.2 S instead of 3.76 S and an isoelectric point of 6.2 pH instead of 5.7 pH.[54] The molecular weight is decreased to 23 000 indicating that an estimated 75 aminoacid residues have been detached from thrombin to yield this product. N terminal acid analysis still registers threonine and isoleucine so the N-terminal ends of both chains are probably intact.

It is possible to make insoluble thrombin derivatives by coupling thrombin chemically to a polymer matrix. However, as in the case of acetylated thrombin, the derivatives display excellent esterase activity but minimal clotting activity. This suggests that it is important to have a particular molecular configuration for thrombin to react with fibrinogen.

Separation of thrombin from plasma usually starts by isolating prothrombin, then activating it to thrombin either by 25% sodium citrate in the presence of factor X or by thromboplastin in the presence of calcium ions and factor V. After activation, the thrombin product is further purified by acetone precipitation and ion exchange chromatography and sometimes in addition by precipitation with ammonium sulphate or ethanol.[47,55]

6.12. FIBRINOGEN

Fibrinogen is discussed fully in another chapter of this book and so this account will be limited to a brief outline. Fibrinogen is the precursor of fibrin, which is the protein constituting the network of the blood clot. All the evidence so far available suggests that fibrinogen is made in the liver. The normal concentration of fibrinogen in the blood is 2.5–3.5 mg/ml but levels of 0.8–1.0 mg/ml are usually adequate for haemostasis.

Deficiency of fibrinogen may be congenital or acquired and if sufficiently severe may result in a haemorrhagic state. The congenital form is very rare and affects the sexes equally. Treatment of bleeding in this condition consists of giving transfusions of fresh plasma or protein concentrates rich in fibrinogen. The acquired form of hypofibrinogenaemia is relatively common and is seen in certain clinical situations e.g. in abruptio placentae in pregnancy, in certain forms of carcinoma and following surgical operations on the lungs or prostate. The reduced fibrinogen concentration in those conditions is thought to be due to consumption of fibrinogen during intravascular clotting, the latter probably being triggered by the entry of tissue thromboplastin into the circulating blood.

Fibrinogen is a large helical protein with a molecular weight between 320 000 and 340 000.[56] Under the electron microscope it appears as three spheres 50–70 Å in diameter connected by strands 15 Å thick.[57] It is thought to exist as a dimer with each half containing three peptide chains of 50 000–65 000 molecular weight. Each half also has a so-called "disulphide knot"[58] which incorporates the N-terminal portions of all three chains. About 36% of the total half-cystine residues are believed to be contained in the "disulphide knots" which are thought to be the active sites of polymerisation. A total of 28–29 disulphide bridges are present and since free thiol groups are not detectable it can be assumed that all the half-cystine residues are involved in disulphide bridges. The carbohydrate entity in fibrinogen is relatively small. Many quantitative determinations have been made and the reported concentrations are mostly in the range of 1 to 5% of the molecular weight.

Fibrinogen is converted into insoluble gels or threads when reacted with various substances including papain, reptilase, vasculokinase and staphylococcal coagulase. In the blood clotting process it is acted upon proteolytically by thrombin. Two of the three chains are attacked releasing the fibrinopeptides A and B (m.w. 50 000–65 000) and forming fibrin monomer. These monomers polymerize to form fibrin and this is stabilised by cross-linking under the influence of factor XIII. The fibrinogen to fibrin reaction may be summarised in the following way:—

Fibrinogen + Thrombin → Fibrin Monomer
+ Fibrinopeptides A and B

Fibrin Monomer ⇌ Aggregates → Fibrin

Unstable Fibrin $\xrightarrow{\text{Factor XIII}}$ Stable Fibrin

In Cohn fractionation of plasma, fibrinogen precipitates with fraction I and this may be further purified by extraction of the precipitate with citrate-ethanol-glycine buffer, followed by precipitation with buffered ethanol or glycine or glycine/ethanol mixtures.[59] Another method of purifying the Cohn fraction I is to stand a solution of it overnight at 0 to 4°C to precipitate out unstable components and then to precipitate fibrinogen from the supernatant with 25% ammonium sulphate.[6] Cryoprecipitation of plasma may also be used to isolate fibrinogen. Here the plasma is frozen at −30°C, thawed at 5°C and the precipitated fibrinogen is removed and washed. A further novel method of isolation involves the use of a chemical complex. Plasma is absorbed with barium sulphate and chromatographed on TEAE-cellulose. The product is mixed with potassium mercuric tetrathiocyanate ($K_2Hg(SCN)_4$) whereupon fibrinogen precipitates as a complex. The washed precipitate is dissolved in a solution of EACA and saline, chromatographed on Sephadex G-25 gel and residual mercuric ions are removed by means of an insoluble chelating agent.

6.13. FACTOR XIII

Factor XIII is the agent necessary for stabilisation of fibrin in clots. Its absence from the blood is usually congenital and results in severe bleeding and delayed wound healing due to the formation of friable clots. This leads to extensive scarring about old wounds. Fortunately only a small amount of factor XIII, 1–2% of the amount present in normal blood, is required for correcting the defect *in vitro* and *in vivo*, so that haemorrhage in factor XIII deficient patients can be easily controlled by a transfusion of fresh or fresh frozen plasma. Moreover it is possible to treat such patients prophylactically by giving a transfusion of plasma once every 2–3 weeks.

Factor XIII has a molecular weight of about 350 000 and can be dissociated into three sub-units each with a molecular weight of

110 000.[61] It is of intermediate stability compared to other clotting factors; it loses about half of its activity after three days at room temperature but can be stored for prolonged periods at $-20°C$. Serum contains about 12% of the plasma level and in normal people the plasma concentration varies from 50% to 200% of the average value. Factor XIII circulates as an inactive molecule but in the process of clotting the thrombin formed converts it into its active enzymic form. Its role in the haemostatic mechanism is to stabilise clots by forming chemical cross-links between the fibrin strands.[62] Present theory suggests that these links are formed by a transpeptidation reaction between the amino groups of lysine and the carbonyl groups of asparagine or glutamine.[63,64]

Clots formed in the absence of this factor can be easily lysed and their solubility in 30% urea or 2% acetic acid is the basis of a diagnostic test for factor XIII deficiency. In the presence of as little as 1% of the normal factor level the clot is stable. Below this level the clot will dissolve completely in a few hours and such a procedure can be easily converted into a quantitative analysis. An alternative method is based upon a measure of enzymic transferase activity. In this test radioactive putrescine is mixed with caseine in the presence of factor XIII and the enzyme activity is measured by the amount of radioactive protein which is produced.[65] This test is suitable only for factor concentrations in excess of 5% average normal level. It must also be remembered that this particular transferase activity can be found in materials other than factor XIII and so false positive results can be obtained. A particular example is with lysed red cells which display transferase activity but have no ability to stabilise fibrin.

Factor XIII is detectable in normal intact platelets but greater quantities are released upon lysis. Factor XIII from platelets is immunologically identical to the plasma factor and in assay systems shows no functional difference. White cells and red cells have no factor XIII activity even after lysis. Concentrates of factor XIII may be prepared by ammonium sulphate precipitation of plasma at controlled levels of ionic strength and pH.[66] The fraction containing factor XIII is heated for three minutes at 56°C to remove heat-labile proteins and precipitated again with ammonium sulphate. Following dialysis the product is chromatographed on DEAE cellulose or Bio-Gel P 200.[67]

6.14. INHIBITORS

Inhibition can occur at many stages in the series of consecutive clotting reactions but three types of inhibition are of particular interest in the study of coagulation defects:—

(i) Inhibitors to factor VIII
(ii) Inhibitors to factor IX
(iii) Inhibitors associated with systemic lupus erythemotosus

In addition to these, there are naturally occurring inhibitors which neutralise the active intermediates of clotting. Of these anti Xa[68] and the several antithrombins[69,70] are best understood. Such substances must be of vital importance to the living organism since they provide a control mechanism by which clotting is contained within the limits required for haemostasis and prevent the dangerous extension of thrombus formation. Little is known about the biochemistry of their action at the present time.

Inhibitors to factors VIII and IX are believed to be antibodies arising from infusions of factor concentrates or plasma. Inhibitors associated with systemic lupus erythemotosus and fibrin degradation products are not clearly defined. Each of these will now be discussed individually.

6.14.1. FACTOR VIII ANTIBODIES

Antibodies to factor VIII arise in 5% of haemophiliacs and are of serious concern since their presence in the blood makes the patient unresponsive to transfusion therapy with factor VIII. In haemophilic patients the inhibitors have been shown to be antibodies and they presumably arise as a consequence of treatment with factor VIII.[20] Antibodies to factor VIII can also arise in previously normal patients with such conditions as rheumatoid arthritis, penicillin allergy and ulcerative colitis. They have also been found in women in the postpartum period. All of these spontaneous non-haemophilic cases are rare; the patients have been previously normal from a blood coagulation viewpoint but after acquiring antibodies they become clinically indistinguishable from a haemophiliac with factor VIII antibodies. In a proportion of cases the antibody may disappear within 2–3 years.

In their kinetic behaviour, factor VIII antibodies appear as a range of types. At one extreme their reaction with factor VIII is adequately expressed by simple second order kinetics and hence by pseudo first order kinetics when antibody is in excess.[71–73] At the other extreme, very complex kinetic behaviour is observed which is exceedingly difficult to present in mathematical form. Some recent work on these kinetics has suggested that an antibody-antigen complex is formed which can have partial factor VIII activity.[74] Factor VIII antibodies are γ-globulins[75] and studies of sub-classification indicate that most of them are γ G globulins with K light chains. However, other types have been identified too and electrophoretic and gel filtration studies suggest that all are heterogeneous to some degree. They do not require to fix complement for their reaction with factor VIII. They are stable in storage and resist heating to 56°C. They also resist the clotting reaction and are present in serum. At present the only practical method of antibody assay is based on clotting tests. Haemagglutination techniques have given positive results but only when exceptionally high antibody concentrations are used.[76] Radioactive immunological assays have so far been unsuccessful largely because factor VIII of adequate purity has not been available.

6.14.2. FACTOR IX ANTIBODIES

Some patients with factor IX deficiency (Christmas disease, or Haemophilia B) develop antibodies following repeated therapeutic transfusions.[77] They are distinctly rarer than factor VIII antibodies but this is at least partly due to IX deficiency being relatively rarer. Antibodies to factor IX have so far been described only in patients with hereditary IX deficiency. They are stable and have properties resembling γ-globulins. Factor IX antibody assays are based on clotting tests but if the antibody is in high concentration it can be detected by haemagglutination techniques. They differ from factor VIII antibodies in being very rapid in action. It is not possible to measure the exact velocity since assay methods are in themselves time-consuming but ten minutes is more than adequate for all *in vitro* reactions to take place.

6.14.3. INHIBITORS IN SYSTEMIC LUPUS ERYTHEMATOSUS

In patients with systemic lupus erythematosus a coagulation inhibitor can rise and lead to a bleeding diathesis. Little is known about this inhibitor. It does not destroy any known clotting factor but it prolongs clotting times in such tests as the kaolin-cephalin time and the one-stage prothrombin time. The present view is that it interferes in some way with the activation of prothrombin. It is stable in storage and resists heating to 56°C for thirty minutes.[78]

REFERENCES

1. R. G. MACFARLANE, *Nature*, **202**, 498 (1964)
2. E. W. DAVIE and O. D. RATNOFF, *Science*, **145**, 1310 (1964)
3. R. BIGGS and R. G. MACFARLANE, Human Blood Coagulation and its Disorders (Blackwell Scientific Publications, Oxford, 1962)
4. R. M. HARDISTY and G. I. C. INGRAM, Bleeding Disorders, Investigation and Management, (Blackwells Scientific Publications, Oxford, 1965)
5. E. W. DAVIE and O. D. RATNOFF, (ed., H. Neurath) The Proteins, 2nd editio n, (Academic Press, New York & London) Vol. III (1965) p. 359
6. R. J. SPEER and H. RIDGEWAY, *Thrombos. Diathes. Haemorrh.* **18**, 259 (1967)
7. R. J. SPEER, H. RIDGEWAY and J. M. HILL, *Thrombos. Diathes. Haemorrh*. **14**, 1 (1965)
8. E. F. MAMMEN and G. L. GRAMMENS, *Thrombos. Diathes. Haemorrh.* **18**, 306 (1967)
9. S. SHERRY, N. AKLJAERIG and A. P. FLETSCHER, *Thrombos. Diathes. Haemorrh.* Supplement. Diffuse Intravascular Clotting. Transactions Conference St. Moritz, Switzerland, September, 1965. (1966) p. 243
10. J. G. G. SCHOENMAKERS, R. MATZE, C. HAANEN and F. ZILLIKEN, *Biochem*. *Biophys. Acta*, **101**, 166 (1965)

11. R. L. ROSENTHAL, *Thrombos. Diathes. Haemorrh.* Supplement. Thrombolytic Activity and Related Phenomena, Transactions Conference Princetown, N.J., U.S.A. September, 1960. (1961) p. 379

12. H. L. NOSSEL, In the Contact Phase of Blood Coagulation page 51. (Blackwell Scientific Publications, Oxford 1964)

13. O. D. RATNOFF and E. W. DAVIE, *Biochemistry*, **1**, 677 (1962)

14. H. S. KINGDON, E. W. DAVIE and O. D. RATNOFF, *Biochemistry* **3**, 166 (1964)

15. K. W. E. DENSON, R. BIGGS and P. M. MANNUCCI, *J. Clin. Path.*, **21**, 160 (1968)

16. H. S. KINGDON and E. W. DAVIE, *Thrombos. Diathes. Haemorrh. Supplement* **17**, 15 (1965)

17. C. R. HARMISON and W. H. SEEGERS, *J. Biol. Chem.*, **237**, 3074 (1962)

18. E. BIDWELL, J. M. BOOTH, G. W. R. DIKE and K. W. E. DENSON, *Brit. J. Haematol.*, **13**, 568 (1967)

19. G. W. R. DIKE, E. BIDWELL and C. R. RIZZA, *Brit. J. Haematol.*, **22**, 469 (1972)

20. C. R. RIZZA and R. BIGGS, *Brit. J. Haematol.*, **24**, 65 (1973)

21. S. SHULMAN, R. H. LANDABURU and W. H. SEEGERS, *Thrombos. Diathes. Haemorrh.*, **4**, 336 (1960)

22. L. KAAS, O. D. RATNOFF and M. A. LEON, *J. Clin. Invest.*, **48**, 351 (1969)

23. D. E. G. AUSTEN, *Brit. J. Haematol.*, **19**, 477 (1970)

24. E. BIDWELL, G. W. R. DIKE and W. FORD, In Human Blood Coagulation. Haemostasis and Thrombosis (ed., R. Biggs). (Blackwell Scientific Publications, Oxford, 1972)

25. J. NEWMAN, A. J. JOHNSTON, M. H. KARPATKIN and S. RUSZKIN, *Brit. J. Haematol.*, **21**, 1 (1971)

26. J. G. POOL, E. J. HERSHGOLD and A. R. PAPPENHAGEN, *Nature*, **203**, 312 (1964)

27. S. E. MICHAEL and G. W. TUNNAH, *Brit. J. Haemtol.*, **12**, 115 (1966)

28. B. ALEXANDER, Thrombosis et Diathesis Haemorrhagica Supplement, Thrombolytic Activity and Related Phenomena. Transaction of Conference Princetown N.J., U.S.A. September 1960. (1961) p. 392

29. C. R. HARMISON, H. SCHRÖER and W. H. SEEGERS, *Thrombos. Diathes. Haemorrh.* **13**, 587 (1965)

30. G. H. TISHKOFF, L. C. WILLIAMS and D. M. BROWN, *J. Biol. Chem.*, **243**, 4151 (1968)

31. E. HÖGENAUER, K. LECHNER and E. DEUTSCH, *Thrombos. Diathes. Haemorph.*, **19**, 304 (1968)

32. H. PRYDZ, *Scand. J. Clin. Lab. Invest.* **17**, Suppl. 84, 78 (1965)

33. H. PRYDZ, *Scand. J. Clin. Lab. Invest.*, **16**, 101 (1964)

34. T. P. TELFER, K. W. E. DENSON and D. R. WRIGHT, *Brit. J. Haematol.*, **2**, 308 (1956)

35. M. P. ESNOUF and W. J. WILLIAMS, *Biochem. J.*, **84**, 62 (1962)

36. J. E. LEVESON and M. P. ESNOUF, *Brit. J. Haematol.* **17**, 173 (1969)

37. M. J. CALDWELL and W. H. SEEGERS, *Thrombos. Diathes. Haemorrh.*, **13**, 373 (1965)

38. C. M. JACKSON, T. F. JOHNSON and D. J. HANAHAN, *Biochemistry*, **7**, 4492 (1968)

39. M. P. ESNOUF and F. JOBIN, *Biochem. J.*, **102**, 660 (1967)

40. P. A. OWREN, *Thrombos. Diathes. Haemorrh.*, Supplement Thrombolytic Activity and Related Phenomena. Transactions of Conference Princetown N.J., U.S.A. September 1960. (1961) p. 387

41. N. AOKI, C. R. HARMISON and W. H. SEEGERS, *Canadian J. Biochem. Phys.*, **41**, 2409 (1963)

42. P. J. BARTON and D. J. HANAHAN, *Biochem. Biophys. Acta*, **133**, 506 (1967)

43. W. H. SEEGERS, *Thrombos. Diathes. Haemorrh.*, **14**, 213 (1965)

44. LANCHANTIN, G. F., HART, D. W., FRIEDMANN, J. A., SAAVEDNA, N. V., and MEHL, J. W. *J. Biol. Chem.* **243**, 5479 (1968)

45. SEEGERS, W. H. Prothrombin. (Harvard University Press, Cambridge, Mass. (1962)
46. S. MAGNUSSON, *Arkiv für Kemi*, **24**, 367 (1965)
47. S. MAGNUSSON, *Arkiv für Kemi*, **24**, 349 (1965)
48. S. S. SHAPIRO and D. F. WAUGH, *Thrombos. Diathes. Haemorrh.*, **16**, 469 (1966)
49. W. H. SEEGERS, E. MARCINIAK, R. K. KIPFER and K. YASIENGA, *Arch. Biochem. Biophys.*, **121**, 372 (1967)
50. E. MARCINIAK and W. H. SEEGERS, *Nature*, **209**, 621 (1966)
51. S. SHERRY and W. TROLL, *J. Biol. Chem.*, **208**, 95 (1954)
52. C. R. HARMISON, R. H. LANDABURU and W. H. SEEGERS, *J. Biol. Chem.*, **236**, 1693 (1961)
53. D. J. BAUGHMAN and D. F. WAUGH, *J. Biol. Chem.*, **242**, 5252 (1967)
54. W. H. SEEGERS, L. McCOY, R. K. KIPFER and G. MURANO, *Arch. Biochem. Biophys.*, **128**, 194 (1968)
55. W. BERG, K. KORSAN-BENGSTEN and J. IGGE, *Thrombos. Diathes. Haemorrh.*, **16**, 501 (1966)
56. E. A. CASPARY and R. A. KEKWICK, *Biochem. J.*, **67**, 41 (1957)
57. N. V. BANG, *Thrombos.* Diathesis Haemorrhagica Supplement. Fibrinogen and Fibrin Turnover of Clotting Factors. Transactions of Gleneagles Conference, Scotland. July 1963 (1964) p. 73
58. B. BLOMBACK, M. BLOMBACK, A. HENSCHEN, B. HESSEL, S. IWANAGA and R. WOODS, *Nature*, **218**, 130 (1968)
59. B. BLOMBACK and M. BLOMBACK, *Arkiv für Kemi*, **10**, 415 (1956)
60. K. LAKI, *Arch. Biochem. Biophys.*, **32**, 317 (1951)
61. A. G. LOEWY, A. DAHLBURG, K. DUNATHAN, R. KRIEL and H. L. WOLFINGER, *J. Biol. Chem.*, **236**, 2634 (1961)
62. K. LAKI and L. LORAND, *Science*, **108**, 280 (1948)
63. S. MATACIC and A. G. LOEWY, *Biochem. Biophys. Res. Comm.*, **30**, 356 (1968)
64. J. J. PISANO, J. S. FINLAYSON and J. S. PEYTON, *Science*, **160**, 892 (1968)
65. A. DVILANSKY, A. F. H. BRITTEN and A. G. LOEWY, *Brit. J. Haematol.*, **18**, 399 (1970)
66. A. G. LOEWY, K. DUNATHAN, R. KRIEL and H. L. WOLFINGER, *J. Biol. Chem.*, **236**, 2625 (1961)
67. L. LORAND and K. KONISHI, *Biochem. Biophys. Acta*, **121**, 177 (1966)
68. R. G. MACFARLANE and B. J. ASH, *Brit. J. Haematol.*, **10**, 217 (1964)
69. V. ABELGAARD, *Scand. J. Clin. Lab. Invest.*, **21**, 89 (1968)
70. P. D. KLEIN and W. H. SEEGERS, *Blood*, **5**, 742 (1950)
71. R. BIGGS and E. BIDWELL, *Brit. J. Haematol.*, **5**, 379 (1959)
72. S. S. SHAPIRO, *J. Clin. Invest.*, **46**, 147 (1967)
73. R. BIGGS, D. E. G. AUSTEN, K. W. E. DENSON, C. R. RIZZA and R. BORRETT, *Brit. J. Haematol.*, **23**, 125 (1972)
74. R. BIGGS, D. E. G. AUSTEN, K. W. E. DENSON, R. BORRETT and C. R. RIZZA, *Brit. J. Haematol.*, **23**, 137 (1972)
75. E. BIDWELL, Annual Review of Medicine, **20**, 63 (1969)
76. H. R. ROBERTS, M. B. SCALES, J. T. MADISON, W. P. WEBSTER and G. D. PENICK, *Blood*, **26**, 805 (1965)
77. R. M. HARDISTY, *Thrombos. Diathes. Haemorrh.*, **8**, 67 (1962)
78. R. BIGGS and K. W. E. DENSON, *Brit. J. Haematol.*, **10**, 198 (1964)

Chapter 7

Complement*

Chester A. Alper

Centre for Blood Research,
800 Huntington Avenue,
Boston, Massachusetts 02115, U.S.A.

7.1. INTRODUCTION

It was almost a century ago that the ability of fresh serum to destroy certain bacteria was noted. This bactericidal capacity was soon resolved into a heat-stable substance, bactericidin, bacteriolysin, or, in modern terms, specific antibody, and a heat-labile substance called alexin by Büchner and complement by Bordet. At the turn of this century, Ehrlich and Morgenroth found similar requirements for the immune lysis of sheep red cells. Since the release of hemoglobin could be quantitated, the sheep cell (E), amboceptor or antibody (A), and complement (C) model system provided a powerful tool for the further study of complement and its actions. It was rapidly established that antibody could combine with the sheep cell in the absence of complement but complement could not exert its lytic action in the absence of antibody.

In 1907, Ferrata discovered that complement was not a single substance but could be resolved into two components by dialysis of fresh serum against water. The insoluble material (euglobulin) when redissolved in buffer or the supernatant (pseudoglobulin) alone did not lyse antibody-sensitized sheep erythrocytes (EA). However, when EA were first exposed to euglobulin, washed, and then incubated with pseudoglobulin, lysis occurred. No hemolysis was observed if the order of exposure was reversed. Thus, the components were named, in order of their reaction sequence, first and second component. Definitive evidence for a third component of complement was obtained by Ritz. In 1912, he noted that the venom of the Indian hooded cobra, *Naja naja*, destroyed the hemolytic activity of fresh serum but did not act on the first or second components. Further evidence for the existence of a third component was obtained when it was observed that yeast and gram-negative bacteria inactivated the same component as did venom.

* The original observations cited in this chapter were aided by U.S. Public Health Service Grants AM 13855 and AM 16392.

A fourth component of complement was discovered in 1926 by its inactivation on treatment of serum with ammonia.

The system remained at four components (designated C'1, C'4, C'2 and C'3) until the late 1950's. It then became apparent that C'1 consisted of three separate proteins (C'1q, r and s) complexed under physiologic conditions. C'3 was also found to be multiple and comprised no less than six distinct proteins. In the presently accepted nomenclature,[1] the complement components are numbered, in order of their sequential reaction, C1, 4, 2, 3, 5, 6, 7, 8, 9. Most of these proteins have been more or less purified as the result of work by Müller-Eberhard, Nelson, Lepow, Mayer and others. It is now clear that certain substances and particles can "by-pass" the first three components, C1, C4 and C2, and can activate C3 and subsequent components by an alternate pathway. Such a pathway was first described by Pillemer and his co-workers who called it the properdin system and it is currently being intensively reinvestigated. In addition to the components of complement themselves, the complement system also contains at least three inhibitors or inactivators at the C1, the C3, the C6, and the properdin Factor B steps. A list of the complement proteins, including synonyms, is given in Table 7.1.

7.2. BIOCHEMISTRY OF THE COMPLEMENT PROTEINS

Many of the complement proteins have been obtained in highly purified form in recent years and the remainder have been at least partially purified. Enormous advances in the purification of proteins in general over the past two decades have made this possible. Standard procedures such as anion and cation exchange chromatography,[2,3] preparative electrophoresis and gel filtration, as well as the more recent and exceedingly promising techniques of affinity chromatography have been used. Many of the proteins in the complement system are extraordinarily labile so that certain methods such as cold ethanol fractionation or ammonium sulfate precipitation are not often suitable in purification.

The physicochemical characteristics of these proteins are outlined in Table 7.1. As they occur in normal serum, their combined concentration is about 300 mg per 100 ml or approximately 4 to 5% of the total serum protein. As can be seen from the Table, several of the proteins (C3, C4, C1q, C$\overline{1}$ inhibitor, GBG) occur at rather high concentration. C3, in fact, is visible on ordinary zone electrophoresis of fresh serum, particularly if Ca^{2+} or Mg^{2+} is added to the electrophoresis buffer to slow its electrophoretic mobility in relation to transferrin and β-lipoprotein.

All of the complement components and related macromolecules thus far studied are glycoproteins and contain between around 3 and 43% carbohydrate. The carbohydrate portion of these molecules may be

TABLE 7.1

Physiochemical properties of human complement and properdin proteins

	Clq	Clr	Cls	C2	C3	C4	C5	C6	C7	C8	C9	Cl inhibitor	C3 inactivator	C6 inactivator	Properdin	GBG	GBGase	CoF-binding protein
Approximate mean normal serum concentration (mg/100 ml)	18	10	3	1	150	40	7	6	6.5	5	0.2	18	1	—	2.5	35	—	<2
Relative electrophoretic mobility	$?\gamma_2$	β	α_2	β_1	$\beta_1\text{-}\beta_2$*	β_1	β_1	β_2	β_2	β_1	α_2	α_2	β	β	$?$†	β_2	α_2	β_2
Approximate molecular weight	400,000	170,000	110,000	117,000	185,000	240,000	200,000	95,000	120,000	150,000	79,000	90,000	100,000	—	185,000†	100,000		
Sedimentation coefficient ($S_{20.w}$)	11	7	4	6	9.5	10	9	6	5	8	4.5	4	5	6	5†	6	3	5
Carbohydrate (%)	15	—	—	—	2.7	14	19	—	—	—	—	42	—	—	9.8†	10.6	—	—
Synonyms	11S component; C'O	—	C1 esterase	—	β_1C−globulin; properdin factor A	β_1E− globulin	β_1F− globulin	—	—	—	—	C1 esterase inhibitor; α_2 neuramingtyco-protein	C3b inactivator conglutinogen activating factor (KAF)	—	—	C3 pro-activator (C3PA); properdin factor B; heat-labile factor (HLF); unknown factor (UF)	C3PAse; factor; D	CVF-binding C3 protein; pro-activator

* Dependent on divalent cation concentration.
† These values are for properdin isolated in activated forms. Native properdin is a β-globulin which has not yet been isolated.

critical for function, as is the case for C9 and the C3 inactivator. The amino acid composition of most of the complement proteins is unremarkable. C1q and GBG are unusual for their high glycine content. C1q is uniquely rich among the complement proteins in hydroxylysine and hydroxyproline[3] and in this respect resembles collagen.

C1 exists in native serum as a macromolecular complex of three separate molecules, C1q, C1r, and C1s, perhaps bonded through Ca^{2+} as a ligand.[4] The complex is dissociable by chelation of the Ca^{2+} with EDTA. Dissociation is also favored by relatively high ionic strength. The C1 complex has a sedimentation coefficient of 18S, whereas the subcomponents are 11S, 7S and 4S, respectively. There is evidence that the complex occurs naturally as 1C1q:2C1r:4C1s.

C1q has a molecular weight of about 400 000 and it consists of five or six similar subunits of 57 000 daltons in non-covalent attachment.[5,6] Since five or six IgG molecules can be bound maximally to one C1q molecule, it may be that there is a combining site for immunoglobulins on each of the subunits. C1q is sufficiently large to visualize by electron microscopy. Such ultrastructural studies suggest that the molecule is a 200 Å diameter disc with five or six subunits arranged around a central core.[7] Another form of the molecule, a 400 Å rod structure, containing the same subunits has also been observed[8] and suggests that C1q may undergo extensive conformational change, perhaps relevant to C1r and C1s activation. C1q with its hydroxylysine, hydroxyproline and high glycine content is collagen-like and it is, therefore, of great interest that its biological activity is destroyed by collagenase.[6]

C1r and C1s have been isolated in both zymogen and activated ($C\overline{1r}$ and $C\overline{1s}$) forms. The zymogen form of C1r can be activated by proteolytic enzymes such as trypsin. $C\overline{1r}$ and $C\overline{1s}$ are esterases, the activities of which can be blocked by diisopropylfluorophosphate. C1s may consist of a single subunit, whereas $C\overline{1s}$ consists of two subunits of 36 000 and 77 000 daltons. It may be that $C\overline{1r}$ cleaves C1s and that the active enzymatic center on $C\overline{1s}$ is on the 36 000 molecular weight fragment. $C\overline{1s}$ cleaves esters containing a positively charged amino acid (N-acetylglycyl-L-lysine methyl ester and tosyl-L-arginine methyl ester, for example) as well as compounds containing an aromatic amino acid, such as acetyl-L-tyrosine ethyl ester or N-carbobenzoxy-L-tyrosine p-nitrophenyl ester.[9] There is evidence that the active center of $C\overline{1s}$ contains an anionic as well as hydrophobic binding site in accord with these specificities.[10]

C4 is a molecule of about 240 000 daltons, which when cleaved by $C\overline{1s}$, yields two fragments, C4a of about 10 000 daltons and C4b of about 230 000 daltons.[11,12] C4b binds to the red cell surface and participates in further complement activation.

C2 has a molecular weight of 117 000. It is acted upon by $C\overline{1s}$ to produce two fragments, C2a (83 000 daltons) and C2b (34 000 daltons).[3] C2a attaches to the antibody-$C\overline{14}$ site. Preliminary electrophoretic evidence suggests that C2 is a tetramer-type molecule.

Studies of C3 structure by SDS polyacrylamide gel electrophoresis in the presence of reducing agent suggest that it consists of two kinds of subunits: an α chain with a molecular weight of 110 000 and a 70 000 molecular weight β chain.[13] Enzymatic cleavage of C3 by $C\overline{42}$ or trypsin appears to cleave a fragment,[14] C3a, with a molecular weight of 6800, from the α chain, leaving C3b (around 180 000 daltons). On interaction of C3 with $EAC\overline{142}$, C3b attaches to the red cell membrane at multiple sites which may be remote from the antibody-$C\overline{142}$ site. Further

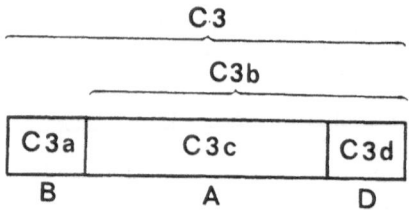

FIG. 7.1. The antigens on the C3 molecule and its fragments.

degradation of C3b occurs in whole serum, partly as the result of the proteolytic action of the C3 inactivator, to C3c (7S) and C3d (around 27 000 daltons). Cell-bound C3b may be attached via its C3d portion so that C3c is released by the action of the C3 inactivator on cell-bound C3b. Antigenic analysis[15] has revealed three distinct determinants on C3 shown diagrammatically in Fig. 7.1. The B antigenic determinant is found on native C3 and anti-B reacts with C3a but not the other fragments. Antibody to the A determinant reacts with C3, C3b and C3c, but not C3a or C3d. The D determinant is found on C3d, C3 and C3b, but not on C3a or C3c.

Preliminary studies of C5 indicate that it, like C3, contains two polypeptide chains.[13] The α chain has a molecular weight of 120 000 and the β chain 80 000. Cleavage of C5 to C5a (10 000–15 000 daltons) and C5b by trypsin or $EAC\overline{1423}$ appears to occur on the α chain.

Structural studies of the remaining complement proteins are incomplete. From the electrophoretic and isoelectric focussing patterns of C6, C7 and GBG, it appears likely that these proteins are tetramers. Indirect evidence from studies of GBG polymorphism suggests that it consists of three kinds of subunits.

C2, C3 and C5 contain available sulfhydryl groups related to the biologic function of these molecules. Oxidation of two free —SH groups in C2 enhances C2 hemolytic activity 13- to 16-fold. The decay of $\overline{\text{EAC14}^{oxy}2}$ is remarkably retarded compared with $\overline{\text{EAC142}}$ and it is probably this effect which is responsible for enhancing hemolytic activity. On reduction of ^{oxy}C2, these effects are reversed.

For fuller discussions of physicochemical aspects of complement proteins, several excellent reviews are available.[3,16]

7.3. REACTIONS OF THE COMPLEMENT SYSTEM

7.3.1. THE CLASSICAL SEQUENCE TO C3 ACTIVATION

Aggregated forms of IgG or IgM, particularly in antigen-antibody complexes, bind to C1q in macromolecular C1. It is presumed that this binding, which occurs in a minimal molar ratio of one IgM or two IgG molecules to one molecule of C1q, produces a conformational change in the C1q such that activation of C1r and C1s as proteolytic enzymes ensues. Of the IgG subclasses, IgG1 and IgG3 are most potent in activating C1, IgG2 is less active and IgG4 is apparently impotent. The activated form of C1s, designated $\overline{\text{C1s}}$, then cleaves C4 and C2 to yield $\overline{\text{C4}}$ and $\overline{\text{C2}}$ which, in turn, form a bimolecular complex, $\overline{\text{C42}}$.[17] This complex (C3 convertase) is capable of cleaving C3 enzymatically into C3a, a fragment of about 6800 daltons and C3b, only slightly smaller than native C3. C3b produced in this fashion on cell surfaces is designated $\overline{\text{C3}}$ and is capable, in conjunction with $\overline{\text{C42}}$, of activating C5. The subsequent sequential activation of C6 and C7 allows the formation of a stable trimolecular complex, $\overline{\text{C567}}$ in the fluid phase and presumably on the cell surface as well. The addition of C8 to red cells sensitized by antibody and the first seven complement components ($\overline{\text{EAC1-7}}$) may result in slow lysis of such cells. Rapid lysis accompanies the addition of C9 to $\overline{\text{EAC1-8}}$. The classical complement sequence is summarized in Fig. 7.2.

From present evidence, the reaction sequence can conveniently be divided into two distinct phases, activation steps consisting of the interaction of the first four components and an attack phase involving the sequential activation of C5–9. It appears that specific limited proteolysis occurs only in the activation steps and in the action of $\overline{\text{C42}}$ in conjunction with $\overline{\text{C3}}$ on C5. A stable fluid phase complex of $\overline{\text{C56789}}$ can be generated and a similar complex forms on the sensitized cell surface as well. Although the mechanism by which the fully activated complement system produces membrane damage and cell lysis is not understood, it may be that this multimolecular complex acts as a detergent or in some other, non-enzymic way alters membrane structure so that damage results. There is no good evidence, at any rate, that the activated terminal components, C8 or C9, possess lipase or proteinase activity.

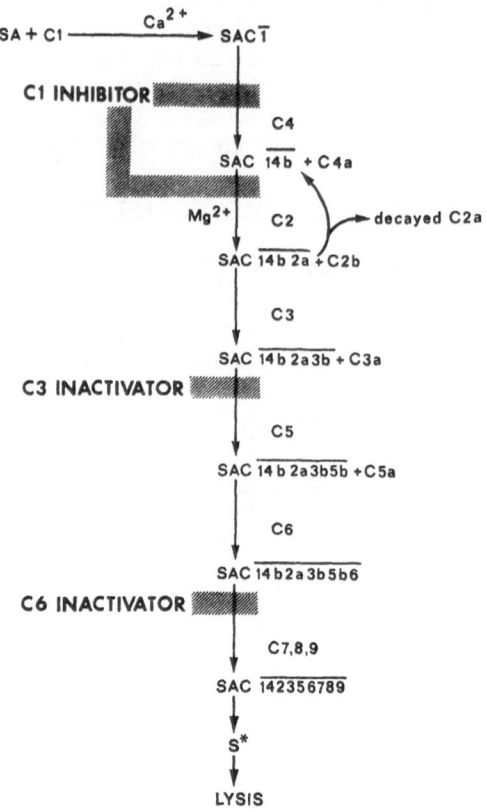

FIG. 7.2. The classical complement sequence.

There are several control steps in the classical complement pathway. The C$\bar{1}$ inhibitor acts by stoichiometrically combining with and inactivating C$\bar{1}$.[18] It has also been demonstrated that activated C2 in the form of EAC$\overline{142}$ or EAC$\overline{42}$ decays spontaneously[19] by the cleavage and discharge of an inactive fragment[20] so that it is no longer able to act on C3. This decay process thus competes with the forward cascade of the complement sequence and acts as a brake on the system. There are several serum proteins designated "decay accelerating factors" which hasten the decay of C$\bar{2}$. The C3 inactivator is a normal serum protein that has no demonstrable effect on native C3 but that proteolytically cleaves C3b and destroys the ultimate hemolytic and immune adherence activities[2] of EAC$\overline{1423}$ or EAC$\overline{43}$ [EAC(1)4(2)3]. Since the C3 inactivator is an enzyme, it is not "consumed", as is the C$\bar{1}$ inhibitor. Although evidence for an inactivator at the C6 step in the complement sequence has been obtained, this inhibitor has not been purified, nor has its exact mechanism of action been elucidated.

By analysis of complement-mediated cytolysis, its requirements and the kinetics of the reactions involved, a theoretical framework for the mechanisms of complement action has been developed. These concepts have allowed the design of tests for the functional measurement of both

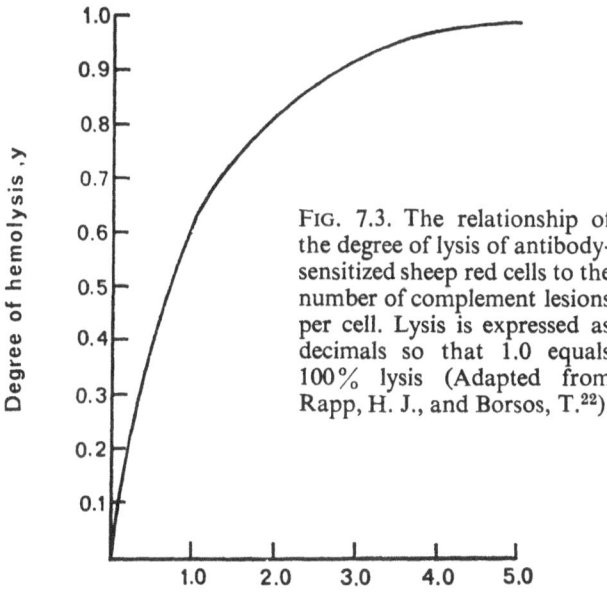

FIG. 7.3. The relationship of the degree of lysis of antibody-sensitized sheep red cells to the number of complement lesions per cell. Lysis is expressed as decimals so that 1.0 equals 100% lysis (Adapted from Rapp, H. J., and Borsos, T.[22]).

Number of lytic sites per cell, z

antibodies and individual complement components on a molecular basis. Of fundamental importance in this regard is the one-hit theory of immune hemolysis[21] which states that lysis of a red cell can occur from the sequential action of complement initiated at a single antigenic site (S) that has reacted with specific antibody. This theory now has considerable experimental support.[22]

Certain assumptions, all of which are consistent with experimental observations, are necessary to the interpretation of immune hemolysis experiments in terms of the one-hit theory. It is assumed that the number of S on a given red cell is large, that the combination of S with specific antibody is random, that the reaction of SA with C1 occurs randomly, and that the subsequent sequential reactions of this site with complement components are similarly random.

From these assumptions, it can be predicted that, in the assay of individual complement components, as the average number of effective sites per cell, z, is increased, the degree of hemolysis, y, as a decimal fraction, increases asymptotically, as shown in Fig. 7.3. This is the Poisson distribution and can be expressed as:

$$y = 1 - e^{-z} \text{ or (more conveniently)}$$

$$z = -\ln (1 - y) \text{ (where } \ln = \log_e).$$

From these considerations, it can be shown that when $z = 1$, or there is an average of one lytic site per cell, 0.632 of the cells will be lysed after the reaction with complement is completed. It follows, therefore, that the number of effective molecules of the complement component being measured can be calculated from the number of cells lysed.

7.3.2. THE PROPERDIN OR ALTERNATE PATHWAY TO C3 ACTIVATION

The properdin system, as originally defined, consisted of properdin itself and two other proteins, designated Factors A and B. Factor A was known to be destroyed by incubation with hydrazine and Factor B by heating serum at 50°C. Properdin itself has been isolated and characterized but only in activated form. It is now clear that Factor A is C3 and that activated Factor A is probably the major cleavage fragment of C3, C3b. Factor B has also been isolated and characterized as glycine-rich beta glycoprotein (GBG),[23] C3 proactivator (C3PA),[24] heat-labile factor (HLF) and unknown factor (UF). An enzyme, variously designated GBGase, C3PAse or Factor D, has recently been found to be part of the system and has the capacity to cleave GBG and destroy Factor B activity. An activated fragment of GBG has the ability, probably through its action on an unidentified intermediary protein(s) to cleave C3 into C3a and C3b. Remarkably, C3b is necessary for the activation of proGBGase and possibly for the activity of GBGase as well.[25] A positive feedback loop is thereby formed by which generation of a product (C3b) activates a system capable of producing further cleavage of C3 and generation of more C3b to serve as both tinder and powder. The C3 inactivator acts as a brake on this process by cleaving C3b in such a fashion that it is no longer able to activate proGBGase and by inactivating GBGase already formed,[26] as shown in Fig. 7.4.

The presence of a positive feedback loop in the form of the alternate pathway has many important implications for our understanding of complement activation *in vitro* and *in vivo*. Any agent capable of altering C3 to a form capable of activating GBGase is thus able to trigger the amplified conversion of C3 to C3b and C3a and the activation of C5–9. Recruitment would also be expected from classical pathway activation by antigen-antibody complexes. Recent evidence by Götze and Müller-Eberhard[27] suggests that properdin acts in a related manner. The activated form of this protein appears to alter C3 (without obvious physicochemical change) so that it can recruit the alternate pathway. The known primary activators of properdin are complex carbohydrates such as inulin, dextran, zymosan (from yeast cell wall) and lipopolysaccharides from gram-negative bacterial endotoxins. In man, most antibodies are probably not capable of activating properdin, although there is the possibility that aggregated IgA and IgE have this capability.

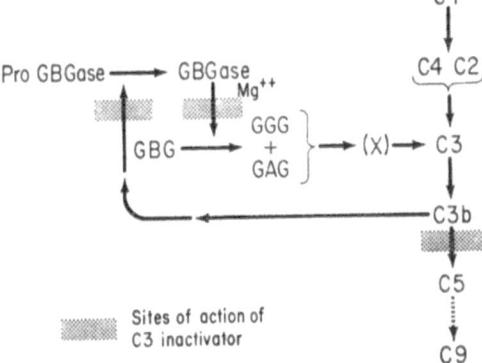

FIG. 7.4. The alternate (properdin) pathway to C3 activation. A possible intermediate(s) to the attack on C3 by one of the two fragments of GBG is indicated by (X). The exact position of properdin in this pathway is uncertain. (By permission, from Alper, C. A., and Rosen, F. S.[53])

The interaction of cobra venom with the complement system has been studied since the turn of the century but, even at this late date, the reaction mechanisms are not completely elucidated. The anti-complementary protein from the venom (CoF or CVF) has been isolated and characterized as 7 S β globulin. This material when added to whole serum in the presence of divalent cation inactivates C3 and later components but is without effect on isolated C3 or isolated C5. CoF added to whole serum in the presence of divalent cation forms a 9 S complex with a normal serum protein (CoF-binding protein). This complex is then capable of cleaving C3 into C3b and C3a, even in the absence of divalent cation. C5 is also attacked by a complex of CoF with a normal serum protein but it is not known whether the same or a different CoF-binding protein is involved as that leading to C3 cleavage. In whole serum, recruitment of the alternate pathway by the C3b generated amplifies the action on C3 and later components but it is not clear whether the proteins of the alternate pathway are directly required for the attack by CoF on C3 and later complement components. The normal function of the CoF-binding protein is also unknown but it certainly must play some physiological role in addition to its interaction with CoF.

There is now considerable evidence that the C3-converting enzymes of the classical and alternate pathways, and probably of certain other pathways as well, whether assembled on surfaces or in the fluid phase, are able to produce a kind of innocent bystander reaction. For example, $EAC\overline{(1)42}$ or $C\overline{(1)42}$ can act on C3 so that there is transient activation of a combining site on $C\overline{3}$ for red cells unsensitized by antibody or early complement components. Zymosan, on interaction with serum, causes the activation of C3 and the assembly of $C\overline{567}$ in the fluid phase which can then attach to unsensitized erythrocytes so that the latter then go on to lysis by the later components, C8 and C9.[28] The latter phenomenon has been termed reactive hemolysis. CoF added to serum induces a similar passive sensitization of red cells leading to hemolysis.

7.4. BIOLOGIC ACTIVITIES OF COMPLEMENT

Complement contributes essentially to the inflammatory response and to host defense against invasion by pathogenic organisms. In certain pathologic states, complement appears to play a role in mediating tissue injury.

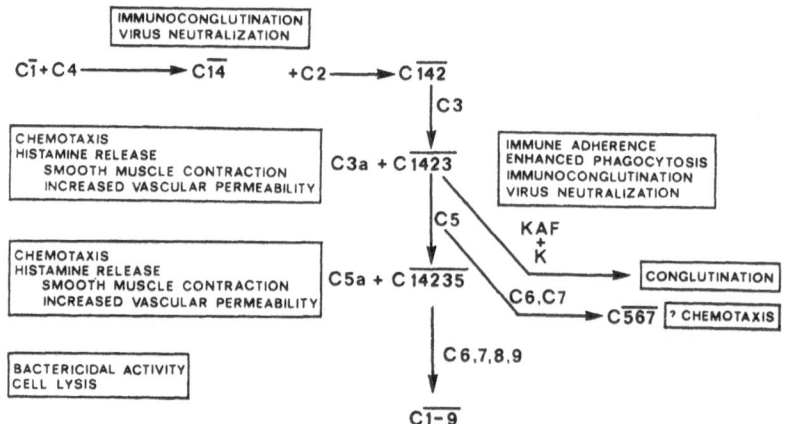

FIG. 7.5. Complement-mediated functions and their elaboration during complement activation.

Although earlier studies of complement emphasized immune cell lysis with its requirement for all components, it is now abundantly established that many of the important biologic activities of the complement system become manifest early in the activation sequence.[29] Figure 7.5 summarizes the known complement-mediated functions and the steps during complement activation at which they are elaborated. These functions are mediated for the most part by cleavage fragments of complement proteins. One function, chemotaxis for polymorpho-nuclear leukocytes, may be carried by a complex of C5b with activated C6 and C7 ($\overline{C567}$) in addition to being a property of the fragments C3a and C5a.

Neutralization of certain viruses may be accomplished by the properdin pathway or by IgM antibody, C1 and C4 or by IgM antibody, C1, C4, C2 and C3. The properdin pathway, the classical pathway, and all of the proteins of the common pathway are probably required for the bactericidal properties of serum for smooth gram-negative bacteria.

Both C3a and C5a are anaphylatoxins, that is to say they both cause contraction of smooth muscle and an increase in vascular permeability. They both cause degranulation of mast cells with attendant histamine release. It appears likely that it is the released histamine which mediates the smooth muscle contraction and vascular permeability change. Although smooth muscle shows a progressively diminishing response to either C3a or C5a (tachyphylaxis), such a muscle will react fully to the other anaphylatoxin on first exposure. Both C3a and C5a are chemotactic for leukocytes. The anaphylatoxin properties of these fragments are rapidly destroyed in whole serum by the action of an enzyme called anaphylatoxin inactivator, probably identical with carboxypeptidase B. There is evidence that another enzyme, chemotactic factor inactivator, destroys the chemotactic activity of C3a, C5a and $\overline{C567}$. As was pointed out earlier, C3a and C5a may be generated by non-complement proteolytic enzymes, such as plasmin, trypsin and tissue and bacterial proteases, as well as by classical or alternate pathway activation. Evidence has been presented that such non-complement enzymes may play a role in the local inflammatory response to non-immune injury.

Activation of the classical sequence through C3 is required for the enhancement by serum of the phagocytosis of antibody-sensitized erythrocytes or smooth pneumococci. For the latter, maximal enhancement also requires proteins of the alternate pathway. Endotoxin particles as a paraffin oil emulsion are phagocytosed only with the aid of the alternate pathway and may be a model for non-antibody properdin-dependent opsonization of gram-negative bacteria. The enhancement of phagocytosis would seem to be one of the most important biologic functions of complement and of particular importance for defense against infection by pyogenic bacteria.

The requirement for C3 in the mobilization of leukocytes has been demonstrated *in vitro* using rabbit femurs. Confirmation of this role for C3 was obtained by noting the absence of a peripheral leukocytosis during bacterial infection in a patient with hereditary deficiency of C3.

Immune adherence[30] is the property of EAC$\overline{(1)}$4$\overline{(2)}$3 to attach to primate red cells, mammalian B lymphocytes, macrophages, monocytes, polymorphonuclear cells and rabbit platelets. Particles other than erythrocytes acted upon by antibody and the first four complement components also exhibit immune adherence. This function of complement is of obvious importance in the clearance from the blood of bacteria and other pathogenic organisms and is a necessary prelude to phagocytosis. It may also be important in the production of humoral antibody to thymus-independent antigens. Because of the normal presence in whole serum of the C3 inactivator which destroys the immune adherence activity of properly sensitized particles, this property is probably short-lived *in vivo*.

EAC($\overline{1}$)4($\overline{2}$)3, after it has been acted upon by the C3 inactivator, exhibits a new property: conglutination, or the ability to be agglutinated by the bovine serum protein, conglutinin (K). In this role, the C3 inactivator has been designated conglutinogen activating factor (KAF).[31] Since conglutinin has been detected only in cow serum and not in the serum of other species, including that of man, the relevance of this function for all species but bovida is questionable. Conglutination must be distinguished from immunoconglutination which is not restricted to cows. Immunoconglutinins also agglutinate EAC($\overline{1}$)4($\overline{2}$)3, but they arise "nonspecifically" after antigenic and reticuloendothelial stimulation of many kinds. They are antibodies to particle-bound C4b and C3b. They presumably potentiate the blood stream clearance of particles, including organisms, sensitized by antibody and the first two to four components of complement.

In dividing the functions of complement as we have done, we tend to separate processes which *in vivo* are highly interdependent and continuous. Chemotaxis, immune adherence and enhancement of phagocytosis clearly form such a continuum.

7.5. SITES OF SYNTHESIS, ONTOGENY, AND PHYLOGENY OF COMPLEMENT

7.5.1. SYNTHESIS

We know very little about the factors controlling synthesis of complement proteins. As is true of virtually all plasma proteins, complement components are synthesized by certain cell types only and the latter differ for the different components.[32] Intact, hemolytically active C1 is produced by cells of the gut columnar epithelium, particularly of the colon in man. Macrophages found in many organs synthesize C4 and C2. Hepatic parenchymal cells are the prime producers of C3, C5 and C$\overline{1}$ inhibitor.

Evidence for production of complement proteins by individual organs, tissues and cells has been of several kinds. Incorporation of ^{14}C-labeled amino acids by tissue in short term culture into proteins reactive with specific antisera is one such line of evidence. Such experiments require extensive controls for proper interpretation. For example, specific incorporation should be inhibited by agents or conditions which block or diminish protein synthesis generally. Ideally, it should be shown that the molecules which have incorporated radiolabeled amino acids are biologically active. If possible, where allotypes exist, the radioactive protein should be shown to be of the same type as the donor of the tissue and different from that of the non-radioactive carrier serum used in making precipitin patterns for autoradiography. Immuno-fluorescence is useful in identifying possible sites of synthesis but cannot alone distinguish synthesis from storage of specific proteins.

Production of functionally active complement proteins by tissues in culture has been elegantly demonstrated using the molecular titrations already described. Low initial concentrations and controls using inhibitors of protein synthesis are necessary for proper interpretation. This method allows the quantitative assessment of synthesis of functionally active molecules. It can be adapted, through the use of red cell-complement intermediates, to the qualitative study of individual complement protein synthesis by single cells in suspension.

An unusual situation permitted the demonstration that the liver was the primary organ of synthesis of C3. The C3 allotype of a recipient in homotransplantation of the liver changed completely to that of the donor.

7.5.2. ONTOGENY

The techniques used to demonstrate complement synthesis *in vitro* have been applied to fetal tissues at various stages of gestation.[33] Information has also been obtained by measuring the concentrations of complement proteins in fetal serum.

C4 and C2 are synthesized by human fetuses at eight weeks of gestation and later; at this stage macrophages in the liver are particularly active in this respect. The CĪ inhibitor is synthesized as early as four weeks of gestation. Fetal liver synthesizes C3 by eight weeks of gestation and C5 by 14 weeks. C5 is also synthesized by fetal lung and intestine. C1 synthesis has been demonstrated in a 19 week-old fetus.

Hemolytic complement and individual proteins of the complement system tend to be lower in concentration at birth than later in life. Cord sera contain a mean of one-half to three-quarters the normal adult levels of C4, C3, C5, properdin and GBG. These levels rise to normal adult levels by three months of age.

Transplacental passage of complement proteins in either direction appears to be negligible. It has been shown that allotypes of C3 and GBG often differ between paired maternal and cord sera. Electro-phoretic differences in C4 similarly may be different. The cord blood of a baby heterozygous for C2 deficiency contained C2 but its homozygous deficient mother's serum did not. Similar observations have been made in complement deficiencies in animals.[34]

7.5.3. PHYLOGENY

The complement system in its entirety or in more rudimentary form is found in many animal classes and is, therefore, rather ancient.[35] The hemolymph of the horseshoe crab and sipunculid worm is not ordinarily lytic for sheep erythrocytes but can be made so by the addition of cobra venom. Starfish hemolymph cannot be made lytic for sheep cells even though a complex is formed with cobra factor.

Complete complement systems seem to occur in both cartilaginous and bony fish, which, like those of mammals, can be inactivated by heat, removal of Ca^{2+} and Mg^{2+} and hydrazine treatment. Hemolytic complement activatable by antibody is also found in the serum of amphibians, reptiles and birds.

7.6. TESTS INVOLVING COMPLEMENT

We will consider three main groups of tests involving complement: (a) those which are designed to test the concentrations of individual proteins in the complement system, (b) those which test for involvement of complement in disease, and (c) those which utilize complement consumption *in vitro* to detect and measure antibodies. The last are commonly referred to as complement fixation tests.

7.6.1. FUNCTIONAL ASSAYS OF COMPLEMENT

The measurement of whole serum hemolytic complement is a useful screening test for the integrity of the complete system. However, decreases in the concentration of individual components to 50% or less of the normal level may have little or no effect in this test.

The test is based on the ability of sheep red cells properly sensitized by rabbit antibody to sheep erythrocytes to be lysed by the complete classical complement sequence. Hemoglobin released by such lysis can be measured spectrophotometrically with great precision and related to the percentage of cells lysed. Figure 7.6 shows the relationship of percent lysis in one hr at 37°C of optimally sensitized sheep cells exposed to various concentrations of fresh guinea pig serum ("complement"). The dose-response curve is clearly sigmoidal and reflects the complexity of steps involved in the reaction. The slope is steepest around 50% lysis and therefore small differences in hemolytic complement activity result in the greatest differences in observed lysis. For this reason, the test is designed to measure hemolytic complement at this endpoint.

The sigmoidal curve of Fig. 7.6 is described by the von Krogh equation:

$$x = \frac{y^{1/n}}{1 - y} \qquad (7.1)$$

which, when converted to log functions becomes:

$$\log x = \log K + \frac{1}{n} \log \frac{y}{1 - y} \qquad (7.2)$$

Thus, if $x =$ volume of added fresh serum and $y =$ per cent lysis expressed as a decimal, and the results are plotted on log-log graph paper, one obtains a straight line whose slope is $1/n$ and whose y intercept is $\log K$. The value of $1/n$ under the conditions chosen is usually 0.2 ± 0.02 and can be checked experimentally. The constant K is the 50% lysis unit of hemolytic complement since at 50% lysis, $y/1 - y$ is 1 $(0.5/0.5)$ so that $1/n \log (y/1 - y) = 0$ and $x = K$. Results are thus defined in terms of CH_{50} units.

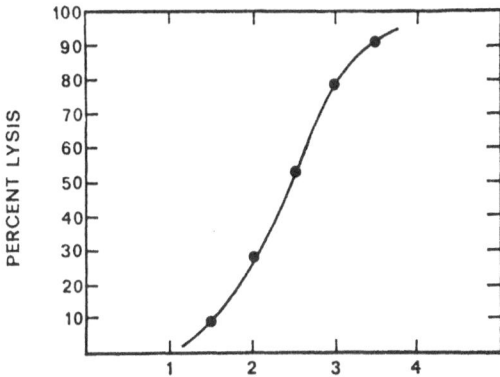

GUINEA PIG COMPLEMENT, 1/500 (ml.)

FIG. 7.6. The relationship of the degree of lysis of antibody-sensitized sheep red cells to increasing amounts of guinea pig serum (complement). (By permission, from Mayer, M. M.[36]).

This test must be performed with optimally sensitized red cells under strictly controlled conditions of temperature, tonicity, pH, Ca^{2+}, Mg^{2+} and protein concentrations. The standard buffer for dilution is gelatin veronal-buffered saline of precise formulation. For details and a fuller analysis of the test, the reader is referred to Mayer[36] and Rapp and Borsos.[22]

Measurement of individual complement components requires solutions of individual complement components purified to the extent that they are free of the activities of all the other complement components. These solutions are used to sensitize sheep red cells optimally treated with antibody and complement components up to the step that is to be measured. These intermediates are then exposed to the test sample and a mixture of all later-acting components is used to produce lysis. Since all of the components except for that being measured are supplied in excess, the percent lysis is related to the concentration of the measured component, as described earlier (section 7.3.1).

In the measurement of the early-acting components, C1, C4 and C2, advantage is taken of the fact that, although Ca^{2+} is required for the C1 step and Mg^{2+} for the C2 step, no divalent cation is needed for later steps. Therefore, the terminal components C3 and C5–9 can be supplied as whole serum in 0.01 M EDTA to chelate divalent cations (C-EDTA) so that early-acting components in the developing reagent do not contribute to the reaction.

As an example of the functional molecular measurement of a complement component, consider the assay of Cl.[22] The cells used for this assay are EAC4, produced with relatively large amounts of IgM hemolytic antibody and C4. Portions of suspensions of these cells are mixed with different dilutions of test solution and incubated for the time needed to reach equilibrium. These cells are exposed to C2 in excess (in other words, adding more C2 would not alter the results) and excess C-EDTA for sufficient time to allow complete reaction. After centrifugation, hemoglobin concentration is determined spectrophotometrically in the supernatant fluid.

Because of the decay of $EAC\overline{142}$, the timing of the incubation of $EAC\overline{14}$ with the test sample in the assay for C2 is critical. Kinetic analysis of the reaction of $EAC\overline{14}$ with C2 to form $EAC\overline{142}$ reveals that formation of the latter reaches a maximum at a time (t_{max}) that varies with temperature and the concentration of $EAC\overline{14}$ but that is independent of C2 input. It is, therefore, possible to relate the functional C2 concentration in a sample to the $EAC\overline{142}$ generated at the t_{max} under specific conditions.

Because the reaction of the $C\overline{1}$ inhibitor with $C\overline{1}$ is stoichiometric the $C\overline{1}$ inhibitor can be assayed functionally by its effect on $EAC\overline{1}$.[37] Other functional assays depend on the inhibition by the $C\overline{1}$ inhibitor of the esterolytic activity of $C\overline{1}$ for such synthetic substrates as N-acetyl-L-tyrosine ethyl ester.[9]

A semi-quantitative assay for the C3 inactivator based on its capacity to render $EAC\overline{(1)4(2)3}$ conglutinable has been developed. Because of the enzymatic nature of the action of the C3 inactivator, its failure to form a durable bond with $EAC\overline{1423}$, and the absence of suitable inhibitors for its activity, more quantitative assays are not yet available.

A fuller description of functional assays for individual complement components is given by Rapp and Borsos.[22] This book not only provides an excellent discussion of the theoretical bases for these tests, but contains a wealth of practical information indispensable to the design and execution of the functional complement assays.

7.6.2. IMMUNOCHEMICAL ASSAYS FOR COMPLEMENT PROTEINS

The measurement of complement proteins as proteins in highly complex mixtures such as human serum requires reagents of very high specificity. Fortunately, antibodies, when properly prepared, satisfy this requirement. Antibodies have now been produced against the majority of the complement proteins and most precipitate with their respective antigens in whole serum. Quantitative immunochemical techniques are therefore available for the quantitation of complement proteins in serum and other biological fluids. For the proper application of these techniques to the measurement of specific components, certain problems and limitations should be borne in mind.

The most generally applicable, reliable, precise, accurate and reproducible technique, in our hands at least, is electroimmunoassay. In this method, antibody is incorporated into agarose gel made up in electrophoresis buffer. Wells are cut in a 1 mm film of this gel and precisely measured volumes of standards and samples are placed in the wells. Electrophoresis is carried out so that rocket-shaped peaks develop to stability and cease to grow. The peak heights (from the proximal edge of the antigen hole to tip) of the samples are then related to those of the standards on graph paper and antigen concentrations in the test samples are thereby established. A certain amount of contaminating antibodies to proteins other than the complement protein measured can be tolerated provided that the desired peak can be distinguished by its character from those of contaminants. By adjusting antibody concentrations, and thereby peak heights, one can deliberately measure two or even three proteins in the same gel with a mixture of highly specific antisera. Practical and theoretical details of the electroimmunoassay technique are discussed by Laurell.[38]

Radial immunodiffusion is a widely used method for the immunochemical measurement of proteins but has certain drawbacks compared with electroimmunoassay. In radial immunodiffusion, antibody-containing gel is prepared as described above and wells are cut and filled in a similar manner. Plates are then simply incubated until stable sharp rings form. The area (or diameter^2) of these rings is then related to a standard curve obtained from the rings produced by the standards in the same gel to yield antigen concentrations in the unknowns.

A comparison of these two methods as applied to the measurement of C3 with the usual anti-C3 (anti-A determinant) is informative. If fresh serum is compared with the same serum stored at 37°C for three days (aged serum) in electroimmunoassay, peak heights are identical or at most 10% higher for the aged serum. However, in radial immunodiffusion the aged serum C3 ring will be considerably larger than the fresh serum ring. The reason for this difference is that the storage conversion product of C3, C3c, is smaller than native C3 (7 S versus 9.5 S) and diffuses more rapidly into the gel during radial immunodiffusion. In electroimmunoassay, on the other hand, antigen is driven into contact with antibody by electrophoresis so that diffusion is minimal and molecular size is unimportant as a determinant in peak height. The slightly higher negative charge on C3c compared with C3 probably accounts for the slightly higher peaks sometimes observed in this test. Conversion of C3 is greater during the relatively long incubation time (at least 48 hrs at room temperature) required by radial immunodiffusion for the optimal development of rings compared with the 4 hrs at a few degrees above 0°C needed for electroimmunoassay as it is usually performed. Incubation at 37°C to hasten stable ring-formation in radial immunodiffusion also hastens conversion of C3. Finally, the dose-response curve in electroimmunoassay is more nearly linear than in radial immunodiffusion. In the latter, concentration is related to the square of the diameter so that, at the least, the relationship is lin-log, leading to lower accuracy.

Both of the techniques described have a threshold of detection of about 0.25 mg per 100 ml (2.5 μg per ml) which can be lowered by "counterstaining" otherwise invisible peaks or rings with a second antibody to the animal antibody. Even further lowering of the threshold can be achieved by the use of radioisotopically labeled antigen or antibody to develop peaks or rings. Thus, virtually all complement proteins can be measured with these methods. For very low concentration proteins, such as properdin, radioimmunoassay is probably most practical.[39]

For the screening of large numbers of samples, a rapid, accurate automated nephelometric method is available.[40] In this technique diluted antiserum is mixed with diluted sample and incubated for 15 min, at which time light scattering by the small antigen-antibody aggregates is read and compared with the behavior of standard antigen solutions. This method, when anti-C3 consisting exclusively of anti-A determinant is used, produces identical values for individual sera tested fresh or aged. At present, the automated nephelometric method is limited (among the complement proteins) to the quantitation of C3, C4, CĪ inhibitor and GBG in serum.

To the extent that antiserum to the B determinant of C3 is used in any immunochemical technique to measure C3, any conversion of C3 will result in a "disappearance" of antigen since this determinant is usually detectable only on the native form of the molecule.

7.6.3. TESTS FOR COMPLEMENT ACTIVATION *IN VIVO*

The measurement of the serum concentration of complement proteins in patients provides some information about the participation of complement in disease. However, this information, by its very static quality, is limited and ambiguous. The serum concentrations of all proteins, including those of the complement system, are determined by

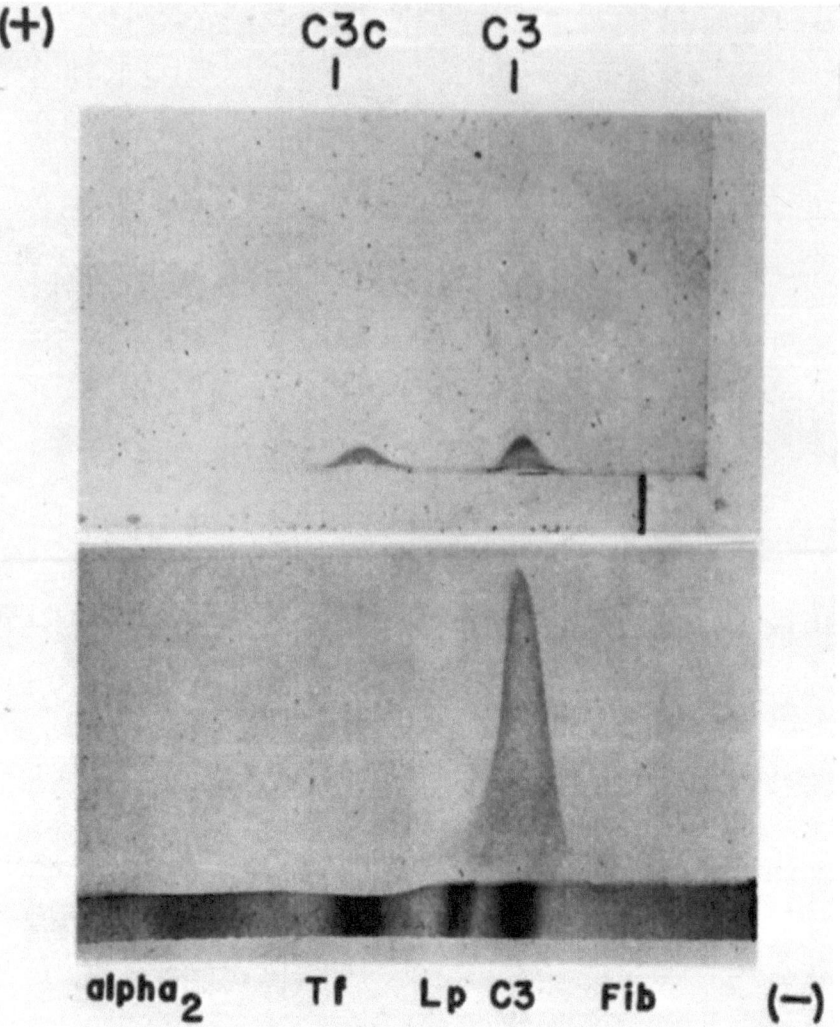

FIG. 7.7. Antigen-antibody crossed electrophoresis with anti-C3. The bottom pattern represents native, unconverted C3 in normal human plasma. The top pattern is from a patient with type II essential hypercatabolism of C3 and shows markedly reduced native C3 and the presence of circulating C3c. For reference a stained gel of separated plasma is shown below these patterns to illustrate the position of α_2-globulins (alpha$_2$), transferrin (Tf), β-lipoprotein (Lp), C3 and fibrinogen (Fib). The anode for the first separation was at the left and for the second electrophoresis into antibody-containing gel was at the top. (By permission, from Alper, C. A., Bloch, K. J., and Rosen, F. S.[51]).

dynamic equilibrium between synthesis and catabolism. A normal serum level may reflect a balance between abnormally accelerated synthesis and catabolism.

From our knowledge of the molecular reactions of complement *in vitro*, we can predict several facets of complement activation *in vivo* and we can examine the tests that have been devised to study the role of complement in disease. Some of these tests are available in special hospital laboratories and others are research procedures.

We know that complement activation *in vitro* is characterized by limited sequential binding of components to antigen-antibody complexes or other activating agents and extensive fluid phase sequential limited proteolytic cleavage of components in the earlier steps and protein-protein associations between later-acting components. An obvious hallmark, then, of complement activation *in vivo* is the elaboration of cleavage products. Whether or not these are detectable in circulating plasma depends on the rapidity with which these fragments are cleared as well as the rate at which they are produced from native complement proteins.[41] As we have noted, these fragments are usually distinguishable from their parent molecules in molecular size, in electrophoretic mobility, and often in antigenic composition.

The two most common techniques for examining specimens for conversion products are immunoelectrophoresis and antigen-antibody crossed electrophoresis (crossed immunoelectrophoresis) (Fig. 7.7). The latter is more sensitive and provides potentially quantitative estimations of conversion. Conditions of collection and storage of specimens are critical since conversion *in vitro* must be avoided. We have found that samples from normal individuals show no evidence of conversion of C3, at least, if (1) the venipuncture is without undue trauma, (2) the first 3–4 ml of blood are discarded, (3) the blood is then collected through the same needle into EDTA, (4) plasma is obtained by centrifugation within 15–30 min and immediately frozen in aliquots at −65° or below. Repeated thawing and freezing of samples induces conversion and is to be avoided. Agarose for electrophoresis should be tested with normal samples obtained as described above to be certain that no conversion results. Under these circumstances it is not necessary to include EDTA in the electrophoresis buffers. Because in whole serum or plasma C3b is further broken down to C3c and C3d, the detection of the D antigen on C3 in C3d by immunoelectrophoresis is another useful test for complement activation *in vivo*.

Deposition of complement proteins (usually along with immunoglobulins) in tissues such as kidney and skin has been taken as evidence for complement activation in disease. Such deposition is detected by the fluorescent antibody technique, using antibodies to individual proteins such as C3, C4 and properdin. From studies in animals, it is clear that such deposition may result either from filtration and trapping of immune complexes within small blood vessels as in the renal glomeruli or from antibody reactive directly with kidney or other tissue antigens. For a full treatment of this subject, the reader is referred to Cochrane and Koffler.[42] It is clear that tissue deposition is not an important direct determinant in the lowering of the serum concentration of complement proteins in disease. In "one-shot" experimental disease, C3 deposition persists after return of the level of this protein to normal. We also know from work *in vitro* that only 10–20% of C3 molecules bind to antigen-antibody complexes or non-sensitized red cells under optimal conditions of activation. This is also true of other complement components.

Metabolic studies with radioisotopically labeled purified complement proteins have yielded important information about complement protein metabolism. Although intuitively one might expect that rapid clearance of labeled complement proteins would attend complement activation *in vivo*, this is not necessarily true.[43] It is only the case if fluid phase conversion products are cleared at a more rapid rate than the native molecule. Such is true of C3b and C4b which are removed from the circulation at rates 5 or more times those of the native proteins. No information is yet available for fragments of other complement proteins.

7.6.4. COMPLEMENT FIXATION TESTS

The use of complement "fixation" for detecting the presence of antigens or antibodies has a very long and productive history. This test can be used with soluble or particulate antigen and is based on the loss of hemolytic activity of complement on incubation with antigen-antibody aggregates.[22] Micromodifications have been developed.[44] Complement fixation is performed in two stages. In the first, the reaction of antigen-antibody in the test sample with a known amount of whole complement is allowed to proceed to completion. In the second step, hemolytic complement is measured to determine the amount consumed. The conditions for the first step are determined by the nature of the antigen and the type of antibody being measured. For example, IgG antibody fixes complement more efficiently at 4°C than at 37°C, whereas the reverse is true of IgM antibody directed against the same antigenic determinants.

Complement fixation tests can be made quantitative. The most elegant example is the $C\overline{1}$ fixation and transfer test ($C\overline{1}$FT) of Borsos and Rapp (1965) which is capable of measuring molecular concentrations of either antigen or antibody. The test exists in two forms, direct and indirect. In brief, in the direct test, particulate antigen-antibody complex is mixed with $C\overline{1}$ to saturate all $C\overline{1}$-fixing sites. The complexes are washed at low ionic strength to remove unbound $C\overline{1}$ but preserve bound $C\overline{1}$. In the transfer step, the washed antigen-antibody-$C\overline{1}$ complex is mixed with EAC4 at physiologic ionic strength which permits the quantitative transfer of the $C\overline{1}$ from the complex to EAC4. In the assay step, EAC$\overline{14}$ are lysed by C2 and C-EDTA and the number of Cl molecules bound is calculated. The indirect test is used to detect antibodies that do not fix complement and employs an additional step in which complement-fixing antibody to the first antibody on the complex is employed.

7.6.5. MISCELLANEOUS

The treponema pallidum immobilization test is highly specific for the infectious agent of syphilis. It requires both specific antibody and complement and is used when ordinary tests yield false positive results.

Tests for paroxysmal nocturnal hemoglobinuria (PNH) depend upon the unusual susceptibility of red cells in this disease to complement lysis. In the Ham test, the test red cells are incubated in acidified (pH $\sim$6.7) fresh serum from an ABO compatible person or from the patient. The minimal complement activation that occurs at somewhat acid pH will produce lysis of PNH cells but not normal erythrocytes. An important control consists of incubating patient's red cells with serum heated at 56°C prior to acidification. With destruction of total hemolytic complement, such serum will not lyse PNH cells. The sugar water test for PNH depends upon minimal complement activation probably attendant upon euglobulin precipitation. In this test whole patient's blood is mixed with 9 volumes of a 10% sucrose solution and incubated.

7.7. COMPLEMENT ABNORMALITIES IN DISEASE

Since the hemolytic activity of complement requires adequate levels of all of the classical components, assessment of this activity has provided a useful test for alterations in the concentration of any or all of the complement proteins. As already mentioned, this activity is best expressed in CH$_{50}$ units. Hemolytic complement may be unaltered despite considerable changes in the concentration of individual components. For example, isolated reductions in C2, C3, C6 or C7 in man, or Cl in human serum in the test tube to half-normal serum concentration results in normal or only slightly reduced hemolytic complement levels. In other words, only one or a few components are present in "limiting" concentration with respect to hemolytic activity whereas others are present in excess.

The most common change in hemolytic complement in disease is an increase in association with a wide variety of inflammatory and necrotic disorders as part of the acute phase plasma protein response. Specific complement proteins which are known to rise in concentration in this circumstance include C4, C2 and C3.

Serum hemolytic complement is lowered in a relatively limited group of acquired diseases such as systemic lupus erythematosus, acute glomerulonephritis, membranoproliferative glomerulonephritis, acute serum sickness and other varieties of immune-complex disorders.[45] It is also lowered in advanced hepatic cirrhosis.

The measurement of serum hemolytic complement for clinical purposes has, in recent years, largely been replaced by the immunochemical assay of C3. In general, this protein is reduced in concentration in those disorders in which hemolytic complement activity is reduced but there is no strict correlation. In systemic lupus erythematosus, for example, depression of C4 concentration may precede and is often more marked than depression of C3. In lupus, many or all complement components are reduced in concentration, probably as the result of complement activation *in vivo*. In acute poststreptococcal glomerulonephritis, there is initially a marked fall in all classical complement components but within two or three days of the onset of symptoms, most of the components return to normal. The concentrations of C3 and C5, however, remain low for three or four weeks. Evidence from metabolic studies with purified ^{125}I-labeled C3 suggests that in the initial phase, the C3 concentration is lowered as part of activation of the complement system, presumably by antigen-antibody complexes. Circulating conversion products of C3 can sometimes be demonstrated during this initial phase.[43] Studies performed two or three days after the onset of symptoms suggest that depressed C3 synthesis is responsible for the prolonged lowering of the level of C3 (and perhaps of C5 as well) after the initial activation phase.

The situation in membranoproliferative or mesangiocapillary glomerulonephritis (also known previously as chronic hypocomplementemic nephritis) is far less clear. In this disorder, particularly as it occurs in children and adolescents, lowered hemolytic complement is a usual finding. This is almost always the result of markedly reduced levels of C3 and C5 with normal levels of most other complement components. Serum concentrations of GBG or properdin Factor B are usually normal in these patients or at most slightly reduced. By immunofluorescent techniques using antibodies to immunoglobulins and C3, the latter proteins are demonstrable on the glomeruli in most patients. The serum of some patients with membranoproliferative glomerulonephritis contains circulating C3d and also may contain a protein termed nephritic factor which, when a normal serum protein, presumably deficient in the nephritic serum, is added, is capable of cleaving C3 *in vitro*. The nature of the nephritic factor is uncertain with some investigators claiming that it is antibody of the IgG3 subclass and others that it is a non-immunoglobulin serum protein. The relevance of the lytic-nephritic interaction with C3 to the lowered C3 levels observed *in vivo* is cast into doubt by metabolic studies with labeled C3 in patients with membranoproliferative glomerulonephritis. The latter studies have shown normal or near-normal plasma disappearance curves of the injected C3 so that the low levels of C3 in some of these patients, at least, including those with nephritic factor, is primarily the result of depressed synthesis. Depressed synthesis of C3 *in vitro* by liver obtained from such patients has also been demonstrated.

It appears likely that activation of the classical complement pathway by antigen-antibody aggregates is responsible for the low levels of many complement components in systemic lupus erythematosus and in other kinds of immune complex disease. This has been confirmed in metabolic studies with labeled complement components and by the immunofluorescent localization of immunoglobulins and C3 in glomeruli and other tissues of such patients. Recently, evidence for the participation of the properdin pathway has been obtained by the demonstration of properdin in the glomeruli of patients with lupus. The mechanisms by which the properdin system is activated remain unclear. An attractive hypothesis is that antigen-antibody complexes lodge in glomeruli, activate the classical pathway and generate chemotaxis for neutrophils. The latter release their lysosomal contents, including proteolytic enzymes, locally, thereby inducing glomerular damage and exposure of basement membrane. The exposed carbohydrate of the glomerular basement membrane may then activate properdin and cause further activation of terminal complement components.

There is some evidence that complement activation may occur in severe disseminated intravascular coagulation with fibrinolysis. In this clinical situation, it may be that the generated plasmin and possibly also thrombin attack C3 directly. Serum levels of the latter protein may be lowered. The lowered C3 serum concentrations observed in patients with advanced hepatic cirrhosis or other hepatocellular disease may result from associated disseminated intravascular coagulation and fibrinolysis or from interference with C3 synthesis in this organ which is its site of synthesis.

We have observed marked elevations of C3 serum concentration in patients with severe biliary obstruction. The mechanisms for this elevation are unknown.

Serum concentrations of C1q may be observed to be low in patients with agammaglobulinemia, particularly in those with severe combined immunodeficiency. Recent metabolic studies with labeled C1q suggest that this lowering is the result of hypercatabolism rather than decreased synthesis.

In certain cases of acquired hemolytic anemia, C3 is detected on erythrocytes by a Coombs antiglobulin reagent specific for this protein. General Coombs reagents vary in their content of anti-C3 and are for the most part, anti-IgG. The presence of C3 on patients' red cells may indicate an anti-erythrocyte antibody which has "fixed complement", activation of C3 by other means with C3 deposition as part of the innocent bystander reaction or an unusual abnormality of the red cell membrane making it more "susceptible" to C3 uptake, as in paroxysmal nocturnal hemoglobinuria.

Total serum hemolytic complement and the levels of individual complement proteins are normal or elevated in patients with rheumatoid arthritis and there is no evidence in serum of complement activation. However, the joint space is relatively sequestered from the circulating plasma and there is now considerable evidence that complement participates locally in rheumatoid joint inflammation.[46] Hemolytic complement is reduced in joint fluid from patients with rheumatoid arthritis, particularly those with rheumatoid factor and with nodules, when compared with joint effusions from patients with other diseases. There is reduction in the relative concentrations in rheumatoid joint effusions of several complement proteins, including C4, C2, C3 and properdin Factor B, and conversion products of the latter two proteins are often found. Chemotactic factors, thought to consist of C5a and $\overline{C567}$, are found in the majority of rheumatoid joint effusions. By immunofluorescence, C3 and C4 have been identified in the lining cells, blood vessels, and intercellular connective tissue of rheumatoid synovial membranes. Incubation of normal leukocytes with joint fluid from patients with seropositive rheumatoid arthritis, but not seronegative disease, developed intracellular inclusions containing IgG, IgM and C3. It is not clear whether complement activation in joints affected by rheumatoid arthritis results from the presence of complexes of IgG antibody with an unknown, possibly viral antigen, of complexes of IgM rheumatoid factor with aggregated γ-globulin, of proteolytic enzymes from leukocytes, or some combination of these factors.

7.8. INHERITED DISORDERS OF THE COMPLEMENT SYSTEM

As was pointed out some 300 years ago by Sir William Harvey, we can learn a great deal about normal physiology from the study of individuals with inherited defects. Such has been true of the inherited disorders of the complement system.

Hereditary angioneurotic edema was known clinically many decades before it was shown to be the result of an inherited defect in the CĪ inhibitor.[47] The tendency to develop episodic attacks of subcutaneous, laryngeal and gastrointestinal swelling is inherited as an autosomal dominant trait. The subcutaneous swellings are painless, non-erythematous and non-pruritic. When the lining of the gastrointestinal tract is affected, there is vomiting, abdominal pain and copious diarrhea; when the respiratory tract is involved, asphyxiation can result. Attacks last 24 to 96 hours. About 80% of affected families have reduced levels of CĪ inhibitor protein whereas in the remaining 20%, this protein is in normal or elevated concentration but does not function normally to inhibit CĪ. The clinical picture in these two varieties of the disease is the same. There are probably three or four different mutant genes which produce dysfunctional proteins.

The defect in inhibition of CĪ results in a proteolytic attack on C4 and C2, the natural substrates of C̄1. Serum levels of these proteins are reduced, particularly during attacks, at which time activated C1 (C̄1) can be demonstrated in patients' serum. A vasoactive peptide with kinin-like activity has been extracted from the plasma of patients with hereditary angioneurotic edema but its origin is unknown. It does not appear to derive from either C4 or C2. The fluid-phase activation of the initial part of the complement sequence is relatively inefficient: levels of C3 are within the normal range and there is only mild elevation in the catabolic rate of C3.

In principle, a simple screening test for hereditary angioneurotic edema consists of the immunochemical estimation of C̄1 inhibitor and C4. If the serum levels of both are normal, the disease is ruled out. If both are low, the patient almost certainly has the disease. If C4 is low but the C̄1 inhibitor level is normal, the electrophoretic mobility of the C̄1 inhibitor should be determined.[34] In most cases with dysfunctional proteins, the mobility will be abnormal. The definitive establishment of the diagnosis requires the demonstration of low or undetectable functional inhibition of C̄1 in the patient's serum.

Hereditary deficiency of C2 has been documented in half a dozen families.[48] Affected persons are homozygotes and heterozygotes have 30 to 50% of the normal level. The level of C2 in serum from homozygotes in different families varies and ranges from none detectable to 3 or 4% of normal. Although abnormalities of complement-mediated functions can be demonstrated in serum from C2-deficient persons, the latter are by and large asymptomatic. Recent reports of systemic lupus erythematosus or lupus-like syndromes in C2-deficient individuals may indicate an increased susceptibility to "collagen" disease. However, a meaningful association is far from proved since the populations screened for C2 deficiency consisted largely of patients with known or suspected lupus or glomerulonephritis.

A single patient homozygous for C3 deficiency has been found.[49] Her clinical history was like that of boys with X-linked agammaglobulinemia; she has had numerous serious infections with pyogenic organisms. The level of C3 in her serum is less than 1/1000th the normal and asymptomatic heterozygotes for the deficiency state have 50% of the normal serum concentration. By means of the inherited structural polymorphism in C3 it has been shown that the deficiency gene (C3⁻) is an allele of normal C3 genes but produces no detectable protein. Complement-mediated functions in the C3-deficient patient's serum are extremely low. Thus, there was no generation of chemotaxis, no bactericidal activity and no enhancement of the phagocytosis of antibody-sensitized bacteria. The addition of purified C3 to her serum restored these functions. There was furthermore no leukocytosis when the patient had severe gram-positive bacterial infection, providing evidence *in vivo* for the importance of the complement system and C3 in particular in the mediation of this function.

Single families with deficiencies of C6[50] or C7 (Boyer, unpublished observations) have recently been found. In homozygous C6 deficiency there is no detectable C6 and heterozygotes have 50% or less of the normal level of this protein. Some C7 is detectable in the serum of the homozygous deficient individual. In neither of these deficiency states are there any important abnormalities of coagulation function. Although affected persons are asymptomatic, hemolytic complement is very low and there is a deficiency of bactericidal activity. Chemotaxis is normal in C6 deficient human serum.

Homozygous deficiency of the C3 inactivator[34] resembles C3 deficiency and agammaglobulinemia in its clinical expression in that the patient with this disorder has a markedly increased susceptibility to infection with pyogenic organisms. Deficient complement-mediated functions in his serum are not restorable by the addition of purified C3. No serum C3 inactivator can be detected and heterozygous family members have approximately half-normal concentrations but no other abnormalities. The absence of the C3 inactivator in the patient results in continuous activation and consumption of the proteins of the alternate pathway *in vivo*. Active GBGase is present at all times, GBG is markedly diminished to not detectable, CoF-binding protein is not detectable, and C3 is markedly diminished, but the properdin level is within the normal range. Circulating C3b is present continuously. It is presumed that C3a is generated *in vivo* since the patient has marked histaminuria and urticaria develops when he takes a shower.

In addition to these disorders which have well-established inheritance there are at least three additional abnormalities of the complement system which may have a genetic basis. Marked deficiency of C1r with a relative diminution in C1s has been described in two kindred. Affected persons have lupus-like syndromes or nephritis alone.

A single patient with low C3, circulating C3c, and a history of life-threatening infections with pyogens has been reported.[52] Studies of her family failed to reveal any abnormality. Complement-mediated functions are grossly abnormal in the patient's serum but were markedly improved by the addition of purified C3. The concentrations of classical and alternate pathway proteins are all normal but an enzyme, designated C3ase, is present in her serum. This enzyme appears to cleave C3 in the presence of Mg^{2+} but there is no apparent production of C3a *in vivo*. A normal serum protein (C3ase co-factor) enhances the activity of C3ase. Metabolic studies indicated that the patient's low serum C3 concentration is the result of both increased catabolism and decreased synthesis.

A familial disorder which may involve a dysfunctional C5 molecule has been reported in several families.[52] Affected persons have eczema and increased susceptibility to infection with staphylococci and gram-negative bacteria during the first year of life. Their serum exhibits decreased enhancement of phagocytosis of yeast by normal peripheral blood leukocytes. The concentration of C5 is normal by immuno-chemical estimation and by hemolytic assay. Inheritance patterns are unclear.

REFERENCES

1. Bull. Wld. Hlth. Org. **39**, 935 (1968)
2. R. A. NELSON, JR., J. A. JENSEN, I. GIGLI and N. TAMURA, *Immunochemistry*, **3**, 111 (1966)
3. H. J. MÜLLER-EBERHARD, *Ann. Rev. Biochem.*, **38**, 389 (1969)
4. I. H. LEPOW, G. B. NAFF, E. W. TODD, J. PENSKY and C. F. HINZ, JR., *J. Exp. Med.*, **117**, 983 (1963)
5. K. YONEMASU and R. M. STROUD, *J. Immunol.*, **107**, 309 (1971)
6. K. B. M. REID, D. M. LOWE and R. R. PORTER, *Bio. Chem. J.*, **130**, 749 (1972)
7. S.-E. SVEHAG and B. BLOTH, *Acta Pathol. Microbiol. Scand.*, **78**, 260 (1970)
8. M. J. POLLEY, *Progr. Immunol.*, **1**, 597 (1971)
9. N. R. COOPER, *Progr. Immunol.*, **1**, 567 (1971)
10. D. BING, *Biochemistry*, **8**, 4503 (1969)
11. R. A. PATRICK, S. B. TAUBMAN and I. H. LEPOW, *Immunochemistry*, **7**, 217 (1970)
12. D. B. BUDZKO and H. J. MÜLLER-EBERHARD, *Immunochemistry*, **7**, 227 (1970)
13. U. NILSSON and J. MAPES, *J. Immunol.*, **111**, 293 (1973) (abstr.)
14. W. DIAS DA SILVA, J. W. EISELE and I. H. LEPOW, *J. Exp. Med.*, **126**, 1027 (1967)
15. C. D. WEST, N. C. DAVIS, J. HERBST, R. SPITZER and J. M. MCCONVILLE, *Fed. Proc.*, **24**, 446 (1965)
16. H. J. MÜLLER-EBERHARD, *Progr. Immunol.*, **1**, 553 (1971)
17. H. J. MÜLLER-EBERHARD, M. J. POLLEY and M. A. CALCOTT, *J. Exp. Med.*, **125**, 359 (1967)
18. I. H. LEPOW, G. B. NAFF and J. PENSKY, (eds. G. E. W. Wolstenholme and J. Knight) "Complement", Ciba Foundation Symposium (J. and A. Churchill, Ltd., London, p. 74. 1968)
19. T. BORSOS, H. J. RAPP and M. M. MAYER, *J. Immunol.*, **87**, 326 (1961)
20. R. M. STROUD, M. M. MAYER, J. A. MILLER, and A. T. MCKENZIE, *Immunochemistry*, **3**, 163 (1966)
21. M. M. MAYER, (eds. M. Heidelberger and O. J. Plescia) "Immunochemical Approaches to Problems in Microbiology," (Rutgers University Press, New Brunswick, N.J., p. 268. 1961)
22. H. J. RAPP and T. BORSOS, "Molecular Basis of Complement Action," (Appleton-Century-Crofts, New York. 1970)
23. T. BOENISCH and C. A. ALPER, *Biochim. Biophys. Acta*, **221**, 529 (1970)
24. O. GÖTZE and H. J. MÜLLER-EBERHARD, *J. Exp. Med.*, **134**, 90s (1971)
25. H. J. MÜLLER-EBERHARD and O. GÖTZE, *J. Exp. Med.*, **135**, 1003 (1972)
26. C. A. ALPER, F. S. ROSEN and P. J. LACHMANN, *Proc. Natl. Acad. Sci. (U.S.)* **69**, 2910 (1972)
27. O. GÖTZE and H. J. MÜLLER-EBERHARD, *J. Immunol.*, **111**, 288 (1973) (abstr.)
28. P. J. LACHMANN and R. A. THOMPSON, *J. Exp. Med.*, **131**, 643 (1970)
29. I. H. LEPOW, *Progr. Immunol.*, **1**, 579 (1971)
30. R. A. NELSON, *Proc. Roy. Soc. Med.*, **49**, 55 (1956)
31. P. J. LACHMANN and H. J. MÜLLER-EBERHARD, *J. Immunol.*, **100**, 691 (1968)
32. H. R. COLTEN, *Transplant Proc.*, in press
33. H. R. COLTEN, *J. Clin. Invest.*, **51**, 725 (1972)
34. C. A. ALPER and F. S. ROSEN, *Adv. Immunol.*, **14**, 251 (1971)
35. F. S. ROSEN, *Transplant. Proc.*, in press

36. M. M. MAYER, (eds., E. A. Kabat and M. M. Mayer) "Experimental Immunochemistry," 2nd ed., (Charles C Thomas, Springfield, Ill., p. 133. 1961)
37. I. GIGLI, S. RUDDY and K. F. AUSTEN, *J. Immunol.*, **100**, 1154 (1968)
38. C.-B. LAURELL, *Scand. J. Clin. Lab. Invest.*, **29**, suppl. 124, 21 (1972)
39. J. O. MINTA, I. GOODKOFSKY and I. H. LEPOW, *Immunochemistry* **10**, 341 (1973)
40. R. F. RITCHIE, C. A. ALPER, J. GRAVES, N. PEARSON and C. LARSON, *Amer. J. Clin. Path.* **59**, 151 (1973)
41. C. A. ALPER, (eds., M. A. Rothschild and T. H. Waldmann) "Plasma Protein Metabolism: Regulation of Synthesis, Distribution and Degradation," (Academic Press, New York, p. 393, 1970)
42. C. G. COCHRANE and D. KOFFLER, *Adv. Immunol.*, **16**, 185 (1973)
43. C. A. ALPER and F. S. ROSEN, *J. Clin. Invest.*, **46**, 2021 (1967)
44. E. WASSERMAN and L. LEVINE, *J. Immunol.*, **87**, 290 (1961)
45. P. H. SCHUR and K. F. AUSTEN, *Ann. Rev. Med.*, **19**, 1 (1968)
46. N. ZVAIFLER, *Adv. Immunol.*, **16**, 265 (1973)
47. V. H. DONALDSON and F. S. ROSEN, *Pediatrics*, **37**, 1017 (1966)
48. M. R. KLEMPERER, H. C. WOODWORTH, F. S. ROSEN and K. F. AUSTEN, *J. Clin. Invest.*, **45**, 880 (1966)
49. C. A. ALPER, H. R. COLTEN, F. S. ROSEN, A. R. RABSON, G. M. MACNAB and J. S. S. GEAR, *Lancet*, ii, 1179 (1972)
50. J. P. LEDDY, M. M. FRANK, T. GAITHER, R. S. HEUSKINVELD, R. T. BRECKENRIDGE and M. R. KLEMPERER, *J. Clin. Invest.*, **52**, 50a (1973)
51. C. A. ALPER, K. J. BLOCH and F. S. ROSEN, *N. Engl. J. Med.*, **288**, 601 (1973)
52. M. E. MILLER and U. R. NILSSON, *N. Engl. J. Med.*, **282**, 354 (1970)
53. C. A. ALPER and F. S. ROSEN, (ed., D. Bergsma) "Birth Defects". (Original Articles Series, The National Foundation, New York, in press.)

Chapter 8

Immunoglobulins

G. T. Stevenson

Tenovus Research Laboratory, General Hospital
Southampton SO9 4XY

8.1. INTRODUCTION

Immunglobulins are a class of structurally related proteins containing antibodies, certain fragments of antibodies, and certain pathological proteins. The earlier, synonymous "γ-globulins" has been largely superseded because it is sometimes not clear whether it refers to a class of structurally related proteins or to an electrophoretic class. The suffix globulin arose from a classification of proteins based upon solubility which is now irrelevant. Within the immunoglobulin class the nomenclature tends to follow recommendations made from time to time by W.H.O. Committees.[1]

Immunoglobulins are synthesized by cells of the lymphoid series, in greatest amounts by plasma cells but altogether by cells with wide ranges of size, ribosomal content, and content of rough endoplasmic reticulum.[2,3] Those exported into the extra-cellular fluid have been readily available for study and are the subject of this article. They have been detected in almost all vertebrates in which they have been sought, but not in invertebrates.[4,5] By definition they mediate the humoral or circulating antibody response. Two additional roles have been postulated for immunoglobulins which remain attached to the surfaces of the secreting cells: the recognition of antigen, and the implementation of cell-mediated immunity. Characterization of these covert immunoglobulins is a major aim of current research.

The functions of antibodies fall under two headings: *combination with antigen* ("antibody activity" is used in this restricted sense) and *adjunctive functions*.

Each antibody molecule shows a high specificity in its combination with antigen, and any single animal is capable of producing antibodies for an indefinitely large number of antigens which it might encounter throughout life. So antibody production is a remarkable example of vertebrates adapting to their environment. Although we have a moderate knowledge of the product, the antibody molecule, and the means whereby it combines with antigen, the mechanisms which call it forth are largely a mystery.

The various adjunctive functions are expressed before or after combination with antigen. Thus before combination some antibodies are transported across the placenta, some not; some are preferentially transported into exocrine secretions; and some show a tendency to fix to various cells. After combination with antigen some antibodies tend to form soluble, some insoluble complexes. Some antibodies in combination with antigen fix complement while others do not. With cellular antigens in the absence of complement some antibodies cause agglutination while others merely coat the cells, and some but not others have an opsonizing effect. This incomplete list illustrates the sort of adjunctive functions which determine much of the physiology and pathology of the immune system.

Although definitive proof is lacking it has become increasingly likely that all variations in the functions of antibodies stem from variations in amino acid sequence, with variations in the folding of a given sequence playing no part.[6,7] For this hypothesis to be true it is necessary but not sufficient that antibody and adjunctive functions be served by a vast array of sequences. Structural studies indicate that this is the case.

The difficulty of investigating structure in the face of such heterogeneity has been lessened by the availability in man and mouse of homogeneous immunoglobulins known as *myeloma proteins*. These are usually products of the neoplastic plasma cells which make up a disseminated tumour called multiple myeloma or plasmacytoma. Sometimes they arise from tumours which are lymphomas rather than plasmacytomas, and sometimes they occur in the absence of an overt neoplasm. The fact that structural features discovered in myeloma proteins have subsequently been found regularly in normal immunoglobulins has led to the belief that a myeloma protein from a single tumour represents a vastly over-produced single member of the normal molecular population. The occasional occurrence of myeloma proteins with large deletions possibly represents an exception to this rule. The rule is consistent with the following important hypotheses. (1) A lymphoid cell committed to secreting immunoglobulins for export is restricted to one type of molecule or to a small group of closely related molecules. (2) Secretion by its descendants is restricted to the same molecule or molecules. (3) Multiple myeloma arises from a malignant transformation of a single such committed lymphoid cell.

8.2. THE FOUR-CHAIN IMMUNOGLOBULIN MOLECULE

A four-chain model for immunoglobulin molecules was proposed by Porter,[8] based largely on studies of rabbit normal immunoglobulins. A complete sequence of a two-chain myeloma protein was reported in 1965[9] and of a four-chain myeloma protein in 1969.[10]

Most immunoglobulin molecules are notably symmetrical units with two pairs of peptide chains, called *heavy* and *light* (Fig. 8.1). The remainder consist of either incomplete or polymeric variants of this structure. Within any single molecule the two heavy chains are identical and the two light chains are identical, but within the immunoglobulins of a single animal there are many varieties of both chains.

The generalizations about the molecule which follow cover the structural findings in man, rabbit and mouse. It seems likely that they will be universally applicable, although some stretching of the size ranges might be required.

The molecular weights of the heavy chains vary between 50 000 and 75 000, and of the light chains between 22 000 and 24 000. Interchain bridging is by non-covalent and disulphide bonds as shown in Fig. 8.1: each light chain being joined to a heavy, and the two heavy chains joined in their carboxy-terminal halves. The non-covalent bonds have been present in all molecules examined. The disulphide bonds have been variable in number and position, and those joining the light to the heavy chains have sometimes been absent.[11-14]

Proceeding along its length each peptide chain is divisible into linear zones ("homology regions") of about 110 amino acid residues,

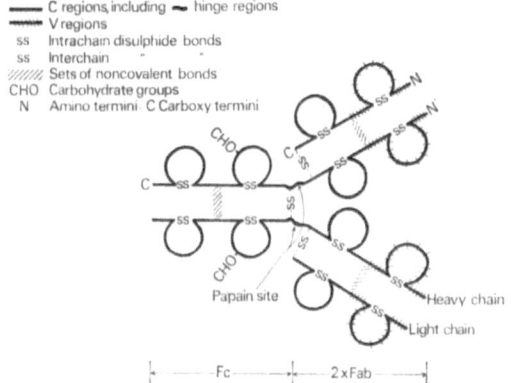

FIG. 8.1. The four-chain immunoglobulin molecule. The model is based on data from mammalian molecules, but it is likely that those of other vertebrates are similar.[5] Known divergences from the model involve the length of the Fc region of the heavy chains, the number and arrangement of the interchain disulphide bonds, and the number and positioning of carbohydrate groups. Various arrangements of the interchain disulphide bonds in human immunoglobulins are described by Milstein and Pink.[14]

with four such regions in a heavy chain (perhaps more in some) and two in a light. Each region contains an intrachain disulphide bond forming a loop of approximately 70 residues. Studies of sequence show the regions to contain further homologies, suggesting that immuno-globulin molecules evolved by gene duplications from a primordial protein of molecular weight approximately 12 000.[15,1]

The carbohydrate content of complete immunoglobulin molecules has been found to vary between 2 and 12%. Commonly a carbohydrate prosthetic group is attached to the heavy chain close to the region indicated in Fig. 8.1, but attachments at other sites have also been described. Immunoglobulins may well obey a general rule that carbohydrate residues are attached to Asx in the tripeptide sequence Asx-X-(thr or ser).

An important feature of the molecule is the arrangement of its chains into a tripartite structure, with functions clearly assignable to each part. The first evidence of this arose from limited digestion with papain (Porter, 1959). There is a zone of susceptibility to papain and some other proteolytic enzymes at about the middle of the heavy chain. In many immunoglobulin molecules cleavage by papain at the site indicated in Fig. 8.1 gives three soluble fragments: two identical Fab and Fc. Physicochemical and antigenic studies suggest that these retain conformations close to those in the parent molecule.[17,18]

Each Fab fragment retains one intact antibody site, having an association constant for antigen the same as in the intact molecule.[19] The "ab" derives in fact from "antigen-binding". The N-terminal half of the heavy chain, contained in Fab, is called Fd.

The Fc fragment was so called because for some immunoglobulins it crystallizes readily. An accumulation of evidence, some to be presented later, suggests that it contains those regions determining all the adjunctive functions.

Pepsin at a pH near 5 cleaves many molecules at a site just C-terminal to the papain site and the inter-heavy chain SS bonds. The Fc domain is broken into small fragments. The Fab-like fragments survive and are designated Fab'. They exist in solution as the disulphide-bonded dimer (Fab')$_2$, which is converted into monomer in the presence of reducing agents.[20]

8.3. DIMENSIONS OF VARIATION

Two given heavy or two given light chains can differ in sequence from one another due to four causes. The resulting four dimensions of variation have been called xenotypic, allotypic, isotypic and idiotypic. Of these only the last appears to be unique to immunoglobulins although isotypic variation is more complex than has been described in other proteins (e.g. iso-enzymes, normal haemoglobins).

Xenotypic variation is that occurring between homologous chains from different animal species. It provides clues to the evolution of immunoglobulins, and to those regions of the peptide chains where amino acid substitutions are best tolerated.[21,22]

Allotypic variation, or genetic polymorphism, is that occurring within a single species due to differences in peptide chains coded by allelic genes. Thus if a given chain occurs within a species as two allotypes x and y, and if the encoding genes are given the same symbols, a single member of that species will have a genotype xx, xy or yy. Similar polymorphisms are found of course in other proteins, but the word "allotype" is restricted to the immunoglobulin literature. A remarkable amount of information has been gathered about allotypes, expecially in man, rabbit and mouse.[23,24,25] Their study has been of great importance in considering the genetic control of antibody variation.

Isotypic variation refers to those readily delineated classes and subclasses which are found in any single individual of a species. It is determined by sequence in the C-terminal half of the light chains and the C-terminal three quarters or more of the heavy chains.* So in a whole molecule the isotype of the heavy chain determines the structure of Fc, and hence the adjunctive functions. The relatively limited number of isotopes (for example, ten currently recognized among the heavy chains in man) has led to these regions of the chains being called "constant" (C). A major difference between the C regions of two chains places them, by definition, in different classes, while a minor difference places them in different subclasses of the one class. What actually constitute major and minor differences are more easily described when considering in the next section the example of human immunoglobulins. Within a single subclass variation in the C region can only be of allotypic origin: the finding of any other variation would define a new subclass.

Idiotypic variation is also encountered in any single individual of a species. It is determined by sequence in the "variable" (V) regions of the chains: the N-terminal region of about 110 amino acid residues in each chain. The number of different idiotypes in an individual is indefinite, but probably *at least* several thousand for both chains. *Thus a given isotype in C regions is associated with many different idiotypes in V regions: a cardinal and unique feature of immunoglobulin populations.* There is little doubt that idiotypes provide the structural variety necessary for the wide range of antibody activities, although C as well as V regions of Fab could participate in antibody sites. (The term "idiotypic" was coined[26] to describe the occurrence of apparently unique antigenic determinants in antibody populations of a single specificity from single animals. Such antigenic determinants can be detected also in individual myeloma proteins, and have been clearly located in the V regions of the chains.) The nature of idiotypic variation will be described in more detail in section 8.6, where we consider the underlying genetic mechanism.

The population of whole immunoglobulin molecules within any single species or individual reflects the isotypic, idiotypic and allotypic

---

* Sets determined by C regions are sometimes called "types" for the light chains, distinguishing them from "classes" for the heavy chains. The latter term will be used for both chains in this review.

variation of its chains. A further dimension of variation is generated by having different pairings available for any given chain. Just how much freedom in pairing is allowed is not clear but it is known that there are restrictions: that is, if the total number of heavy chain variants in an individual is m, and of the light chain variants n, then the number of whole molecules which he can form is less than m × n. This also is discussed in more detail in section 8.6.

We now illustrate the above concepts by examining the best studied immunoglobulin population, that of man. It might be helpful first to refer to Fig. 8.2 where much of our discussion about structure and organization is summarized. There will be occasion to refer in section 8.6 to allotypes of rabbit and mouse immunoglobulins, but otherwise space does not allow a consideration of other species, for which a

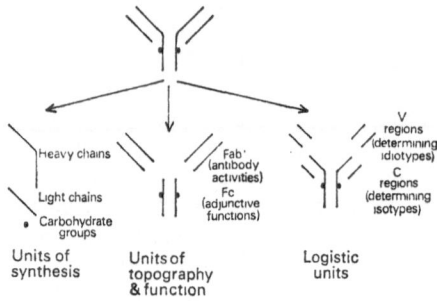

FIG. 8.2. Dissections of the immunoglobulin molecule. That into logistic units is at present conceptual only. The other dissections are readily performed in the test tube.

recent review can be consulted.[5] In general immunoglobulins of other mammals are similar to those of man, although correspondences are not precise and difficulties arise in correlating the nomenclatures. Less isotypic variation has been found in more primitive species.

8.4. VARIETIES OF HUMAN IMMUNOGLOBULINS

8.4.1. PEPTIDE CHAINS

The recognized variants of human heavy and light chains are set out in Table 8.1. Classes are designated by Greek letters and subclasses within these by appending an Arabic numeral: thus $\gamma 1$, $\gamma 2$, etc. Allotypes are designated in a variety of ways. For the origins of the nomenclature the reader is referred to the original W.H.O. publication (1964).

Isotypic variation should soon be definable strictly in terms of amino acid sequence, and it is anticipated on the basis of efforts to date[27] that determination of the isotypes of unknown samples will be possible by simple chemical tests. Pending these events we rely chiefly on the original

TABLE 8.1

Peptide chains of human immunoglobulins

	Isotype		Markers of allotypes*	Idiotypes
	Classes (Molec. wts.)	Subclasses		
Heavy chains	γ (53 000)	$\gamma1$	a, x, z, f	(Numbers indefinitely large. Probably allotable to a limited number of subgroups within each set.)
		$\gamma2$	n	
		$\gamma3$	g, b⁰, b¹, b³, b⁴, b⁵	
		$\gamma4$	4a, 4b	
	α (64 000)	$\alpha1$	N.R.	
		$\alpha2$	Am₂	
	μ (68 000)	$\mu1$	N.R.	Possibly one shared set
		$\mu2$	N.R.	
	δ (70 000)	N.R.	N.R.	
	ϵ (75 000)	N.R.	N.R.	
Light chains	κ	N.R.	1, 2, 3.	Unique set
	λ	λ, Oz+ † λ, Oz−	N.R. N.R.	Unique set

The data in Tables 10.1, 10.2 and 10.3 are largely from the reviews by Cohen and Milstein[34] and Milstein and Pink.[14]

N.R. = None recognized.

* Antigenic determinants by which the allotypic determinants are recognised. Others have been described. Note that separate sets exist for each subclass or undivided class. Some of the determinants in a single set are mutually exclusive, while some occur commonly in certain combinations: for details see Natvig and Kunkel.[35] Determinants on the γ chains are known collectively as Gm, those on the κ chains as Inv. A Committee on Nomenclature[1] has recommended the use of numbers to designate the various Gm determinants, but this has not been widely adopted.

† Oz+ and Oz− probably reflect isotypic rather than allotypic differences.[36]

tool, antigen structure, both to define and to determine isotypes. Different classes of chains are antigenically distinct, *e.g.* an antiserum raised in rabbits to human γ chains does not react with human μ chains. Within a single class chains belonging to different subclasses show a partial antigenic overlap. Thus a rabbit antiserum to human γ1 chains will react with γ2, but after absorption with the latter will still display some reaction with γ1. Difficulty in applying these rules arises from variation in antisera and the need to exclude allotypic and idiotypic determinants from consideration.

The difficulty in stating precisely and consistently the difference between classes and subclasses, the trouble that this hierarchy causes when comparing the immunoglobulins of different species, and the fact that any unique isotype however defined seems to be associated with one cistron (section 8.6), all suggest that the class-subclass distinction should be abandoned. Instead all isotypic differences could be taken to define classes, which would vary in the closeness of their relationship.

Allotypes also are delineated usually by antigenic structure. Two main sources of antibodies have been used: human sera, and rabbit antisera to myeloma proteins in which activity directed to non-allotypic determinants has been removed. Anti-allotypic activity in human sera is part of the complex anti-immunoglobulin activity apparently present at low levels even normally.[28] The anti-allotypic component appears to occur either fortuitously as part of the general activity directed towards the individual's own immunoglobulin, or following active immunization, incidental to blood transfusion or transplacental passage, with a foreign allotype.[29] Rheumatoid arthritis stands out in its high incidence of high concentrations of antibodies to autologous immunoglobulin (the antibodies are often called "rheumatoid factor"), and some rheumatoid sera have been useful in characterizing allotypic determinants. Indeed it was with such sera that the Gm group of determinants was first defined.[30] Other human sera with useful anti-allotype activity come from a variety of sources, including apparently healthy blood donors.

In some cases there is a one-one relation between allotypes and the antigenic determinants by which they are recognized (Table 8.1). Thus the γ4 chains exist as allotypic variants 4a and 4b, while the γ2 chains exist as n and (reflecting the absence to date of an antiserum specific for the allelic chains) n(−). In the case of γ1 chains, however, allotypic variation occurs at more than one site on the chain so that allotypes of whole chains must be defined by the *sets* of determinants: for example the three common allotypes among Caucasians have the sets za, zax and f.

All allotypic variations recognized among human immunoglobulins involve the constant regions. However experience in the rabbit and mouse indicates that unrecognized allotypes in man could involve the variable regions[31,32] (see section 8.6).

TABLE 8.2
Classes of human immunoglobulins

Class	Heavy chain	Light chain	Formulae of commonly occurring species ($L = \kappa$ or λ)	Molec. wt.	Carbohydrate content, %	Concentration in normal plasma, mg/ml.
IgG	γ	κ or λ	$\gamma_2 L_2$	150 000	2.9	8–17
IgA	α		$(\alpha_2 L_2)_{1,2 \text{ or } 3}$, T* $(\alpha_2 L_2)_2$ T*	175 000–500 000 400 000	7.5	1.4–4.2
IgM	μ		$(\mu_2 L_2)_5$	900 000	12	0.5–1.9
IgD	δ		$\delta_2 L_2$	180 000		0.003–0.4
IgE	ϵ		$\epsilon_2 L_2$	196 000	11	<0.001

* The predominant species in external secretions. T is "secretory" or "transport" piece.

In the $\gamma 1$ chains sequence data suggest that f is associated with arginine and z with lysine at position 214 (in the constant part of Fd), while the presence or absence of a is associated with a substitution of two amino acids in Fc. In the κ chains Inv(1) is associated with leucine and Inv(3) with valine at position 191[33]—a remarkably subtle difference to be picked up by antigenic characterization. The position with regard to the determinant Inv(2), usually associated with Inv(1), is obscure.

8.4.2. WHOLE MOLECULES

Whole immunoglobulin molecules are divided into classes and sub-classes on the basis of isotypic variation in their heavy chains (Table 8.2). For classes, Roman letters which correspond to the Greek letter of the heavy chain are used adjectivally to qualify the noun immuno-globulin: thus immunoglobulin G, abbreviation IgG, contains γ chains. Arabic numerals are appended to indicate the subclass, *e.g.* IgG3 containing $\gamma 3$ chains. The convention can be expanded to en-compass the light chains as well, so that the molecule $\gamma_2 \kappa_2$ can be designated IgGK. However we sometimes do not know whether the light chain present is κ or λ, and preparations from normal immuno-globulin populations usually contain both. So it is convenient in general not to specify the light chain, and "IgG" implies $\gamma_2 \kappa_2$ and/or $\gamma_2 \lambda_2$.

The practice of basing immunoglobulin classes simply on the heavy chains arose because it is the C regions of the latter which determine most of the physicochemical and antigenic differences among the parent molecules. As the same C regions determine adjunctive functions it follows that there is a correlation between these functions and immuno-globulin class (Table 8.3).

However there is probably no correlation between antibody activities and class. There are many studies[38] where a given antibody activity has been sought and found in several classes, and it is probably true to say that if a certain activity can be expressed at all then it can be expressed in any of the classes. Certainly some activities have been much more frequently associated with one class than others, for example the association of anti-I with IgMK in the cold haemagglutination syndromes.[39] But such examples could arise from the mode of pres-entation of the antigen and/or the method of detecting the antibody, rather than from any limitations of activity in the non-participating classes.

In summary: members of a given immunoglobulin class are so grouped on the basis of the C regions in their heavy chains, have similar physicochemical and antigenic properties, manifest similar adjunctive functions, and have no distinctive antibody functions. We now consider some properties of each of the classes of human immunoglobulins individually.

TABLE 8.3

Adjunctive functions of human immunoglobulins

Class	Subclass	Fixation of complement	Transport across placenta	Transport into external secretions	Sensitization of skin	
					Human (Homocytotropism*)	Other species (Heterocytotropism†)
IgG	1	+	+‡	−	−	+
	2	++				−
	3	+				+
	4	−				+
IgA	1 and 2	−	−	+	−	−
IgM	1 and 2	++	−	−	−	−
IgD		−	−	−	−	−
IgE			−		+	−

* Would be regarded as synonymous with "reaginic activity" by most authors.
† The species has usually been the guinea pig, and the sensitization detected by passive cutaneous anaphylaxis (PCA) or reverse PCA. Heterocytotropism has been demonstrated for IgE in monkey skin.[37]
‡ No information for the individual subclasses.

Immunoglobulin G

IgG is the predominant immunoglobulin of mammalian sera. Many older descriptions of "γ-globulin" or "7S γ-globulin" can be taken to refer to it, and it has been the archetype for studies of physicochemical properties and antibody activities. Some statements which follow derive from experiments on IgGs of species other than man, but experience suggests that the principal conclusions apply to all mammalian IgGs.

IgG is synthesized in man at the rate of about 30 mg/kg/day.[40] In general it appears to make up the bulk of antibody populations occurring late in primary immunizations or throughout the response to booster immunizations.[41] The subclasses IgG1, 2 and 4 have notably smaller catabolic rates than do other immunoglobulins: their half-lives in adult man average about 21 days, compared with about 7 days for IgG3 and between 2 and 7 days for the other immunoglobulins[42] (see Table 8.4). The IgGs appear to be unique in that their catabolic rate is

TABLE 8.4
Metabolism of human immunoglobulins

Class and subclass	Circulating pool, mg/kg	Approx. % Intravascular	T½, days	Approx. synthetic rate, mg/kg/day
IgG1, 2 and 4	500	50	21	32
IgG3			7	
IgA	90	40	6	25
IgM	36	76	5.1	7
IgD	1		2.8	0.4
IgE	<0.02	50	2.4	<0.02

Data from Waldmann and Strober[42]

higher the higher their extracellular concentration. This applies both to normal and abnormal IgGs and is independent of subclass: in ten cases of multiple myeloma a raised concentration of any subclass was associated with increased catabolism of all the subclasses.[43] The molecular site determining the catabolic rate appears to be in Fc. Thus injected Fc fragment has a survival time comparable to the whole molecule and much longer than those of Fab and light chains,[44] and an increased concentration of Fc increases the catabolism of IgG.[45] The catabolic rate can in fact be regarded as another adjunctive function. The mechanisms underlying this regulated catabolism are not known, although an interesting suggestion has been made by Brambell.[46]

Some adjunctive functions additional to catabolic rate can be seen from Table 8.3 to vary in the different subclasses, furnishing further argument for abolishing the subclass-class distinction. Among other functions IgG antibodies have generally been found to be precipitating, although a proportion of rabbit IgG antibodies to heterologous serum albumins were found by Christian[47] to be non-precipitating. With erythrocytic antigens IgG antibodies are non-agglutinating,[48] but can induce complement-mediated lysis and have an opsonizing effect (which complement might enhance). Opsonization requires an intact Fc.[49] So also does full complement-fixation. although partial fixation has been described with (Fab')$_2$ fragments, and the position is complicated by the possible existence of several combining sites serving different components of complement.[50]

There are receptors on monocytes and macrophages, and perhaps neutrophils, specific for the Fc region of IgG1 and IgG3. These cause aggregates of antigen and IgG1 or IgG3 antibody to adhere to these cells, furnishing a plausible explanation for the opsonizing effect.[51,52]

Placental transfer of IgG antibodies is a vital means of conferring passive immunity on the neonate in certain mammals, including man and other primates, rabbit and guinea pig.[53] In these animals absorption of antibodies from maternal milk is negligible (although secretory IgA in the milk might help defend the infantile gut). This was the first adjunctive function to be clearly localized to Fc. Brambell, Oakley and Porter[54] showed that Fc and whole IgG, but not Fab, are readily transmitted across the rabbit placenta.

Human IgG as isolated from plasma by chromatography, electrophoresis or precipitation with ammonium sulphate is entirely monomeric with an extrapolated sedimentation coefficient ($S^0_{20,w}$) of 6.6 S.[55] However the occurrence of a myeloma IgG as a noncovalently bonded aggregate of molecular weight about 800 000[56] suggests that a minute amount of normal IgG could be polymeric. IgG for therapeutic use is usually prepared by the cold ethanol process,[57] which yields up to 25% in the form of low-order aggregates ($S_{20,w} = 9.5$ to 12 S). Upon injection such aggregates can fix complement and cause tissue damage,[58] so that they are sometimes removed by prolonged centrifugation before therapeutic administration.

Molecules of IgG have a range of net charges, reflected in a range of electrophoretic mobilities, isoelectric points, and chromatographic behaviour on ion-exchange resins. On electrophoresis the bulk migrate in the γ zone, giving by the moving boundary method a broad peak with a mobility at the apex of -1.1×10^{-5} cm^2 V^{-1} sec^{-1}, in barbiturate of pH 8.6 and ionic strength 0.1; the corresponding isoelectric point is 7.3.[59] However there is an overall spread of mobilities from the slow γ right up to the α_2 zone, well illustrated by immunoelectrophoresis (Fig. 8.3). The subclass IgG4 has an average mobility higher than those of the other subclasses.[60]

An examination of Fig. 8.3 reveals the overlap in mobilities among the main immunoglobulin classes and emphasizes the difficulty of obtaining pure IgG in high yield by electrophoresis or ion-exchange chromatography. Preparations yielded by these techniques are pure only at the cost of sacrificing the more acidic components.

Zone electrophoresis in sieving gels in the presence of urea resolves the isolated heavy and light chains of IgG into discrete bands, giving 10 bands for the light chains[61,61a] and up to 18 for the heavy.[62] Feinstein[63] showed that there is a difference of one between the net charges of adjacent components of the light chains. Awdeh et al.[64] report that they have been able to resolve whole IgG into discrete bands by isoelectric focussing in polyacrylamide gel.

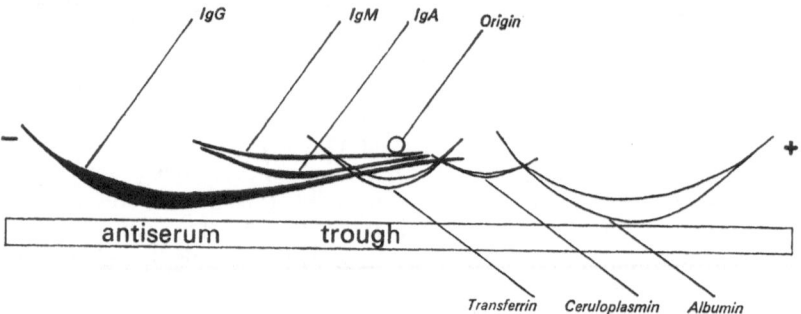

FIG. 8.3. Electrophoretic distribution of the predominant human immunoglobulins, demonstrated by immunoelectrophoresis in agar gel. The technique involves two stages: (1) serum at the origin is subjected to an electrostatic field which causes the proteins to migrate at varying rates; (2) the field is removed and antiserum is inserted into a trough parallel to the electrophoretic migration. Antibodies from the length of the trough and antigens from their sites in the electrophoretic path diffuse towards each other and precipitate where suitable concentrations meet. Precipitates formed by transferrin (β_1 mobility), ceruloplasmin (α_2 mobility) and albumin are indicated as electrophoretic markers. The relative mobilities and broad electrophoretic spreads of the immunoglobulins are apparent.

Hydrodynamic and electron microscopic studies of IgG tend to corroborate the tripartite structure suggested by limited proteolysis. From calculations of frictional ratios and intrinsic viscosities Noelken et al.[18] concluded that the Fab and Fc fragments of rabbit IgGs resembled typical globular proteins, whereas the parent molecules were less compact. The best electron microscopic pictures available to date (reviewed by Green[65]) have been of rabbit IgG bound to antigens. They have suggested a Y-shaped molecule, the three limbs representing the Fc and two Fab zones (Fig. 8.4). The angle between the Fab limbs in the antigen-antibody complexes varied between 10° and 180°, with no preferred angle apparent. Feinstein and Rowe (1965)[65a] obtained compact dimensions for unbound rabbit and human IgGs, and considered that the inter-Fab angle was probably closed in this state. A review of more recent pictures, however, led Green[65] to conclude that there is no good electron microscopic evidence of a change in conformation accompanying combination with antigen.

Studies of optical rotatory dispersion and circular dichroism (reviewed by Dorrington and Tanford[66]) have been particularly useful in characterizing conformational changes, such as upon exposure to denaturants[67] or upon recombination of isolated γ and light chains (Stevenson and Dorrington, 1970[67a]). Between 220 and 270 nm the spectra of optical rotatory dispersion of all human IgGs, although

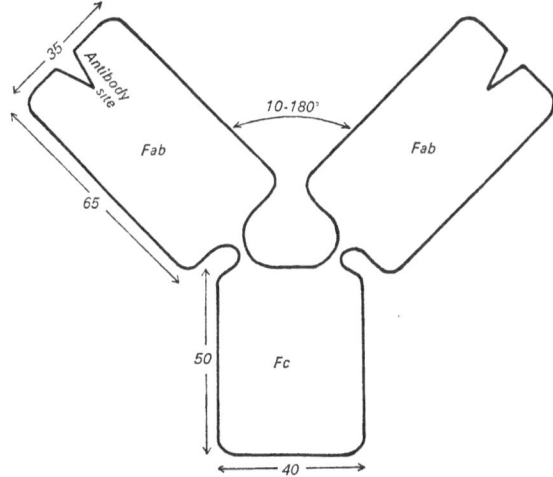

FIG. 8.4. Molecular envelope of rabbit IgG, deduced from electron micrographs. Dimensions are in Å. Modified from Green.[65]

showing quantitative differences, have been qualitatively similar: with a small amount of rotation throughout the range, Cotton effect minima near 230 and 225 nm, and a small negative Cotton effect near 240 nm. Other immunoglobulins have been broadly similar. The findings have been held to reflect a paucity of α-helix and the presence of β-structure, but Dorrington and Tanford emphasize the difficulties in such interpretations.

A number of crystallographic studies of human myeloma IgGs are at present under way. A major technical difficulty has been the obtaining of suitable crystals. The completion of these studies will of course represent an enormous advance in our understanding of immunoglobulins.

Immunoglobulin M

IgM, formerly called "γ-macroglobulin" or, on account of its extrapolated sedimentation coefficient, "19 S γ-globulin", occurs predominantly as the pentamer $(\mu_2 L_2)_5$. This form is implied unless stated otherwise, although small amounts of IgM in both normal and pathological sera have been identified in the forms of both larger polymers and monomeric $\mu_2 L_2$ units.[68,69,70] In primitive vertebrates such as the sea lamprey and dogfish only one class of immunoglobulin has been found, whose heavy chains resemble mammalian μ chains in size, electrophoretic mobility and amino acid composition. This analogue of IgM occurs in both monomeric and polymeric forms.[5]

Probably because of its high molecular weight IgM is largely confined to the intravascular compartment: 76% compared with 50% or less for the other major classes.[71] This distribution, its efficiency in agglutination and complement fixation,[72] and the manifestations of IgM deficiency[73] all suggest that it plays an important role in protecting the bloodstream against microbial invasion. In experimental immunizations of mammals IgM is characteristically the first class of antibody to appear, is supplanted by IgG later in the primary response, and may not be apparent in secondary responses.[41] However IgM is not invariably absent from secondary responses;[74] and it appears to form the bulk of "natural" antibodies to erythrocytes and bacteria,[75] where the antibody response is possibly evoked by repeated stimuli from widely distributed antigenic determinants.

Recent findings in the rabbit suggest that IgM is the major immunoglobulin to be found on the surfaces of lymphocytes.[76]

In the presence of thiols, e.g. 0.1 M mercaptoethanol, IgM dissociates into its five $\mu_2 L_2$ subunits with sedimentation coefficients of about 7S. The subunits are therefore held together in the pentamer by interchain disulphide bonds, probably occurring between C-terminal half-cystines on the μ chains.[77,78] Possibly an additional joining or J chain, not structurally related to the μ or L chains, is involved[78a]. Note that this dissociation does not require the additional presence of an unfolding agent, in contrast to the dissociation of basic 4-chain immunoglobulin units into heavy and light chains.

The susceptibility of the pentameric structure to thiols is paralleled by an apparent susceptibility of the antibody activity. This thiol-sensitivity is often used as a crude means of differentiating antibody activities due to IgG and IgM, the former being resistant. However it is not the antigen-antibody combination which is prevented by the thiol, but certain adjunctive functions, such as haemagglutination and haemolysis, which seem to require the full pentameric structure.[79] A further objection to the use of thiol-sensitivity to identify activity due to IgM is that some activities of IgA are similarly sensitive.[80]

Fragments of IgM analogous to Fab and Fc of IgG have been obtained by similar proteolytic procedures.[81,82] Digestion by papain in the absence of thiol has been found to yield a pentameric Fc of 320 000 daltons in which the five fragments are linked by disulphide bonds.[83]

Although an antibody valency of 10 is suggested by structural studies, several descriptions of anti-hapten IgM have indicated values of 5.

There is a suggestion however that a wide distribution of association constants might have resulted in sites with low affinities not being counted. Merler et al.[84] have described an IgM antibody to salmonellal lipopolysaccharide with a valency of 10.

Recent electron microscopic studies reveal IgM as a cyclic pentamer with a central ring made up of Fc regions from which, spider-like, Fab arms project.[5,85]

Immunoglobulin A

This class was described by Heremans et al. (1959), who correlated the arc attributed to "β_{2A}-globulin" on immunoelectrophoresis (Fig. 8.3) with the occurrence of a class of myeloma proteins having electrophoretic mobilities above the average and distinctive antigenic determinants. Chodirker and Tomasi[86] observed that IgA forms the major immunoglobulin in seromucous secretions. It was thereby identified as the antibody class associated with the long recognized local immunity shown by various mucosal surfaces, such as of the gut, respiratory tract, and female genital tract (reviewed by Tomasi and Bienenstock).[87]

Plasma IgA makes up some 15% of the plasma immunoglobulin. More than 80% occurs in the form $\alpha_2 L_2$, with the remainder as polymers having sedimentation coefficients in the vicinity of 10, 13 and 15 S. The polymers, as in the case of IgM, fall into monomers on exposure to a thiol, and so are maintained only by inter-monomer disulphide bonds, which, again analogously to IgM, might occur between C-terminal half-cystines on the α chains.[78,88,89] The polymers can exhibit haemagglutinin activity, abolished by thiols.[80]

The allotypic form $Am_2(+)$ of the minor subclass IgA2 is of considerable interest in lacking a heavy-light disulphide bond, with the light chain half-cystines usually participating in this bond forming instead a light-light bond.[90,91] Such an arrangement of interchain bonds seems to apply to most or all of mouse IgA.[78]

In seromucous secretions IgA occurs predominantly as a unit called "secretory IgA". This has a sedimentation coefficient of 11 S and consists probably of two IgA monomers linked to a non-immunoglobulin peptide chain called "secretory piece" or "transport piece".[87] Accompanying it are monomers and polymers identical with plasma IgA.

Tomasi and Bienenstock emphasize that secretory IgA is to be found only in "external" secretions: lacrimal, salivary, gastrointestinal, nasal and tracheobronchial, genitourinary, mammary. In these the ratios of IgG:IgA vary but are generally, and sometimes considerably, less than 1. In internal secretions—synovial, amniotic, pleural and peritoneal, cerebrospinal fluid, aqueous humour—secretory IgA does not occur and the ratios IgG:IgA are much the same as in plasma.

Secretory piece has a molecular weight of 58 000 and appears to bear no antigenic relationship to immunoglobulins. It is probably linked to the IgA units by both noncovalent and disulphide bonds (Fig. 8.5).

When free secretory piece is added in vitro to a mixture of immuno-globulins it complexes specifically with IgA, although it might complex with IgM in the absence of IgA.[92] In patients with agammaglobulinaemia it has been found free in saliva,[93] and there are reports of its being found partly in the free state in secretions from normal subjects.

Immunofluorescent studies, summarized by Tomasi and Bienenstock[87] indicate that much of the IgA of external secretions is probably synthesized locally, since IgA is the predominant immunoglobulin to be found within lymphoid cells of the respiratory and gut mucosae. There is similar but less emphatic evidence that secretory piece is synthesized locally by epithelial cells. Whether the two components are

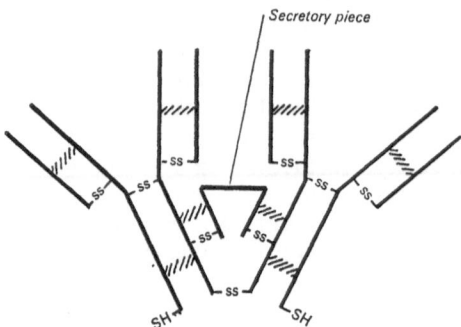

FIG. 8.5. A likely structure for secretory IgA. Two four-chain units are attached to secretory piece by both disulphide and non-covalent bonds, and to each other by a disulphide bond linking C-terminal half-cystines on the α chains. Modified from Tomasi and Bienenstock.[87]

joined before or after delivery into the secretion is not known, and as a corollary we cannot say whether secretory piece facilitates transport of immunoglobulin across the epithelial barrier.

One useful property which secretory piece does appear to confer on IgA is an increased resistance to digestion by proteolytic enzymes.[94]

IgA is clearly the major immunoglobulin concerned with the protection of mucous surfaces. In infections such as influenza and cholera where the parasite wreaks its damage locally and seldom spreads systemically, the local rather than the systemic antibody response has a diagnostic and prognostic significance. Interest has now been revived in older work (*e.g.* Fazekas de St. Groth[95]), showing that local (*e.g.* intranasal) immunization can be much more effective than systemic in eliciting an antibody response in the local mucus.

Immunoglobulin D

Rowe and Fahey,[96] on investigating an unusual myeloma protein, were able to assign it to a previously unknown class, IgD, constituting on average less than 1 % of the total plasma immunoglobulin. Although IgD has several times been shown to manifest antibody activities no distinctive adjunctive function has been found for it. With a half-life of 2–3 days it shares with IgE the shortest survival time among the immunoglobulins.

Immunoglobulin E

In 1966 Ishizaka *et al.*[97] were able to present evidence for a new class of immunoglobulins among which skin-sensitizing (reaginic, homocyto-tropic*) activity resides. By a happy coincidence Johansson and Bennich[98] were characterizing at the same time an atypical myeloma protein and its counterpart in normal serum. The two classes of immunoglobulins thus defined were soon shown to be one, which was named IgE. In healthy subjects it represents less than 0.01 % of the plasma immunoglobulin. It has been reviewed by Ishizaka and Ishizaka.[100]

The physicochemical properties of this class have been deduced largely from the two myeloma proteins presently available, although the similar electrophoretic, chromatographic and antigenic properties of normal IgE suggest that the myeloma proteins are truly representative. The molecular weight of ϵ chains (75 000) suggests that these have 5 or 6 rather than 4 homology regions.

The known adjunctive function can be described as an ability to fix to mast cells in tissues (which is unrelated to antibody specificity), and, upon subsequent exposure to specific antigen, to cause local anaphylaxis mediated by the release of histamine, serotonin, bradykinin and slow reacting substance. This sequence of events is exemplified in the classical Prausnitz-Küstner test for reaginic activity, and appears to underlie the associated clinical syndromes (atopic allergies) which include hay fever, allergic asthma and urticaria. In these conditions an increased plasma IgE has frequently been found. In accord with the geographic allocation of adjunctive functions in other classes, a Prausnitz-Küstner reaction could be inhibited by either an IgE myeloma protein or its Fc fragment, which presumably competed for the cellular receptors.[101] Heating at 56°, known to destroy the reaginic activity of antisera, did not abolish the antibody activity of IgE but did alter its antigenic structure.[102]

No physiological role has been assigned yet to IgE. However on the basis of IgE-producing cells being found near mucosal surfaces and of increased plasma concentrations occurring in metazoan infestations it has been suggested that local anaphylaxis might be useful in dislodging helminths.

* The development of the subject of antibody-mediated sensitivity, with its complex and varying manifestations in different species, has led to a redundant and imprecise terminology. Definitions of allergy, anaphylaxis, atopy, reagin, homocytotropism, etc., can be found in the book by Humphrey and White,[99] but they are not universally accepted.

Immunoglobulins of low molecular weight

Under this heading can be grouped any species with molecular weights appreciably less than 150 000, that of the archetypal 4-chain molecule. Such immunoglobulins have been said to make up several per cent of the normal plasma population in man.[103] They may logically be divided into (1) complete 4-chain units in which one or other of the chains is unusually small, (2) incomplete units. The latter in turn could represent either breakdown products or incomplete syntheses.

Small 4-chain molecules ($S_{20,w} \simeq 5.8$ S) have been described in Australian lung fish[104] and ducks[5] and might occur in rabbits.[105] They have not been detected in man. The low molecular weight reflects a small heavy chain, perhaps with three instead of four homology regions.

The best known incomplete units are free light chains, the normal analogues of Bence Jones proteins and like the latter a synthetic spill-over.[106,107] About 5 mg/day is excreted in urine as an electrophoretically heterogeneous mixture of κ and λ. This can be compared with the Bence Jones protein of myeloma, antigenically and electrophoretically homogeneous, requiring an excretory rate of about 500 mg/day to be detectable by the conventional heat test, and sometimes reaching a rate of 50 g/day.

Also in normal urine are Fc and Fc′ (a smaller proteolytic fragment) of γ chains, which are catabolic products.[108,109] Fab has not been detected in normal human serum or urine. Nor have free heavy chains, nor heavy chain fragments akin to those found in heavy chain disease (see section 8.8). However Franěk and Říha[110] have described in the serum of colostrum-deprived piglets a 5 S immunoglobulin whose tryptic peptide map suggested the presence of only heavy chains.

There have been intermittent claims and denials of significant antibody activity among material of molecular weight 50 000 or less. Current concepts would predict it only in Fab-like fragments.

8.5. ANTIBODY ACTIVITY

It has of course been impractical to raise antibodies at will in man, so the discussion which follows is derived mainly from experiments in animals, particularly the immunologists' favourite the rabbit.

There is no strict definition of antibody. We understand by the term an immunoglobulin with its antibody site having a significantly high affinity for the considered antigenic determinant, but there is no agreement as to what "significantly high" means in terms of association constant. Further, lacking a detailed topographical knowledge of immunoglobulin molecules, it is impossible to say in any given case whether a unique area properly called the antibody site is involved in the binding of antigen. These difficulties have been particularly irksome in assessing the significance of background antibody activity in un-immunized animals, and of antibody-like activities in myeloma proteins.

8.5.1. THE NATURE OF ANTIBODY POPULATIONS

This situation is clearer following an experimental immunization with an antigen x, so great is the typical increase in both titre and as-sociation constant of x-binding activity. The antibody activity can be examined in whole serum or other extracellular fluid, in a purified immunoglobulin population, or most usefully in a purified anti-x population. Details of the preparation of purified antibody populations are to be found in standard texts, such as Williams and Chase.[111] One group of methods for preparing anti-hapten populations entails (1) precipitating the antibodies from serum with macromolecular antigen having hapten-like determinants on its surface; (2) dissolving the washed precipitate with concentrated hapten, which displaces macromolecular antigen from the antibody sites and thereby disrupts the insoluble lattices; and (3) removing the macromolecular antigen and then the hapten by suitable chromatographic procedures. Unfortunately, convenient methods are often inherently selective for such factors as class of immunoglobulin and antibody association constant.

Purified antibody populations vary greatly in their homogeneity, whether judged by the isotypes present, physicochemical properties or affinity for antigen. At one end of the spectrum they exhibit a physico-chemical heterogeneity almost as great as that of the whole immuno-globulin population (e.g. Velick et al.),[112] at the other a homogeneity resembling that of myeloma proteins (e.g. Braun et al.).[113] As might be expected homogeneity is favoured by obtaining antibody to a single determinant from a single animal, but even then heterogeneity can be manifest.[38] Occasionally homogeneous antibodies turn up in man, sometimes in pathological conditions of which the most frequent seems to be the cold haemagglutination syndrome,[39] sometimes following experimental immunization.[114]

The fact that antibody of a single specificity from a single animal can be quite heterogeneous implies that the lymphoid system possesses in general many solutions to the problem of providing antibody of a given specificity. This is also strikingly illustrated by examining the antibody responses of an inbred group of animals immunized with the same antigen. Frequently in the purified antibodies from single animals some V regions occur in sufficient concentration to permit delineation of predominant idiotypic antigenic determinants by gel diffusion studies. Such determinants have differed from one animal to another. Indeed this fact first led Oudin[26] to coin the term "idiotypic". Thus antibodies to Proteus vulgaris from 19 different rabbits were found by Kekus and Gell[114a] all to possess apparently unique determinants; and Daugharty et al.[115] report a similar finding with antibodies to a single determinant, azobenzoate. These facts confront us again with the magnitude of antibody variability: for we must take into account not only a large number of antigenic determinants but also the large number of different antibodies which can be directed against each determinant.

Previous discussion has given some indications of the classes of immunoglobulins to be expected in antibody populations under varying circumstances. Thus plasma antibodies early in an immune response are likely to be largely IgM and later largely IgG, while antibody activity in mucus will usually reside predominantly in secretory IgA. Antibody passively acquired by the human foetus (as in haemolytic disease of the newborn) is IgG, but antibody actively secreted in response to intrauterine infection includes both IgM and IgG.[116]

Apart from the switch from IgM to IgG the change in antibody populations most consistently reported with increasing time of immunization is an increasing affinity for antigen.[117] Thus continuing exposure to small amounts of antigen seems to favour expansion of those clones producing antibodies of higher affinity.

8.5.2. THE ANTIBODY SITE

As already discussed the antibody site is in Fab and survives intact the fragmentation into Fab and Fc. Much effort has been expended to no avail in trying to obtain proteolytic fragments of Fab which retain an intact site.

There is much evidence to suggest that the site is shared between the heavy and light chains. The necessary sequence variation occurs in the N-terminal region of each. Despite some considerable differences in detail most workers have agreed with the first reports[118,119] that there is little antibody activity in isolated heavy or light chains, and that appreciable activity is regained only upon allowing chains from the same antibody population to recombine. Finally the results of affinity-labelling experiments, in which areas at or near the antibody site tend to be labelled by a chemically reactive group attached to a hapten, have implicated both chains.[15]

Evidence from a variety of experiments suggests that a typical antibody site has in round terms a depth of 10 Å and an area of up to 500 Å^2. This implies that it occupies something like 1% of the immunoglobulin surface. Early experiments of Landsteiner and van der Scheer,[120] still often quoted, revealed that the antibodies evoked by a complex haptenic determinant encompassed only a limited area of it, thereby suggesting sites of dimensions considerably smaller than those above. The high negative charge on this hapten might have caused difficulty in accommodating it within one site, because subsequent work (e.g. Levine)[121] has shown that other haptens of a comparable size can undoubtedly be accommodated.

Studies of polysaccharide and protein antigens (summarized by Kabat)[122] also suggest sites larger than the early estimates. For example the human antibodies to dextran studied by Kabat's group appear to have sites ranging in size up to an area, possibly about 500 Å², complementary with a hexasaccharide.

An interesting approach to ascertaining the depth of an antibody site has been described by Hsia and Piette.[123] They examined the reactions of rabbit IgG antibodies to the dinitrophenyl determinant with a series of haptens in which a spin-label was separated by a varying distance from a dinitrophenyl group. If the labelled group, which in the hapten-antibody complex would be outermost, abutted against the antibody its electron spin resonance was altered. The results led to an estimate of site depth of about 10 Å. A variety of other studies suggest a comparable figure, although there may exist appreciably shallower sites.[124]

There is some evidence that antibody sites are smaller on IgM than on IgG. For example Kaplan and Kabat,[125] studying human antibodies to blood group substance A, found that small haptens were more efficient in inhibiting the combination of IgM antibodies with homologous antigen than the combination of IgG antibodies. It is difficult to assess the significance of such results. As pointed out the antibody site changes with time of immunization even without a single class. So how much do the findings with IgM reflect a temporal change independent of class, how much a characteristic of the class, and how much other factors? Our present understanding of immunoglobulin structure— with light chains and V regions of the heavy chains shared among different classes—suggests that fundamental differences in antibody sites related to class are unlikely.

8.5.3. THE REACTION WITH ANTIGEN

A large number of studies with haptens have shown that both steric and chemical complementarities are involved in the antigen-antibody reaction (reviewed by Karush).[126] The importance of chemical complementarity is suggested by observations such as the decrease in hapten-antibody association when isosteric uncharged NO_2 is substituted for a COO^- of the homologous hapten.

There is little doubt that the fit between antibody and antigen is most critical in the deepest part of the antibody site, a fact which has led to the corresponding part of the antigenic determinant being called the "immunodominant group". Thus Beiser et al.[127] studied the reactions of rabbit antibodies to the galactosido-azophenyl-tyrosyl determinant (in which the sugar residue was outermost on the immunogen and hence innermost when in the antibody site) with a variety of related haptens. A minor change in the sugar affected the combination with antibody much more than did more radical changes peripherally in the determinant.

Estimates of association constants and kinetics in antigen-antibody reactions have also, because of technical considerations, been largely confined to situations in which the antigen is a hapten. Heterogeneity of the antibody is a major problem and requires the use of various assumptions and approximations: Chapter 13 by Eisen in Davis *et al.*[128] can usefully be consulted. The "average association constant" quoted for antibodies is usually the value of $\frac{[\text{antigen-antibody}]}{[\text{antigen}][\text{antibody}]}$ when half the antibody sites are occupied.

Average association constants found in hapten-(purified antibody) systems have ranged between 2×10^3 and $>10^9 \, M^{-1}$.[129,112] The difficulty of interpretation posed by low association constants (antibody or not?) has been referred to.

The forward reaction (hapten + antibody → hapten-antibody) is very rapid: for a dinitrophenyl hapten Froese[130] reported the rate to vary within the range $1.4–20 \times 10^7 \, M^{-1} \, sec^{-1}$. There was a much wider variation in rates of the reverse reaction. So it is likely that variations in the association constants of antibodies of a given specificity reflect mainly different rates of dissociation of the antigen-antibody complexes.

Early in the history of immunology the "avidity" of an antibody was defined loosely as its tendency to form stable complexes with its homologous macromolecular antigen. So a more precise definition might have it as the reciprocal of the dissociation rate of the antigen-antibody complexes. However imprecision in the use of the term, particularly a confusion with association constant, persists.

There is no good evidence that a conformational change in antibody accompanies the binding of hapten so as to provide a better fit. For example Cathou and Haber[131] detected no difference in optical rotatory dispersion at 220–260 nm following binding.

8.6. CONTROL OF VARIATION

Proper functioning of the immune response requires that antibody of appropriate specificity and adjunctive function appear at the right time. The former property depends upon three variables, the two V regions and chain pairing, the latter simply upon the C region of the heavy chain. So the science of immunology must ultimately include descriptions of the mechanisms controlling all these variables. We are a long way indeed from such a goal.

It seems likely that, once committed to produce antibody of a given specificity, a lymphoid cell and its progeny remain so committed. But even the fundamental point as to whether commitment of a cell line occurs before or after exposure to the antigen remains undecided. The currently popular *selective hypothesis* (*e.g.* Burnet)[132] states that commitment occurs in a quasi-random fashion independently of the antigen, which merely selects and stimulates into division those cells with antibody sites, in the form of surface receptors, happening to fit its determinants. This simple picture might be modified in view of recent findings about cellular cooperation in the immune response. No hypothesis has been offered about the control of adjunctive functions. Lacking such fundamental knowledge of immunological pathways it is difficult even to frame decisive questions about immunoglobulin variation.

8.6.1. GENETIC CONTROL OF V AND C REGIONS

Neglecting the problem of *selection* of V and C regions for given functions still leaves us with the following questions about their genetic control. What mechanism generates the great number of V regions? How do multiple V regions come to be associated with single C regions within the molecular population?

Information relating to these unanswered questions is accumulating from two main sources, sequence studies of myeloma proteins and the inheritance of allotypes. A vast mass of evidence can be summarized in the following eleven propositions. Further details and discussions are available in reviews by Lennox and Cohn,[132a] Kelus and Gell,[24] Herzenberg et al.,[25] Natvig and Kunkel,[23] Edelman and Gall,[10] Wu and Kabat[133] and, recommended as an overall summary, Milstein and Pink.[14]

(1) *The gene complexes encoding heavy and light chains are not linked to each other, and neither is on the X-chromosome.* This holds for both species, man and rabbit, in which genetic polymorphisms of both heavy and light chains have been well studied.

(2) *Each different C region, whether defined as a class or a subclass, appears to be associated with a single gene* (or "cistron", in the sense of a stretch of DNA encoding a single peptide chain). Thus there is one set of allotypes for each different C region, inherited according to simple Mendelian laws and exhibited by all molecules carrying this C region. A simple example is provided by the κ chains of man: in normal subjects homozygous for Inv(3) all the κ chains have Val at position 191, and in Inv(1, 3) heterozygotes approximately equal numbers have Val as have Leu (Terry et al. 1969[133a]).

Further evidence for this proposition comes from a human pedigree showing an apparent crossover between the loci for the $\gamma1$ and $\gamma3$ subclasses, consistent with single adjacent genes for each locus.[134]

In both man and mouse the loci for all heavy chain isotypes appear to be closely linked. There is a preliminary report that the loci for κ and λ C regions in the rabbit are not linked.[135]

A difficulty is presented by two γ chain subclasses in the mouse exhibiting some shared allotypic markers in parallel. The review by Herzenberg et al.[25] should be consulted.

(3) *Different V regions appear to be encoded by different genes*, rather than to result from some mechanism such as variable translation of a limited number of genes. The evidence comes from sequence studies. If

two V regions from a given class of chain are aligned* and translated back to DNA sequences, then a conversion from one to the other is seen to require the same changes (predominance of single base substitutions, transversion twice as common as transition, and a high mutation rate for G) as pertain for say, two haemoglobins from different species, where the proteins are encoded by different genes related by the evolutionary process.[135a] The sequences in Fig. 8.6 illustrate this point.

It seems an inescapable corollary of propositions (2) and (3) that *at least two genes, one for V and one for C, encode each peptide chain in immunoglobulin molecules.*[136] No precedent is known for such a situation. Nor is there any knowledge of the mechanism which might fuse V and C, although there are indications that it must operate at the level of DNA rather than of mRNA or peptide chains: (a) the peptide chains appear to be synthesized in a smooth sequence from N to C terminus;[137,138] (b) deletions found in the proteins of heavy chain disease (section 8.7) involve adjacent stretches of V and C regions, suggesting a fusion of V and C genes prior to the deletion.[139]

The different genes for the V regions might all be carried in the germ line, or might arise from a small number of germ-line genes by some mutational process during somatic development. Both possibilities have been vigorously argued.[143,144,135a]

(4) *Different classes of light chains possess different sets of V regions, while the different classes of heavy chains probably share one set of V regions.* The first part of this proposition follows clearly from the available sequences of κ and λ chains in both man and mouse. The sharing of a single V set by all heavy chains is suggested by four lines of evidence. (a) The sequences available for γ and μ chains from human myeloma proteins (*e.g.* Press and Hogg).[145] (b) Cases of human myeloma in which the serum contains both IgM and IgG homogeneous proteins, apparently differing only in the C region of their heavy chains.[146,147] (c) The sharing of idiotypic determinants between rabbit IgM and IgG antibodies, produced in sequence to the one antigen.[148] (d) The sharing and parallel expression of V region allotypic determinants by γ, μ, α and ϵ chains in the rabbit (*e.g.* Kindt and Todd).[149]

(5) *Within a set of V regions there are subsets (called "subgroups") such that any two regions within a subgroup show greater resemblances than any two regions from different subgroups.* This phenomenon is best defined for human κ chains, showing three subgroups (Fig. 8.6). Data summarized by Hood and Talmage[135a] showed that V regions within a subgroup were identical for 73–87% of their sequences, whereas those from different subgroups had 51–68% identity. The subgroups are also

---

* Because of minor differences in length aligning requires the introduction of gaps and insertions to maintain homology. This introduces some difficulty in the numbering of residues. Numberings used in this review are from sequences currently used as prototypes, but understandably these change from time to time.

	10			
Protein Roy, subgroup I	Asp-Ile-Glu-Met-Thr-Gln-Ser-Pro-Ser-Ser-Leu-Ser-Ser-Ala-Ser-Val-Gly-Asp-Arg-Val-			
Protein Aq, subgroup I	Asp-Ile-Gln-Met-Thr-Gln-Ser-Pro-Ser-Ser-Leu-Ser-Ser-Ala-Ser-Val-Gly-Asp-Arg-Val-			
Protein Ti, subgroup III	Glu-Ile-Val-Leu-Thr-Gln-Ser-Pro-Gly-Thr-Leu-Ser-Leu-Ser-Pro-Gly-Asp-Arg-Val-			

	20	30	30a	31	40
Roy	Thr-Ile-Thr-Cys-Gln-Ala-Ser-Gln-Asp-Ile-Ser——Ile-Phe-Leu-Asn-Trp-Tyr-Gln-Gln-Lys-Pro-Gly-Lys-				
Aq	Thr-Ile-Thr-Cys-Gln-Ala-Ser-Gln-Asp-Ile-Asn——His-Tyr-Leu-Asn-Trp-Tyr-Gln-Gln-Gly-Pro-Lys-Lys-				
Ti	Thr-Leu-Ser-Cys-Arg-Ala-Ser-Gln-Ser-Val-Ser-Asn-Ser-Phe-Leu-Ala-Trp-Tyr-Gln-Lys-Pro-Gly-Gln-				

	60	70
Roy	Ala-Pro-Lys-Leu-Leu-Ile-Tyr-Asp-Ala-Ser-Lys-Leu-Glu-Ala-Gly-Val-Pro-Ser-Arg-Phe-Ser-Gly-Phe-Gly-	
Aq	Ala-Pro-Lys-Leu-Ile-Tyr-Asp-Ala-Ser-Asn-Leu-Glu-Thr-Gly-Val-Pro-Ser-Arg-Phe-Ser-Gly-Ser-Gly-	
Ti	Ala-Pro-Arg-Leu-Ile-Met-Tyr-Val-Ala-Ser-Arg-Ala-Thr-Gly-Ile-Pro-Asp-Arg-Phe-Ser-Gly-Ser-Gly-	

	80	90
Roy	Ser-Gly-Thr-Asp-Phe-Thr-Phe-Thr-Ile-Ser-Ser-Leu-Gln-Pro-Glu-Asp-Ile-Ala-Thr-Tyr-Tyr-Cys-Gln-Gln-	
Aq	Phe-Gly-Thr-Asp-Phe-Thr-Phe-Thr-Ile-Ser-Ser-Leu-Gln-Pro-Glu-Asp-Ile-Ala-Thr-Tyr-Tyr-Cys-Gln-Gln-	
Ti	Ser-Gly-Thr-Asp-Phe-Thr-Leu-Thr-Ile-Ser-Arg-Leu-Glu-Pro-Glu-Asp-Phe-Ala-Val-Tyr-Tyr-Cys-Gln-Gln-	

	100	108
Roy	Phe-Asp-Asn-Leu-Pro-Leu-Thr-Phe-Gly-Gly-Gly-Thr-Lys-Val-Asp-Thr-Lys-Arg	
Aq	Tyr-Asp-Thr-Leu-Pro-Arg-Thr-Phe-Gly-Gln-Gly-Thr-Lys-Leu-Glu-Ile-Lys-Arg	
Ti	Tyr-Gly-Ser-Ser-Pro-Ser-Thr-Phe-Gly-Gln-Gly-Thr-Lys-Val-Glu-Leu-Lys-Arg	

FIG. 8.6. V regions of three human κ chains, illustrating homologies occurring within a subgroup and between subgroups. All the chains were from Bence Jones proteins. The names assigned to individual proteins derive usually from the name of the patient. Data from Hilschmann,[140] Titani et al.,[141] Suter et al.,[142] Milstein and Pink.[14]

distinguishable by differences in length. V regions of other chains are similarly divisible: thus at the time of writing five subgroups have been defined for human λ chains and four for human heavy chains.

At least two and possibly three of the κ chain subgroups are present regularly in the immunoglobulins of individual normal men, so they cannot be allelic in nature (Milstein et al., 1969[149a]). A point of great theoretical interest is that they appear to set a *minimum* number for the germ-line genes encoding the given set of V regions, for it is difficult to imagine any somatic mutational mechanism regularly yielding the observed inter-subgroup differences.

(6) *Within the best studied sets of V regions there are areas ("hot-spots") of high variability.* This is well brought out in the data for human light chains summarized by Wu and Kabat.[133] Whether one considers a whole set or merely a subgroup three stretches of high variability are discernible: residues 24–34, 50–56 and 89–97. The first and third of these begin after the invariant $\frac{1}{2}$-Cys residues of the intrachain SS loop, and so must be adjacent in the intact molecule. Clearly such a patch of high variability is a good candidate for a contribution by the light chain to the antibody site.

A hypervariable region has also been described in residues 31–37 of human heavy chains, the $\frac{1}{2}$-Cys of the intrachain SS being at position 22.[150]

(7) *The V region of a set, or at least of a subgroup, can share allotypic characteristics which are inherited as though controlled by a single locus.* This puzzling observation is best documented for the V regions of rabbit heavy chains. The *a* locus determines three allotypic variants (a1, a2, a3) which appear to be expressed in parallel, according to Mendelian laws, in at least 80% of the heavy chains.[151] These variants are defined by antisera raised in the same species. The lack of rabbit myeloma proteins severely impedes their further characterization, but Porter's group have tackled the problem by sequencing the vastly heterogeneous normal heavy chains with conventional techniques, and defining from the data "average" or "predominant" sequences. The results suggest that the allotypic variants differ in multiple associated residues scattered throughout the V region—for example in some 13 residues in the first 34.[31] (These 13 must represent relatively invariant residues in the V region. A reminder that one is coping simultaneously with allotypic and idiotypic variation gives an idea of the difficulties.) Such differences are reminiscent of those occurring among V subgroups, which we have seen are not expressed as alleles, and are much more extensive than known allotypic differences in C regions. So one possibility which must be entertained is that the *a* locus is a regulatory gene drastically affecting the quantitative expression of V subgroups. But it would be surprising to find such a regulatory gene tightly linked to the structural genes for C regions, as described under the next proposition. If on the other hand the *a* locus represents structural V

genes we have a strong argument in favour of only a limited number of such genes being transmitted in the germ line.

More recently allotypic variation has been described in the V regions of mouse κ chains.[32] At least 10% of the chains were involved. Further details will be of great interest.

(8) *Given that there are separate genes for the V and C regions, then at least for rabbit γ chains the linkage of V and C appears to be strong in germinal transmission, and strong or absolute during somatic development.* In addition to the V alleles discussed above, two sets of C alleles are known for rabbit γ chains: A11–12 associated with a Met-Thr interchange in the hinge region, and A14–15 associated with a Thr-Ala interchange in Fc. Either of these sets can be used in conjunction with the a1–3 set to study the linkage of V and C genes. Taking first the question of somatic linkage, Kindt *et al.*[152] found that individual γ chains from doubly heterozygous rabbits all bore an *a* allele inherited from the same parent as was the A11–12 allele: so the V and C regions were apparently always encoded from the one chromosome, with no evidence of crossing over between these chromosomes after numerous mitoses. A similar finding was reported by Landucci Tosi *et al.*[153]

Breeding experiments have shown no V–C crossovers during germinal transmission apart from one apparent recombinant among 151 progeny of double backcross matings.[154]

(9) *Within a single immunoglobulin molecule the V regions of the heavy and light chains show no more resemblance than do the V regions of heavy and light chains from different molecules.* This generalization must be regarded with some reserve as it derives from data on only two myeloma proteins.[155,156]

(10) *In a subject heterozygous at an immunoglobulin locus the immunoglobulin secreted by a single cell derives from only one of the alleles at that locus.* This phenomenon ("allelic exclusion") has been defined in myeloma proteins, where the product of a large monoclonal line can be readily examined, and in normal immunoglobulin-producing cells by immunofluorescent and other techniques. For example in rabbits it has been demonstrated for allotypes of both the heavy chain V regions and the light chain C regions.[157,158] Allelic exclusion is well known for proteins whose structural genes are on the X chromosome in female mammals, but no example apart from immunoglobulins is known to involve proteins encoded from autosomes: haemoglobulins, for example, do not exhibit allelic exclusion.

(11) *There is evidence that the immunoglobulin secreted by a cell line can exhibit a change in the C region of its heavy chain without other alteration.* This statement has been put in a general form but is in fact based on preliminary knowledge of only one switch, from IgM to IgG. During the course of immunization a switch in the predominant antibody from IgM to IgG is well known. This can take place within colonies apparently derived from single cells,[159] and at the time of

switching single cells might produce both classes.[160] A notable report from Oudin and Michel[148] indicated that the sequential IgM and IgG antibody produced by an individual rabbit can exhibit common idiotypic determinants: i.e. that the V regions of their chains are similar or identical. Reason for suspecting that such antibodies might indeed differ only in the C regions of their heavy chains comes from the probability of such being the case for the IgM and IgG secreted simultaneously by some human myelomas, each presumably representing a single neoplastic clone.[146,147] In the case described by Wang et al. the IgM and IgG were produced in different cells. Confirmation and generalization of these findings would obviously have profound implications.

No existing theory encompasses propositions (1)–(11). They provide a rich basis for discussion and argument, as a glance at the literature will confirm. It is likely that some undescribed genetic mechanism is operative, with further experiment needed to define it.

8.6.2. SPECIFICITY OF CHAIN PAIRING

The discovery that immunoglobulins are multichain proteins, with the antibody site probably shared between the heavy and light chains, appeared to go some way towards explaining the provision of sufficient structural variety for a seemingly indefinite numer of antibody specificities: because variety could be generated by different heavy-light pairings as well as by variation within each chain. It was found indeed that virtually any heavy and light chain could pair.[161] However it has become apparent that the pairings allowed in vivo are considerably restricted, so that m heavy chains and n light chains can form considerably fewer than m × n different whole molecules. The experiments leading to this conclusion were of the following type.

Two myeloma proteins, $H_A L_A$ and $H_B L_B$, are split into their constituent chains by reduction and alkylation, transfer into a dissociating solution, and mass separation of the chains. The mixing of solutions of H and L followed by transfer to a near-physiological pH and ionic strength will now produce $H_2 L_2$ recombinants, homologous or heterologous according to the solutions used. The combination depends upon non-covalent interchain bonds, the re-forming of interchain disulphide bonds having been prevented by alkylation of the thiol groups. Grey and Mannik[162] performed experiments in which homologous and heterologous light chains were allowed to compete for a molar deficiency of heavy chain: $H_A + (L_A + L_B) \rightarrow H_A L_A + H_B L_B + L_A + L_B$. In this and subsequent work[163] a large number of experiments revealed that in only about 10% of cases does L_B compete on approximately equal terms with L_A for union with H_A. So the occurrence of an H-L pair in vivo implies that this heavy and light chain have a much higher than average affinity for each other.

Isotypes are one determinant of specificity in the heavy-light interaction. For example all myeloma proteins of subclass IgG4 reported to date contain κ chains, and a γ4 chain was found by Grey and Mannik to have little tendency to combine with a λ chain. However known isotypic differences cannot account for all the preferential interactions, as was shown by the recombinants formed among five IgG1K proteins examined by Grey and Mannik. Hence the V regions probably help dictate the specificity of chain pairing.

It is not surprising that heavy and light chains vary in their affinities for each other, but the occurrence *in vivo* only of high affinity pairings is unexplained. There is nothing to suggest that low affinity pairings give a notably unstable molecule. Thus hybrid pairings of rabbit heavy and light chains appeared stable at neutral pH, even in the absence of inter-chain disulphide bonds.[164] There might be some somatic evolutionary process favouring only those cells which happen to produce pairings of high affinity, although it is not apparent how the discrimination could be made. The alternative must be some intracellular process selecting pairs, but again there is no hint as to how this might operate.

8.7. ABNORMAL IMMUNOGLOBULINS

The term abnormal immunoglobulins is used loosely to denote any change in the immunoglobulin population not explicable in terms of a physiological immune response. The presentation can be as a physico-chemical abnormality, for example a serum electrophoretic pattern of the type seen in myeloma; or as an abnormal antibody specificity, seen in the various autoantibodies; or as an abnormal sequel to antibody-antigen reaction, seen in antibody-mediated hypersensitivities. Clearly the underlying cause can reside either in the immune system itself or in the nature of the stimulus presented to it. Frequently it cannot be located.

Here we must restrict discussion to the *homogeneous immunoglobulins* exemplified by myeloma proteins and conforming to the same criteria of homogeneity. These are often called "monoclonal" immunoglobulins, a tendentious but perhaps accurate term, or "paraproteins", which at least is brief. They appear to arise usually, but not always, from frankly neoplastic lymphoid tissue. Whether they are abnormal intrinsically or merely with regard to their presenting concentrations is a problem to which categorical answers cannot be given, largely because of the variation in normal immunoglobulins and the difficulty of characterizing trace constituents. However, as pointed out before, the fact that structural features turning up in myeloma proteins have so often been found subsequently among normal immunoglobulins suggests that the abnormality is usually one of over-production. Likely exceptions are proteins with gross deletions in their primary structure, such as occur in human heavy chain disease (see below) and some mouse plasma-cytomas.[165]

Homogeneous immunoglobulins are generally detected upon electrophoresis of the serum proteins (Fig. 8.7). Most clinical biochemists use electrophoresis in paper, cellulose acetate or agar, giving as a rule a clear distinction between the diffuse normal proteins and any superimposed homogeneous component. But some cases occur in which the distinction cannot be made confidently, especially it would seem when

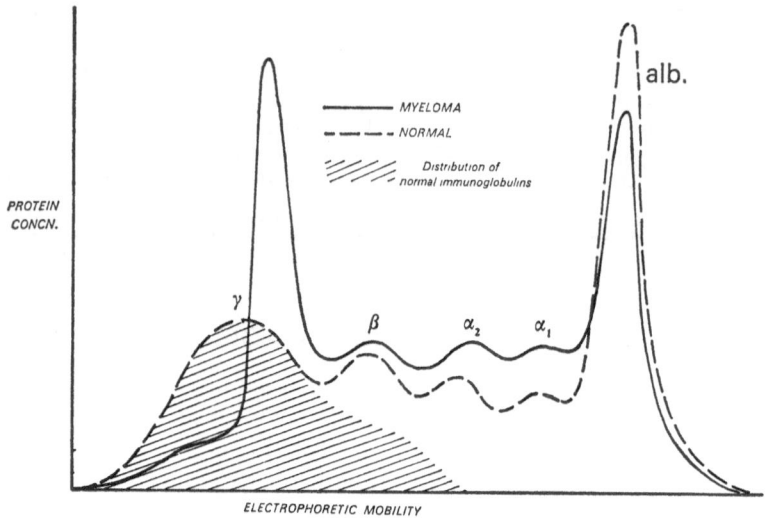

FIG. 8.7. Electrophoretic pattern from a case of myeloma, showing a homogeneous immunoglobulin of fast $\gamma(\gamma1)$ mobility. In the presence of such a myeloma protein the other immunoglobulins are usually markedly reduced, and here this is apparent from the paucity of protein of $\gamma2$ mobility. Other abnormalities apparent are the raised α-globulins and decreased albumin, common non-specific changes in disease. Peaks due to a homogeneous immunoglobulin can occur anywhere along the hatched normal distribution.

a strong antibody response has happened to raise the immunoglobulin concentration over a narrow electrophoretic range. More discriminating electrophoretic techniques, such as in polyacrylamide gel, often bring out minor degrees of heterogeneity even in myeloma proteins, and so paradoxically can give less clear pictures that the simpler methods.

It is clear then that further examination of an immunoglobulin, preferably in a purified state, is required to establish whether it is as homogeneous as a typical myeloma protein. Antigenic characterization should reveal only one isotype each for the heavy and light chains. If allotypic characterization is available allelic exclusion should be sought.

Other useful characterizations can include electrophoresis of the separated peptide chains and analytical centrifugation. Many anomalies in the literature have undoubtedly sprung from too easy an acceptance of a distinction between homogeneous and heterogeneous (monoclonal and polyclonal) based on a simple electrophoretic examination.

Clinical conditions associated with homogeneous immunoglobulins are summarized in Table 8.5. Useful reviews are available from Snapper et al.,[166] Martin,[167] Osserman and Takatsuki,[168] and Michaux and Heremans.[169] A provocative discussion of myeloma is provided by Hobbs.[170] The lymphomas are clearly named inconsistently on a mixed

TABLE 8.5.
Homogeneous immunoglobulins in man

A. *Arising from neoplastic tissue* (*lymphomas*)
  1. Myeloma (plasmacytoma)
  2. Macroglobulinaemia (of Waldenström)
  3. Heavy chain diseases
  4. Atypical cases

B. *Apparently not arising from neoplastic tissue*
  1. With no other detectable abnormality
  2. As an occasional concomitant of:
      Various neoplasms
      Chronic infection
      Hepatic fibrosis
      and others
  3. As a regular concomitant of:
      Cold haemagglutination syndrome
      Papular mucinosis

histological and biochemical basis but no reform of the nomenclature can be expected at this stage.

8.7.1. HOMOGENEOUS IMMUNOGLOBULINS ARISING FROM NEOPLASTIC TISSUE

About 80% of patients diagnosed histologically as having myeloma show an obvious serum myeloma protein. Of these about two thirds are IgG, one third IgA, while accounting for less than 2% are the rare instances of IgD, IgE or two coincident proteins. The myeloma protein sometimes exceeds in concentration the sum of all other proteins present and occasionally reaches 8–10 g/100 ml. Critical examination of a typical IgG or IgA protein will usually reveal heterogeneity. In the former this consists generally of a small degree of charge heterogeneity, and is observed also in mouse myeloma IgG, where it is due to an increasing negative charge acquired after synthesis.[171] The heterogeneity of IgA myeloma proteins arises from a variable degree of polymerization.

About 50% of the patients have Bence Jones protein detectable in their unconcentrated urine by a heat test, in about one third of these without an accompanying serum myeloma protein. The Bence Jones proteins are free light chains, identical to those of the serum protein if present, synthesized in excess of the heavy chains.

The heat test for Bence Jones protein depends upon its having two unusual thermosolubility characteristics: it is precipitated at a relatively low temperature (often beginning at 45°), and the heat-denatured protein then shows an unusually high solubility, especially as the temperature is raised towards boiling point. It is often said to dissolve on boiling and reprecipitate on cooling, but such obligingly obvious behaviour is rare. The critical dependence of the initial precipitation upon pH is described in detail by Putman et al.,[171] and the test suggested in Table 8.6 is based largely upon their work. The test is useful in that

TABLE 8.6.
Heat test for Bence Jones protein

1. Screen urine samples using a general protein precipitant, e.g. trichloro-acetic acid to 8%. In the absence of frank turbidity the test is negative.
2. To 8 ml urine add 2 ml 2 M sodium acetate buffer, pH 4.9. Clarify by centrifuging.
3. Heat at 55 ± 2° for 30 min. If no precipitation occurs the test is negative.
4. Suspend the precipitate in a 1 in 10 dilution of the acetate buffer. Place in a boiling water bath for 5 min. A positive result is provided by any of the following:
   (a) The precipitate obviously dissolving.
   (b) The supernatant solution becoming cloudy as it cools.
   (c) In the absence of (a) and/or (b), 2 ml of supernatant, cleared by centrifuging when cool, plus 0.5 ml trichloroacetic acid, giving a turbidity greater than that given by a solution of plasma albumin at 0.1 mg/ml.

a positive result with unconcentrated urine, indicative of more than about 0.4 mg/ml, is exceedingly rare in the absence of either myeloma or macroglobulinaemia. More sensitive tests reveal Bence Jones protein more frequently, but with an increasingly uncertain implication, until at a level of about 5 mg per day it is found in normal urine.[172]

An uncertain but small proportion of patients with myeloma exhibit a third protein abnormality, amyloidosis. It might be particularly common in cases with overt Bence Jones proteinuria in the absence of a serum myeloma protein. From the other viewpoint myeloma might underly many cases labelled primary amyloidosis. Like so many substances embedded in connective tissue amyloid has proved extremely refractory to extraction and characterization, but a recent report suggests that its protein component is related to immunoglobulin light chains.[173]

Most series contain a few patients with myeloma who exhibit none of the above protein abnormalities. If the histology is typical it is likely that the malignant plasma cells are exporting protein of some sort which has gone undetected.

The lymphomas secreting immunoglobulins G, A, D and E are reasonably lumped together under the one heading of myelomas, as they all exhibit broadly similar features: multiple plasmacytic tumours causing osteolytic lesions, spreading diffusely throughout the bone marrow, and accompanied often by anaemia, renal failure and recurrent infection.

However the lymphoma secreting IgM is in its typical form quite different. The predominant cell is lymphocytic rather than plasmacytic. It is diffusely distributed but tends especially to involve lymph nodes, liver and spleen and rarely causes osteolytic lesions. In the absence of the homogeneous IgM it would be regarded as an unexceptional lymphosarcoma. Sometimes there are manifestations attributable to the high plasma viscosity caused by the high concentration of the IgM: retinopathy, a haemorrhagic tendency and Raynaud's phenomenon. The urine gives a positive heat test for Bence Jones protein in perhaps 10% of cases, although the gross proteinuria sometimes seen in myeloma is not described.

Heavy chain diseases are a rare, histologically disparate group characterized by the appearance in extracellular fluids of protein related to free heavy chains. The few proteins well characterized have revealed an interesting internal deletion.

The first of the group to be recognized was γ chain disease.[174] Only about a dozen cases have since been identified. The histology resembles that of macroglobulinaemia. The serum protein has a molecular weight of 50 000–80 000, antigenically resembles Fc of IgG, and is cleaved into identical halves by sequential treatment with a thiol and denaturant. Sometimes it appears in appreciable quantities in the urine. It does not exhibit the thermosolubility properties of Bence Jones protein. The simultaneous secretion of light chains by tumour has not been observed. Three proteins studied in detail (summarized by Franklin and Frangione)[175] have been revealed as heavy chains with a large internal deletion in Fd: after a short N-terminal sequence a deletion of 100–200 residues involved both V and C regions, to be followed by a resumption of normal γ chain sequence up to the C-terminus.

α-Chain disease[176] seems to occur mainly in underprivileged populations in the Middle East. It presents as a diffuse, mostly plasmacytic, lymphoma of the small gut with an accompanying malabsorption. The serum shows a protein about two thirds the molecular weight of α chains, antigenically resembling Fc of IgA. A deletion analogous to that found in γ chain disease could therefore be present. The serum proteins have been polydisperse due to variable polymerization. Small amounts have been found in the urine. Simultaneous secretion of light chains has not been found.

One case of μ-chain disease has been reported.[177] A tumour with the histology of chronic lymphocytic leukaemia was secreting μ-chain-like material of molecular weight 55 000 as well as free κ chains.

It will readily be appreciated that many cases of protein-secreting lymphoma do not fit the typical descriptions above. Thus occasionally a patient appears with a histological picture of myeloma but a biochemical picture of macroglobulinaemia, or vice versa. Protein variants also turn up sporadically, such as polymerized IgG, monomeric (7 S) IgM, quadrimeric light chains persisting in the serum, and fragmented or deleted proteins other than those described.

Some attention is now being given to lymphomas with surface immunoglobulins but no appreciable export. Some cases of chronic lymphocytic leukaemia and Burkitt's tumour are in this category.[178] The homogeneity and other characteristics of these immunoglobulins remain to be assessed.

8.7.2. HOMOGENEOUS IMMUNOGLOBULINS APPARENTLY NOT ARISING FROM NEOPLASTIC TISSUE

The association of homogeneous immunoglobulins with no detectable abnormality or with the occasional diseases listed under heading B.2 in Table 8.5 is difficult to assess. Conflicting accounts have come from different centres and the criteria for homogeneity have varied. However experimental findings in rabbits[113] leave little doubt that under certain circumstances normal lymphoid tissue can yield an antibody population apparently as homogeneous as a myeloma protein.

Homogeneous immunoglobulins are described occasionally in apparently healthy subjects, especially in the elderly and in close relatives of patients with macroglobulinaemia.[180] Such subjects must be observed carefully for the appearance of a protein-secreting lymphoma. As regards their immunoglobulins ominous signs would be the appearance of overt Bence Jones proteinuria, depletion of the normal immunoglobulins, or a rising concentration of the homogeneous component.

Descriptions of patients with various diseases associated exceptionally with homogeneous immunoglobulins are to be found in the review by Michaux and Heremans.[169] These patients should also of course be observed for the appearance of a lymphoma.

The idiopathic cold haemagglutination syndrome is associated with an antibody (cold agglutinin) of anti-I specificity, class IgMK, and electrophoretically homogeneous.[39] Otherwise the protein varies from one patient to another. Frequently there is lymphocytic hyperplasia, sometimes infiltrative, so that the syndrome shades into macroglobulinaemia of Waldenström.

Papular mucinosis is a very rare, chronic, apparently benign skin disorder associated with a highly basic homogeneous IgG to which an antibody activity has not yet been assigned.[181]

8.8. CONCLUSION

The study of immunoglobulins has advanced rapidly over the last fifteen years, but there is now a pause while two difficult problems are confronted: the mechanism of their genetic control, and the nature of immunoglobulins on cell surfaces. The surface immunoglobulin proving particularly intractable is the hypothetical IgX on the surface of T lymphocytes, the thymus-processed line effecting cell-mediated immunity. Always a potential source of solutions to such problems are observations arising in clinical medicine.

REFERENCES

1. Committees on Nomenclature for Human Immunoglobulins. *Bull. Wld. Hlth. Org.* (1964) **30,** 447; (1965) **33,** 721; (1966) **35,** 953; (1968) **38,** 151; (1969) **41,** 975
2. T. N. HARRIS, K. HUMMELER and S. HARRIS, *J. exp. Med.,* **123,** 161 (1966)
3. S. AVRAMEAS and E. H. LEDUC, *J. exp. Med.,* **131,** 1137 (1970)
4. R. T. SMITH, P. A. MIESCHER and R. A. GOOD, (eds.) *Phylogeny of Immunity.* University of Florida Press, Gainesville, Fla. (1966)
5. H. M. GREY, *Advances Immun.,* **10,** 51 (1969)
6. E. HABER, *Proc. natn. Acad. Sci., U.S.A.,* **52,** 1099 (1964)
7. M. E. NOELKEN and C. TANFORD, *J. biol. Chem.,* **239,** 1828 (1964)
8. R. R. PORTER, in *Basic Aspects of Neoplastic Disease,* p. 177. eds., A. Gellhorn and E. Hirschberg, New York, Columbia U.P. (1962)
9. N. HILSCHMANN and L. C. CRAIG, *Proc. natn. Acad. Sci., U.S.A.,* **53,** 1403 (1965)
10. G. M. EDELMAN and W. E. GALL, *Ann. Rev. Biochem.,* **38,** 415 (1969)
11. G. M. EDELMAN, *J. Am. chem. Soc.,* **81,** 3155 (1959)
12. J. B. FLEISCHMAN, R. H. PAIN and R. R. PORTER, *Archs. Biochem.* Suppl. 1, 174–80 (1962)
13. C. A. ABEL and H. M. GREY, *Biochemistry,* (*Wash.*), **7,** 2682 (1968)
14. C. MILSTEIN and J. R. PINK, *Progr. Biophys.,* **21,** 209 (1970)
15. S. J. SINGER and R. F. DOOLITTLE, *Science, N.Y.,* **153,** 13 (1966)
16. R. L. HILL, R. DELANEY, R. E. FELLOWS Jr., *et al. Proc. natn. Acad. Sci., U.S.A.,* **56,** 1762 (1966)
17. J. W. GOODMAN and D. GROSS, *J. Immun.,* **90,** 865 (1963)
18. M. E. NOELKEN, C. A. NELSON, C. E. BUCKLEY *et al., J. biol. Chem.,* **240,** 218 (1965)
19. J. FUJIO and F. KARUSH, *Biochemistry,* (*Wash.*), **5,** 1856 (1966)
20. A. NISONOFF, F. C. WISSLER, L. N. LIPMAN and D. L. WOERNLEY, *Arch. Biochem. Biophys.,* **89,** 230 (1960)
21. E. A. KABAT, *Proc. natn. Acad. Sci., U.S.A.,* **58,** 229 (1967)
22. L. HOOD, K. EICHMANN, H. LACHLAND, R. M. KRAUSE and J. J. OHMS, *Nature, Lond.,* **228,** 1040 (1970)
23. J. B. NATVIG, H. G. KUNKEL, W. J. YOUNT and J. C. NIELSEN, *J. exp. Med.,* **128,** 763 (1968)

24. A. S. KELUS and P. G. H. GELL, *Progr. Allergy*, **11**, 141 (1967)
25. L. A. HERZENBERG, H. O. MCDEVITT and LEONORE A. HERZENBERG, *Ann. Rev. Genetics.* **2**, 209 (1968)
26. J. OUDIN, *Proc. R. Soc.*, **166**, 207 (1966)
27. B. FRANGIONE, C. MILSTEIN and E. C. FRANKLIN, *Nature, Lond.*, **221**, 149 (1969)
28. H. G. KUNKEL and E. M. TAN, *Advances Immun.*, **4**, 351 (1964)
29. J. C. ALLEN and H. G. KUNKEL, *J. clin. Invest.*, **45**, 29 (1966)
30. R. GRUBB, *Acta path. microbiol. Scand.*, **39**, 195 (1956)
31. J. M. WILKINSON, *Biochem. J.*, **112**, 173 (1969)
32. G. M. EDELMAN and P. D. GOTTLIEB, *Proc. natn. Acad. Sci.*, *U.S.A.*, **67**, 1192 (1970)
33. C. BAGLIONI, ZONTA L. ALESCIO, D. CIOLI *et al.*, *Science, N.Y.*, **152**, 1517 (1966)
34. S. COHEN and C. MILSTEIN, *Advances Immun.*, **7**, 1 (1967)
35. J. B. NATVIG and H. G. KUNKEL, *Ser. Haemat.*, **1**, 66 (1968)
36. D. EIN, *Proc. natn. Acad. Sci.*, *U.S.A.*, **60**, 982 (1968)
37. R. W. NEWCOMB and K. ISHIZAKA, *J. Immun.*, **105**, 85 (1970)
38. N. R. KLINMAN, J. H. ROCKEY and F. KARUSH, *Int. J. Immunochem.*, **2**, 51 (1965)
39. M. HARBOE, *Ser. Haematologica*, **4**, 65 (1965)
40. T. FREEMAN, *Ser. Haemat.*, **4**, 76 (1965)
41. J. W. UHR and M. S. FINKELSTEIN, *J. exp. Med.*, **117**, 457 (1963)
42. T. A. WALDMANN and W. STROBER, *Progr. Allergy*, **13**, 1 (1969)
43. W. D. TERRY and T. A. WALDMAN, *J. clin. Invest.*, **49**, 673 (1970)
44. H. L. SPIEGELBERG and W. O. WEIGLE, *J. exp. Med.*, **121**, 323
45. J. L. FAHEY and A. G. ROBINSON, *J. exp. Med.*, **118**, 845 (1963)
46. F. W. R. BRAMBELL, *Lancet*, **2**, 1087 (1966)
47. C. L. CHRISTIAN, *Immunology*, **18**, 457 (1970)
48. J. V. DACIE, The Haemolytic Anaemias, Part 2. 2nd Edit. London, Churchill. (1969)
49. P. G. QUIE, R. P. MESSNER and R. C. WILLIAMS Jr., *J. exp. Med.*, **128**, 553 (1968)
50. H. J. MÜLLER-EBERHARD, *Advances Immun.*, **8**, 11 (1968)
51. H. HUBER, S. D. DOUGLAS and H. H. FUDENBERG, *Immunology*, **17**, 7 (1969)
52. R. P. MESSNER and J. JELINEK, *J. clin. Invest.*, **49**, 2165 (1970)
53. F. W. R. BRAMBELL, *Biol. Rev.*, **33**, 448 (1958)
54. F. W. BRAMBELL, W. A. HAMMINGS, C. L. OAKLEY and R. R. PORTER, *Proc. R. Soc.*, **151**, 478 (1960)
55. J. R. CANN, *J. Am. chem. Soc.*, **75**, 4213 (1953)
56. S. KOCHWA, E. SMITH, M. BROWNELL *et al.*, *Biochemistry*, (*Wash.*), **5**, 277 (1966)
57. E. J. COHN, L. E. STRONG, W. L. HUGHES, D. J. MULFORD, J. N. ASHWORTH, M. MELIN and H. L. TAYLOR, *J. Am. chem. Soc.*, **68**, 459 (1946)
58. K. ISHIZAKA, T. ISHIZAKA and J. BANOVITZ, *J. Immun.*, **94**, 825 (1965)
59. R. A. PHELPS and F. W. PUTNAM, in *The Plasma Proteins*, Vol. I. ed. by F. W. Putnam. New York, Academic Press. (1958)
60. H. M. GREY and H. G. KUNKEL, *J. exp. Med.*, **120**, 253 (1964)
61. S. COHEN and R. R. PORTER, *Advances Immun.*, **4**, 287 (1964)
61a. S. COHEN and R. R. PORTER *Biochem. J.*, **90**, 278 (1964)
62. J. SJÖQUIST and M. H. VAUGHAN Jr., *J. molec. Biol.*, **20**, 527 (1966)
63. A. FEINSTEIN, *Nature, Lond.*, **210**, 135 (1966)
64. Z. L. AWDEH, A. R. WILLIAMSON and B. A. ASKONAS, *Nature*, **219**, 66 (1968)
65. N. M. GREEN, *Advances Immun.*, **11**, 1 (1969)
65a. A. FEINSTEIN, and A. J. ROWE, *Nature, Lond.*, **205**, 147 (1965)
66. K. J. DORRINGTON and C. TANFORD, *Advances Immun.*, **12**, 333 (1971)

67. R. E. CATHOU and T. C. WERNER, *Biochemistry, N.Y.*, **9**, 3149 (1970)
67a. G. T. STEVENSON and K. J. DORRINGTON, *Biochem. J.*, **118**, 703 (1970)
68. E. C. FRANKLIN, *J. Immun.*, **85**, 138 (1960)
69. N. F. ROTHFIELD, B. FRANGIONE and E. C. FRANKLIN, *J. clin. Invest.*, **44**, 62 (1965)
70. A. HUNTER, A. FEINSTEIN and R. R. COOMBS, *Immunology*, **15**, 381 (1968)
71. W. F. BARTH, R. D. WOCHNER, T. A. WALDMANN and J. H. FAHEY, *J. clin. Invest.*, **43**, 1036 (1964)
72. T. BOROS and H. J. RAPP, *J. Immun.*, **95**, 559 (1965)
73. J. R. HOBBS, R. D. MILNER and P. J. WATT, *Brit. Med. J.*, **4**, 583 (1967)
74. R. J. PORTER, *Proc. Soc. exp. Biol. Med.*, **121**, 107 (1966)
75. S. V. BOYDEN, *Advances Immun.*, **5**, 1 (1966)
76. B. PERNIS, L. FORNI and L. AMANTE, *J. exp. Med.*, **132**, 1001 (1970)
77. F. MILLER and H. METZGER, *J. biol. Chem.*, **240**, 2740 (1965)
78. C. A. ABEL and H. M. GREY, *Science, N.Y.*, **156**, 1609 (1967)
78a. S. L. MORRISON and M. E. KOSHLAND, *Proc. natn. Acad. Sci., U.S.A.*, **69**, 124 (1972)
79. P. C. CHAN and H. F. DEUTSCH, *J. Immun.*, **85**, 37 (1960)
80. J. H. ROCKEY and H. G. KUNKEL, *Proc. Soc. exp. Biol. Med.*, **110**, 101 (1962)
81. F. MILLER and H. METZGER, *J. biol. Chem.*, **241**, 1732 (1966)
82. C. MIHAESCO and M. SELIGMANN, *J. exp. Med.*, **127**, 431 (1968)
83. K. ONOUE, T. KISHIMOTO and Y. YAMAMURA, *J. Immun.*, **100**, 238 (1968)
84. E. MERLER, L. KARLIN and S. MATSUMOTO, *J. biol. Chem.*, **243**, 386 (1968)
85. S.-E. SVEHAG, B. BLOTH and M. SELIGMANN, *J. exp. Med.*, **130**, 691 (1969)
86. W. B. CHODIRKER and T. B. TOMASI, *Science, N.Y.*, **142**, 1080 (1963)
87. T. B. TOMASI and J. BIENENSTOCK, *Advances Immun.*, **9**, 1 (1968)
88. J. F. HEREMANS, M. P. HEREMANS and H. E. SCHULTZE, *Clinica. chim. Acta*, **4**, 96 (1959)
89. T. B. TOMASI, E. M. TAN, A. SOLOMON and R. A. PRENDERGAST, *J. exp. Med.*, **121**, 101 (1965)
90. H. M. GREY, C. A. ABEL, W. J. YOUNT and H. G. KUNKEL, *J. exp. Med.*, **128**, 1223 (1968)
91. L. M. JERRY, H. G. KUNKEL and H. M. GREY, *Proc. natn. Acad. Sci., U.S.A.*, **65**, 557 (1970)
92. R. A. THOMPSON, *Nature, Lond.*, **226**, 946 (1970)
93. M. A. SOUTH, M. D. COOPER, F. A. WOLLHEIM, R. HONG and R. A. GOOD, *J. exp. Med.*, **123**, 615 (1966)
94. W. R. BROWN, R. W. NEWCOMB and K. ISHIZAKA, *J. clin. Invest.*, **49**, 1374 (1970)
95. S. FAZEKAS DE ST. GROTH, *Aust. J. exp. Biol. med. Sci.*, **29**, 339 (1951)
96. D. S. ROWE and J. L. FAHEY, *J. exp. Med.*, **121**, 185 (1965)
97. K. ISHIZAKA, T. ISHIZAKA and M. M. HORNBROOK, *J. Immun.*, **97**, 75 (1966)
98. S. G. JOHANSSON and H. BENNICH, *Nobel Symposium*, **3**, 193 (1967)
99. H. J. HUMPREY and R. G. WHITE, *Immunology for Students of Medicine*. 3rd edit. Blackwell Scientific Publications, Oxford. (1970)
100. K. ISHIZAKA and T. ISHIZAKA, *J. Allergy*, **42**, 330 (1968)
101. D. R. STANWORTH, J. H. HUMPREY, H. BENNICH et al., *Lancet.*, **2**, 17 (1968)
102. K. ISHIZAKA, T. ISHIZAKA and A. E. MENZEL, *J. Immun.*, **99**, 610 (1967)
103. J. PORATH and N. UI, *Biochim. Biophys. Acta*, **90**, 324 (1964)
104. J. J. MARCHALONIS, *Transplantation Proceedings*, **2**, 3 (1970)
105. M. W. STEWARD, C. W. TODD, T. J. KINDT et al., *Immunochemistry*, **6**, 649 (1969)
106. G. T. STEVENSON, *J. clin. Invest.*, **41**, 1190 (1962)
107. I. BERGGARD and G. M. EDELMAN, *Proc. natl. Acad. Sci. U.S.*, **49**, 330 (1963)
108. M. W. TURNER and D. S. ROWE, *Nature, Lond.*, **210**, 130 (1966)

109. J. H. VAUGHAN, R. F. JACOX and B. A. GREY, *J. clin. Invest.*, **46**, 266 (1967)
110. F. FRANĔK and I. ŘÍHA, *Immunochemistry*, **1**, 49 (1964)
111. C. A. WILLIAMS and M. W. CHASE, eds., *Methods in Immunology and Immunochemistry*, Vol. I. Academic Press, New York and London, (1967). pp. 335–385
112. S. F. VELICK, C. W. PARKER and H. N. EISEN, *Proc. natn. Acad. Sci., U.S.A.*, **46**, 1470 (1960)
113. D. J. BRAUN, K. EICHMANN and R. M. KRAUSE, *J. exp. Med.*, **129**, 809 (1969)
114. W. J. YOUNT, M. M. DONNER, H. G. KUNKEL *et al.*, *J. exp. Med.*, **127**, 633 (1968)
114a. A. S. KELUS and P. G. H. GELL, *J. exp. Med.*, **127**, 215 (1968)
115. H. DAUGHARTY, J. E. HOPPER, A. B. MACDONALD *et al.*, *J. exp. Med.*, **130**, 1047 (1969)
116. J. ŠTERZL and A. M. SILVERSTEIN, *Advances Immun.*, **6**, 337 (1967)
117. H. N. EISEN and G. W. SISKIND, *Biochemistry, (Wash.)*, **3**, 996 (1964)
118. F. FRANĔK and R. S. NEZLIN, *Folia microbiol. Praha.*, **8**, 128 (1963)
119. G. M. EDELMAN, D. E. OLINS and J. A. GALLY, *Proc. natn. Acad. Sci., U.S.A.*, **50**, 753 (1963)
120. K. LANDSTEINER and J. VAN DER SCHEER, *J. exp. Med.*, **67**, 709 (1938)
121. B. B. LEVINE, *J. exp. Med.*, **117**, 161 (1963)
122. E. A. KABAT, *J. Immun.*, **97**, 1 (1966)
123. J. C. HSIA and L. H. PIETTE, *Arch. Biochem.*, **129**, 296 (1969)
124. C. WARNER and V. SCHUMAKER, *Biochem. biophys. Res. Commun.*, **41**, 225 (1970)
125. M. E. KAPLAN and E. A. KABAT, *J. exp. Med.*, **123**, 1061 (1966)
126. F. KARUSH, *Advances Immun.*, **2**, 1 (1962)
127. S. M. BEISER, G. C. BURKE and S. W. TANENBAWM, *J. molec. Biol.*, **2**, 125 (1960)
128. B. D. DAVIS, R. DULBECCO, H. N. EISEN, H. S. GINSBERG and W. B. WOOD Jr., *Microbiology.* Harper and Row, New York. (1967)
129. C. MORENO and E. A. KABAT, *J. exp. Med.*, **129**, 871 (1969)
130. A. FROESE, *Immunochemistry*, **5**, 253 (1968)
131. R. E. CATHOU and E. HABER, *Biochemistry, (Wash.)*, **6**, 513 (1967)
132. F. M. BURNET, in *Cellular Immunology*, Cambridge University Press, London. (1969)
132a. E. LENNOX and M. COHN, *Ann. Rev. Biochem.*, **36**, 365 (1967)
133. T. T. WU and E. A. KABAT, *J. exp. Med.*, **132**, 211 (1970)
133a. W. D. TERRY, L. E. HOOD and A. G. STEINBERG, *Proc. natn. Acad. Sci., U.S.A.* **63**, 71 (1969)
134. H. G. KUNKEL, J. B. NATVIG and F. G. JOSLIN, *Proc. natn. Acad. Sci., U.S.A.*, **62**, 144 (1969)
135. R. G. MAGE, G. O. YOUNG, J. REJNEK, R. A. REISFELD and E. APPELLA, in *Collog. Protides of the Biological Fluids.* 17—in press. (1969)
135a. L. HOOD and D. W. TALMAGE, *Science, N.Y.*, **168**, 325 (1970)
136. W. J. DREYER and J. C. BENNETT, *Proc. natn. Acad. Sci., U.S.A.*, **54**, 864 (1965)
137. J. B. FLEISCHMAN, *Biochemistry, (Wash.)*, **6**, 1311 (1967)
138. P. M. KNOPF, R. M. PARKHOUSE and E. S. LENNOX, *Proc. natn. Acad. Sci., U.S.A.*, **58**, 2288 (1967)
139. W. D. TERRY and J. OHMS, *Proc. natn. Acad. Sci., U.S.A.*, **66**, 558 (1970)
140. N. HILSCHMANN, *Hoppe-Seyler's Z. physiol. Chem.*, **348**, 1077 (1967)
141. K. TITANI, T. SHINODA and F. W. PUTNAM, *J. biol. Chem.*, **244**, 3550 (1969)
142. L. SUTER, H. BARNIKOL, S. WATANABE and N. HILSCHMANN, *Hoppe-Seyler's Z. physiol. Chem.*, **350**, 275 (1969)
143. O. SMITHIES, *Science, N.Y.*, **157**, 267 (1967)
144. J. A. GALLY and G. M. EDELMAN, *Nature, Lond.*, **227**, 341 (1970)
145. E. M. PRESS and N. M. HOGG, *Biochem. J.*, **117**, 641 (1970)
146. G. M. PENN, H. G. KUNKEL and H. M. GREY, *Fed. Proc.*, **29**, 258 (1970)

147. A. C. WANG, K. S. WILSON, J. E. HOPPER et al., Proc. natn. Acad. Sci., U.S.A., 66, 337 (1970)
148. J. OUDIN and M. MICHEL, J. exp. Med., 130, 619 (1969)
149. T. J. KINDT and C. W. TODD, J. exp. Med., 130, 859 (1969)
149a. C. MILSTEIN, C. P. MILSTEIN and A. FEINSTEIN, Nature, Lond., 221, 151 (1969)
150. J. D. CAPRA, Nature New Biology, 230, 61 (1971)
151. S. DRAY, G. O. YOUNG and A. NISONOFF, Nature, Lond., 199, 52 (1963)
152. T. J. KINDT, W. J. MANDY and C. W. TODD, Biochemistry, (Wash.), 9, 2028 (1970)
153. S. L. TOSI, S. DUBISLI and R. G. MAGE, J. Immun., 104, 647 (1970)
154. R. J. MAGE, G. O. YOUNG-COOPER and C. ALEXANDER, Nature New Biology, 230, 63 (1971)
155. G. M. EDELMAN, B. A. CURRINGHAM, W. E. GALL et al., Proc. natn. Acad. Sci., U.S.A., 63, 78 (1969)
156. H. KOHLER, A. SHIMIZU, C. PAUL and F. W. PUTNAM, Science, N.Y., 169, 56 (1970)
157. B. PERNIS, G. CHIAPPINO, A. S. KELUS et al., J. exp. Med., 122, 853 (1965)
158. J. J. CEBRA, J. E. COLBERG and S. DRAY, J. exp. Med., 123, 547 (1966)
159. J. ŠTERZL and A. NORDIN, Folia Microbiol., 16, 1 (1971)
160. G. J. NOSSAL, A. SZENBERG, G. L. ADA et al., J. exp. Med., 119, 485 (1964)
161. D. E. OLINS and G. M. EDELMAN, J. exp. Med., 119, 789 (1964)
162. H. M. GREY and M. MANNIK, J. exp. Med., 122, 619 (1965)
163. M. MANNICK, Biochemistry, (Wash.), 6, 134 (1967)
164. O. A. ROHOLT, G. RADZIMSKI and D. PRESSMAN, J. exp. Med., 125, 191 (1967)
165. J. F. MUSHINSKI, J. Immun., 106, 41 (1971)
166. SNAPPER, I., TURNER, L. B. and MOSCOVITZ, H. L. (1953). Multiple Myeloma. Grune and Stratton, New York.
167. MARTIN, N. H. (1960). Quart. J. Med. 29, 179.
168. OSSERMAM, E. F. and TAKATSUKI, K. (1963). Medicine (Balt). 42, 357.
169. J. L. MICHAUX and J. F. HERENANS, Amer. J. Med., 46, 562 (1969)
170. J. R. HOBBS, Brit. Med. J., 2, 67 (1971)
171. Z. L. AWDEH, A. R. WILLIAMSON and B. A. ZSKONAS, Biochem. J., 116, 241 (1970)
171a. F. W. PUTNAM, C. W. EASLEY, L. T. LYNN, A. E. RITCHIE and R. A. PHELPS, Arch. Biochem. Biophys., 83, 115 (1959)
172. G. T. STEVENSON, J. clin. Invest., 39, 1192 (1960)
173. G. G. GLENNER, W. TERRY, M. HARADA, C. ISERSKY and D. PAGE, Science, N.Y., 172, 1150 (1971)
174. E. C. FRANKLIN, J. LOWENSTEIN, B. BIGELOW and M. MELTZER, Amer. J. Med., 37, 332 (1964)
175. E. C. FRANKLIN and B. FRANGIONE, Proc. natn. Acad. Sci., U.S.A., 68, 187 (1971)
176. M. SELIGMANN, E. MINAESCO, D. HUREZ et al., J. clin. Invest., 48, 2374 (1969)
177. J. BUXBAUM, E. C. FRANKLIN and M. D. SCHARFF, Science, N.Y., 169, 770 (1970)
178. E. KLEIN, T. ESKELAND, M. INOUE et al., Expl. Cell Res., 62, 133 (1970)
179. J. WALDENSTRÖM, Acta med. scandinav., 170, (suppl. 367), 110 (1961)
180. M. SELIGMANN, F. DANON, C. MIHAESCO and H. H. FUDENBERG, Amer. J. Med., 43, 66 (1967)
181. K. JAMES, H. FUDENBERG, W. L. EPSTEIN et al., Clin. Exp. Immun., 2. 153 (1967)

Chapter 9

Plasma Albumin
Aspects of its Chemical Behaviour and Structure

Geoffrey Franglen

Department of Chemical Pathology
St. George's Hospital Medical School
Hyde Park Corner, London SW1

Plasma albumin is easy to prepare in bulk in a highly purified and apparently native state. Because of this, it has been used extensively during the past fifty years as a model protein in a wide variety of studies directed towards an understanding either of the properties of albumin as such or, more often, of the physico-chemical behaviour of soluble proteins in general. In these experiments the purity and nativeness of the albumin samples have usually been established by such means as the demonstration of a single, symmetrical peak on electrophoretic analysis, by the absence of other proteins on immuno-analysis, or by solubility and viscosity estimations. It has been realised only comparatively recently that preparation and storage of plasma albumin is practically always accompanied by significant formation of polymeric forms; it is difficult to maintain albumin samples in a completely monomeric state, even after considerable care in preparation. This must be taken into account when assessing work, especially early work, on albumin.

The molecular weight of albumin has been determined by methods common in the physico-chemical study of proteins, *e.g.* X-ray crystallography,[1,2] light scattering,[3,4] osmotic pressure,[5] sedimentation velocity and diffusion.[6-12] The published figures vary considerably from, for example, 61 500 found by Charlwood to 78 300 found by Polson. A value of about 69 000 was commonly accepted, since it was often obtained by methods considered to be reliable and by recognised masters in the field, such as Oncley *et al.*[6]; this value is the one usually quoted in text-books, even in those as definitive as that by Schultze and Heremans.[13] Nevertheless, the uncertainty of so much of this work led Phelps and Putnam[14] to comment:

"Altogether, the fact that the molecular weight of that much studied model protein, . . . serum albumin, cannot be stated much better than as being within the range of 65 000–70 000 is a serious reflection on the present status of methods for the determination of molecular weights of proteins".

In the case of albumin, one of the main reasons for this is concerned with effects of its preparation, as mentioned above, not only with errors of methodology. Nevertheless, these do play a part; for example, many physico-chemical determinations of molecular weight depend on the accuracy of the estimation of the partial specific volume of the molecule, and this, in part, depends on the accuracy of the estimation of its degree of hydration. Charlwood[7] noted that an error of 1% in the partial specific volume leads to an error of at least 3% in the final result. With albumin, a more significant error can occur when allowance is not made for the presence of polymeric molecules in the sample. Hoch and Morris[15] showed that electrophoretically-prepared albumin contains at least two components, and later Pedersen[16] showed by gel-filtration that commercial samples of purified albumin contain high proportions of polymeric molecules. This was also demonstrated by starch-gel electrophoresis.[17] It is not certain whether these forms normally exist *in vivo* or not; like Andersson,[18] the author considers that they are largely an artefact of preparation. Samples of human plasma albumin, fraction AP3 prepared by the method of Kekwick and Mackay,[19] were found to contain 2–3% of polymeric forms when fresh human plasma was used as the starting material and when the final product was not subjected to freeze-drying; when dried, even with the greatest care, the proportion rose to 5–8% (Franglen and Sharrard, unpublished results). The use of low pH levels and organic solvents during preparation increases the proportion of polymeric albumins in the final sample. As an extreme example, apparently native albumin can be prepared by precipitation of serum proteins with trichloracetic acid followed by extraction of the albumin with organic solvents;[20,21] such albumin contains up to 30% of polymeric forms (Franglen and Swaniker, unpublished results).

In determination of the molecular weight of albumin the effect of polymeric forms in the sample must be taken into account. Alternatively, only monomeric albumin could be used, or the method could be one which is not affected by the presence of polymers. Champagne[10] in an ultracentrifugal study allowed for the presence of 8% dimer-albumin in her sample by a beautifully elegant use of Newtonian iteration and obtained a molecular weight for monomeric albumin of 66 400. Squire *et al.*[12] separated monomeric albumin from other polymers by gel-filtration and obtained a molecular weight of 66 700 by the method of sedimentation. Estimations of the minimum molecular weight of a protein can be obtained from its amino acid composition, using the method of Delaage.[22] Albumin is a particularly good subject for such an approach, since it is free of any structural lipid or carbo-hydrate[23] and the polymeric forms are all simple aggregates of the monomer.[24] The most accurate analyses of the amino acid compositions of bovine and human albumins are those published by Spahr and Edsall;[25] these are given in Table 9.1. When the technique of Delaage

TABLE 9.1
Amino acid composition of albumins[25]

| Amino acid | Human | | Bovine | |
	g/100 g dry protein	residues/ 66 000 g protein	g/100 g dry protein	residues/ 66 000 g protein
Gly	1.02	12	1.38	16
Ala	6.64	62	4.97	46
Val	6.23	41–42	5.47	36–37
Leu	10.30	60	10.54	62
Ile	1.38	8	2.40	14
Pro	3.79	26	4.40	30
Phe	6.73	30	5.95	27
Tyr	4.34	17–18	5.00	20
Try	0.29	1	0.57	2
Ser	2.52	19	3.43	26
Thr	3.77	24 25	5.25	34
Cys	5.70	37	5.59	36
Met	1.13	6	0.75	4
Arg	5.63	24	5.30	22–23
His	3.33	16	3.50	17
Lys	12.02	62	11.98	62
Asp	8.89	51	9.46	54
Glu	15.28	78–79	15.17	77–78
Amide NH_3	0.84	34–35	0.79	32–33
Total	98.99	608–613	101.11	617–621

is applied to these results, molecular weights of 66 400–66 800 are obtained for both human and bovine variants. It thus appears from these three quite different approaches that the most likely value for the molecular weight of plasma albumin is about 66 500.

Other forms of heterogeneity can be found in purified albumin, even in the most carefully prepared samples. This heterogeneity can be divided into two types, the first of which is well-established and is due to a number of different factors. It might be termed "macrohetero-geneity" in contrast to a second type, the "microheterogeneity" of Foster.[26]

The earliest evidence implying the presence of macroheterogeneity in purified albumin was that normal samples contain less than one reactive thiol group per molecule.[27] To put it another way, up to 70% of the molecules in a sample can contain one reactive thiol group, whilst the remainder do not. In the latter the reactive group is blocked by cysteine or glutathione.[28,18] Blocking can occur fairly readily this way, and, for example, a serum sample left to stand at room temperature will show a gradual decrease in free cysteine, whilst the albumin prepared from it will show a concomitant fall in free thiol content. It would seem likely that each albumin molecule initially contains a free thiol group which may be blocked later in its life cycle. Some proteolytic enzymes, such as cathepsins, are unable to attack native albumin molecules as long as they contain a free thiol group;[29,30] the author has suggested that blocking of this group may be an important step in the normal catabolism of albumin.[31] A consequence of the presence of the single thiol group is that albumin molecules can be converted to dimers through reaction with mercuric ions or suitable organic mercurials, and Hughes[32,33] and Hughes and Dintzis[34] used this to separate albumin containing the free thiol group. On removal of the mercurial, a sample is obtained containing one mole of free thiol/mole of albumin; this is commonly known as "mercaptalbumin".

Macroheterogeneity through the presence of polymeric forms of albumin has been discussed above; it is probably largely an artefact of preparation, although a marked increase in formation of dimer-albumin has been noted in one family of bisalbuminaemics[35] and in certain pathological conditions.[36] In view of the marked reactivity of the thiol group, it might be thought that the most common form of polymerization, dimer formation, would occur through disulphide bond formation between two molecules of mercaptalbumin. This is only partly true; the author has found that, when dimer-albumin prepared by gel-filtration is subjected to polyacrylamide-gel electrophoresis, it runs as a doublet band, suggesting that there are different ways of linking the molecules in pairs to give two classes of dimers with slightly different molecular volumes of rotation. On treatment with a reducing agent, such as β-mercapto-ethanol, only one of the doublet bands is broken down to monomer-albumin, whilst the other remains apparently unaltered; this confirms the observation of Hartley, Sober and Petersen[37] that about half of dimer-albumin prepared by chromatography on DEAE-cellulose can be transformed to monomer when treated with disulphide-bond reagents. It is not known what linkage is involved in the formation of non-reducible dimer-albumin; it is not broken by lowering the pH level nor by the action of guanidine (Franglen, unpublished observations).

Heat-denaturation can be used to demonstrate the macroheterogeneity of albumin. This has been the subject of detailed work by Czech workers who proposed the presence of up to nine different albumin classes in normal samples, each with a different rate of heat-denaturation,[38-40] it should be noted, however, that in these studies no allowance was made for any effects due to the presence of polymeric forms. A similar, earlier study was carried out by Mackay and Martin[41] on Lister human albumin, fraction AP2; these workers found that the sample was converted into a mixture of slightly aggregated and un-aggregated molecules by heating at 58°C at pH 6.5. The two components could be separated by zone-electrophoresis and had a proportion of about 1:1 when the reaction was complete. The majority of the reaction occurred within 1 hr of heating; there was only a slight increase in the proportion of aggregated material on prolonging the time to 24 hr. This Laboratory has since been able to show that this reaction is not related to the presence of the free thiol group, since blocking with iodo-acetamide does not inhibit the reaction. It is not due to initiation of aggregation by polymeric forms, since the reaction occurs in monomer-albumin prepared by gel-filtration. Variation in fatty acid associated with the sample or defatting it by the method of Chen[42] slightly affects the temperature at which the reaction takes place; it does not alter the final proportion of the two components. Although the electrophoretic mobility of the unaggregated component was the same as that of native, unheated albumin, it was found to be partially denatured in that its content of α-helix had been reduced by about 50%. The components do not form an equilibrium mixture; when the aggregated component was removed by gel-filtration, the unaggregated monomeric molecules were unaffected by further heating at 58°C and only aggregated when the temperature was raised to about 61°C (Franglen and Taylor, unpublished results). The author considers that these observations can be regarded as indicating that normal samples may contain two classes of monomeric albumin with a ratio to each other of about 1:1, differing slightly in their mode of heat-denaturation. It is not clear in what lies the difference between the two albumins; it could be due to subtle differences in their primary structure.

Support for this hypothesis of the normal presence of two different monomeric albumin classes can be deduced from the phenomenon of bisalbuminaemia. This is a genetically determined condition in which a normal albumin is accompanied by another with a manifestly different electrophoretic mobility. Reviews have been published of the types of bisalbuminaemia discovered in different families; there are at least eight distinguishable variants.[43,44] Studies on such families have shown that there is a close genetic linkage between the loci for human plasma albumin and the Group Specific Component (Gc).[45,46] Although no full studies have been carried out so far, it is likely that the difference between the pairs of albumins found in bisalbuminaemics lies in an alteration of the primary amino acid sequence, similar to those found in abnormal haemoglobins.[47] In the ten bisalbuminaemic families the author has examined, the ratio of the aberrant to the normal albumin has always been very close to 1:1; this has been repeatedly noted in other families, for example, by Knedel,[48] Tarnoky and Lestas,[49]

Adams[50] and Bell, Nicholson and Thompson.[51] There are two possible hypotheses to account for this. Some form of genetic alteration in the bisalbuminaemic has occurred, resulting in the production of an aberrant albumin in addition to an apparently normal type. It could be postulated that this also generates a control which ensures that the two albumins are made in about equal amounts, even from family to unrelated family. Alternatively, and more simply, it could be postulated that two albumin classes of identical electrophoretic mobility are normally present in plasma with a ratio maintained at about 1:1 by a control which is normally present; in the bisalbuminaemic the genetic alteration affects one of the albumins in its primary structure, but does not affect the control of their ratio.

Macroheterogeneity of albumin preparations has also been demonstrated by chromatography on various supports, such as DEAE-cellulose[37] and hydroxylapatite;[52] this has been found to be due to the presence of polymeric forms or of the free thiol group in a particular sub-fraction, or to variation in adsorbed fatty acids. More recently, Salaman and Williamson[53] demonstrated the heterogeneity of purified bovine albumin by iso-electric focusing in polyacrylamide-gel in the presence of urea. The sample was resolved into two main groups, each consisting primarily of three components; it was suggested that the sub-fractions obtained in this way may have minor differences in primary structure.

A more subtle form of heterogeneity of albumin samples, "microheterogeneity", has been proposed by Foster on the evidence of a long series of experiments; he gave a full account of his conclusions in a lecture to the Chemical Society.[26] Even the purest preparations of albumin show electrophoretic heterogeneity in the neighbourhood of pH 4.[54-56] Aoki and Foster[57,58] suggested that this arose from the presence of two isomeric forms of albumin which they called 'N' (for native or normal) and 'F' (for fast, since this form has the higher electrophoretic mobility near pH 4). The two forms exist in a pH-dependent equilibrium with the F-form being favoured at lower pH levels. Conversion from one form to the other is completely reversible, but with normal preparations the complete transition of one form to the other is extended over the pH-range from about 3.4 to 4.8.[59] Sogami and Foster[60] considered that, if normal albumin was truly homogeneous, the pH-dependence of the N–F transition should be very sharp, and they suggested that preparations consist of a continuum of molecules which, whilst being very similar, differ in the pH-range in which they undergo the N–F transition. Albumin can be precipitated from solution by high concentrations of potassium chloride over the pH-range of 3.5 to 4.8.[61] By choosing appropriate shifts of pH within this range it is possible to precipitate different fractions of albumin with potassium chloride, and these fractions were found to have different and much sharper N–F transition curves than the original preparation. Foster and his colleagues concluded that they had isolated sub-fractions of albumin, and had demonstrated the microheterogeneity which they had postulated. They were not able to elucidate the nature of the dissimilarities between the sub-fractions; Foster[26] suggested that it might be due to subtle differences in secondary and tertiary structures of the molecules, possibly through random distributions of the amide groups amongst the glutamyl and aspartyl residues, or through alternative pairings of disulphide bonds between the cysteinyl residues.

These suggestions by Foster have far-reaching implications in relation to present theories on the synthesis of proteins. If differences between sub-fractions of albumin were due to variations in chemical bonds in molecules of the same primary structure, it would imply that genetic mechanisms do not have exclusive control over the formation of tertiary structure, and presumably that solution or environmental forces within the cell play a significant role in the folding of the newly-synthesized peptide chain. It would have to be postulated that these forces could be satisfied by a number of different configurations of the polypeptide chain, as long as they were of sufficiently similar energy levels. It would seem that this is the only simple way to permit molecules of identical primary structure to give rise naturally to the multiplicity of forms of albumin needed to satisfy Foster's hypothesis.

Albumin binds many types of small molecules, both normal metabolites and many foreign compounds. Certain properties of albumin are altered by its association with some of these compounds, particularly when they are of a lipophilic nature. McMenamy and Lee[62] investigated the microheterogeneity of albumin samples before and after defatting the protein by the method of Goodman.[63] As a result of this treatment, the pH-ranges at which sub-fractions underwent the complete N–F transition was reduced to about 4.25–4.40 in all cases, irrespective of the values of extent of the ranges found originally in the untreated samples and sub-fractions. The site of indole binding in albumin is sensitive to minor changes in tertiary structure; microdifferences in sub-fractions of albumin, if they did involve the tertiary structure, should be demonstrable through variation in ligand associations of this compound. Whilst such differences were found between sub-fractions of un-defatted albumin, all disappeared after the protein had been defatted.

Foster's observations thus can be explained in terms of the effect of variation in small molecules adsorbed to different molecules of albumin, and careful and complete extraction of these from the sample results in the disappearance of microheterogeneity. Nevertheless, Foster still claims that microheterogeneity exists as a structural entity of covalent origin; he has published another series of experiments to demonstrate that defatted albumin shows signs of heterogeneity in solubility and electrophoretic behaviour.[64–67] These can be criticised for possible inefficiency in defatting of the albumin. In their study McMenamy *et al.* used the prolonged, cold heptane-acetic acid extraction procedure of Goodman under rigorously controlled conditions to ensure that the protein was efficiently freed from adsorbed material and, at the same time, was not significantly denatured. Foster and his colleagues initially used their own method of defatting which consists of storing the albumin solution at pH 2.7 at 2°C for 60 hours, followed by centrifugation to separate adsorbed lipid.[68] Later they used the method of Chen[42] in which the lipids are released from the protein by acid treatment and are absorbed into charcoal. Whilst both these latter methods are effective in removing gross amounts of lipids from albumin, it is doubtful whether they can be as efficient as the Goodman procedure in extracting the final traces of adsorbed materials; it should be noted, though, that Andersson[69] showed that mercaptalbumin becomes completely homogeneous with regard to pH-solubility properties even as a result of defatting by acid-charcoal treatment. If it can be shown by one method of extraction that defatting of the sample results in disappearance of microheterogeneity, any subsequent demonstration of "microheterogeneity" after other extraction procedures is more a reflection of the inefficiency of the latter methods rather than a proof of structural differences in the molecules of the sample.

As noted in Table 9.1 the albumin molecule contains about 600 amino acid residues. In all species examined, albumins have been found to contain one N-terminal and one C-terminal residue (Tables 9.2 and 9.3). Rupture of all 16–18 disulphide bonds in the molecule does not release polypeptide sub-units; reduction and alkylation in the presence of detergent[3] or the performic acid oxidation of albumin[70] lead to no reduction in the molecular weight of the molecule. The 600 residues are thus shown to be formed into a single polypeptide chain. In the case of the N-terminal sequence, the early work of Thompson[71] and Ikenaka[72] was extended considerably by Peters and his colleagues.[73,74] These workers prepared peptides from bovine albumin by brief peptic digestion. One of these, 24 residues in length, was considered to be derived from the N-terminal end of the polypeptide chain

TABLE 9.2
N-Terminal sequences of albumin from various animals

Man[74]	H₂N-Asp-Ala-His-Lys-Ser-Glu-Val-Ala-His-Arg-Phe-Lys-Asp-Leu-Gly-Glu-Glu- Asn- Phe-Lys-Ala-Leu-Val-Leu-
Horse, donkey, mule[69(a)]	H₂N-Asp-Thr-
Pig[76]	H₂H-Asp-Thr-Tyr-Lys-Ser-Glu-Ile-Ala-His-Arg-Phe-Lys-Asx-Leu-Gly-Glx-(Glx,Tyr)-Phe-Lys-Leu-Gly-Leu-Val-Gly-Ser-Arg-
Bull[74,76]	H₂N-Asp-Thr-His-Lys-Ser-Glu-Ile-Ala-His-Arg-Phe-Lys-Asp-Leu-Gly-Glu-Glu-His-Phe-Lys-Gly-Leu-Val-Leu-
Duck, chicken, turkey[69b]	H₂N-Asp-

TABLE 9.3
C-Terminal sequences of albumin from various animals

Man[75]	-Gly-Val-Ala-Leu-CO_2H
Horse, donkey, mule[69(a)]	-Leu-Ala-CO_2H
Dog, rabbit[69(c)]	-Leu-CO_2H
Sheep, goat[69(c)]	-Ala-CO_2H
Pig[76]	-Gly-Ile-Leu-Ala-CO_2H
Bull[76]	-(Glu,Ala,Cys,Cys,Glu)-Phe-Ala-Val-Glu- Gly-Pro-Lys-Leu-Val-Val-Ser-Thr-Gln-Thr- Ala-Leu-Ala-CO_2H
Duck, chicken[69(b)]	-(Ser,Thr,Gly,Val,Leu)-Ala-CO_2H
Turkey[69(b)]	-(Thr,Gly,Ala,Leu)-Val-CO_2H

and was analyzed in detail; it was also found that this peptide bound two copper ions through its histidyl residues in positions 3 and 18. The C-terminal sequence of human albumin was disputed initially. Both White, Shields and Robbins[75] and Ikenaka[72] observed the release from albumin of leucine, alanine, valine and glycine in that order by carboxypeptidase. White *et al.* assumed that this represented the sequence from the C-terminal end of the polypeptide chain, but Ikenaka proposed a different arrangement as a result of further studies. Nevertheless, the sequence proposed by White *et al.* was preferred by Peters *et al.* in their work on peptides produced by peptic digestion of albumin. These workers isolated a peptide containing 77 residues which was assumed to represent that part of the polypeptide chain containing the C terminal end, since it was found to contain the sequence (Gly, Ser, Val)-(Ala,Thr)-Leu-Ala.CO_2H in good agreement with the sequence proposed by White *et al.* This peptide has been analyzed in considerably more detail by Low[76] and confirmed as being from the C-terminal portion of the polypeptide chain.

The work on internal sequences has mostly been confined to bovine albumin, probably because it is easily available commercially in high purity. In this field the most extensive work has been carried out by Low[76] who has determined over two-thirds of the primary structure of the molecule. She also made a comparison of these results with her similar work on porcine albumin, and noted a difference of about 20% between the two. Of the internal sequences, the most interesting is that containing the free reactive thiol group associated with the one unpaired cysteinyl residue of the molecule; this was first determined by Witter and Tuppy.[77] The sequence in this region of the polypeptide chain was found to be identical for both human and bovine albumins over at least nine residues, having the structure -Leu-Gln-Asp-Glu-Gln-Glu-Cys-Pro-Phe-. In that the thiol is the most reactive group in this sequence, it is interesting to note that a prolyl residue is next to the cysteinyl concerned, indicating that this group is sited on the polypeptide chain where it bends sharply.

Only a small number of "finger-print" maps have been published of the peptides produced by enzyme hydrolysis of albumin, and these have been concerned mostly with the examination of variants in bisalbumin-aemia, e.g. Earle, Hutt, Schmid and Gitlin.[47] Schultze and Here-mans[18] noticed that those published show a far smaller number of peptides than might be expected from the molecule's size and amino acid composition. They suggested that this might indicate the presence of a number of repeated amino acid sequences, perhaps related to the sub-unit structure proposed by Foster.[23] From his own work the author considers that it is more likely to be due to incomplete digestion; even in the most carefully controlled conditions a certain amount of un-digested protein is left. Whilst admittedly incomplete, the extensive analyses by Low[76] do not suggest the presence of significant amounts of repeated sequences.

The secondary structure has been examined by optical rotatory dispersion.[78,79] Callaghan and Martin[80] measured the specific rotation of human albumin between 350 nm and 600 nm, and found the mole-cule to have between 40–50% of its amino acid residues in the right-handed α-helical conformation. Jirgensons[81] extended these measure-ments into the far-ultraviolet region of the spectrum, and found posi-tive and negative Cotton effects at 233 nm and 198 nm respectively, giving values for the α-helical content of 64% and 49%.

Ruddiford and Jennings[82] reviewed the various physico-chemical studies which showed that human and bovine albumin are globular proteins in the native state. Oncley et al.[6] first described albumin as a prolate ellipsoid of length 140 Å and of diameter 38 Å; later the same Laboratory[83] concluded from dielectric dispersion studies that the human albumin molecule would be better represented either as a prolate ellipsoid of length 147 Å and diameter 37 Å or as a generalized ellipsoid with axial lengths of 144, 48 and 24 Å. Other workers using low-angle X-ray scattering techniques obtained similar dimensions.[84,85] In contrast, Riley and Oster,[86] also using low-angle X-ray scattering tech-niques, concluded that the bovine molecule was not far from spherical, and that their data could be best accommodated by assuming the molecule had a height and diameter of 45 Å and 49 Å. A spherical form for the human molecule was assumed by Loeb and Scheraga[9] in the interpretation of their viscosity, sedimentation and diffusion data.

Weber[87] attempted to obtain a more intimate idea of the general molecular structure of albumin through studies on the polarization of the fluorescence of the molecule when conjugated with dimethyl-aminonaphthalene sulphonyl chloride. The degree of polarization was constant over the pH-range 4–9, but outside this it fell rapidly. This was in complete contrast to ovalbumin whose degree of polarization re-mained stable over the whole of the pH-range studied. Weber[88] con-cluded that at low pH-levels the albumin molecule dissociated com-pletely into two separate sub-units of roughly equal molecular weights;

it should be remembered that, at that time, it had not been demonstrated that the albumin molecule contains only one polypeptide chain. His conclusion was strongly contested, particularly by Pedersen;[89] other workers using light-scattering methods[90,91] and osmotic-pressure measurements[92] demonstrated the constancy of the molecular weight from neutrality to pH 3.3.

Harrington, Johnson and Ottewill[93] extended this range by investigating the behaviour of bovine albumin down to pH 1.9, and measuring the changes which occur in its sedimentation coefficient, diffusion rate, intrinsic viscosity and the polarization of fluorescent conjugates. They concluded that the molecule exists at acid pH-levels as two compact, reasonably symmetrical sub-units joined by a flexible peptide segment. Under neutral conditions these sub-units are held together by secondary forces so that their relative rotation is not possible and the molecule forms a compact globule. Direct evidence for this hypothesis was provided by electron microscopy of bovine albumin.[94] At about pH 8.0 albumin molecules were shown to exist in the form of two round particles closely associated with one another, giving the appearance of doublets. These doublets were probably single molecules, since their size and volume corresponded well with accepted values. In another similar study Slayter[95] suggested that the Chatterjees' doublet did not represent the neutral form of albumin, but that artefacts of technique had partially ruptured the tertiary structure of the molecules, causing them to resemble the open form of the Harrington model. Supporting chemical evidence for the Harrington hypothesis was provided, however by Adkins and Foster.[96,97] Digestion of bovine albumin by subtilisin at pH 8.9 in the presence of detergent liberated two major fragments with molecular weights roughly half that of the intact molecule. Although the amino acid compositions of the two fragments were distinctly different, their sum accounted for nearly all the amino acid content of the molecule. It appeared that the fragments had been formed by cleavage of the molecule at a central point in the polypeptide chain and thus constituted the two Harrington sub-units.

Harrington et al. concluded that the albumin molecule underwent rapid and reversible expansion as the pH-level was lowered from 4 to 2. This has already been suggested by Tanford[98] from titration studies, and by Yang and Foster[99] from studies on the viscosity and optical rotational behaviour of bovine albumin. Two distinct plateaux occur in the viscosity/pH plot as the pH-level is lowered over this range, and Tanford, Buzzell, Rands and Swanson[100] deduced that the acid expansion of the molecule occurs in two steps. In the progress from the compact neutral globule the molecule first attains an intermediate stage of swelling (the expandable form), which is stable over a short pH-range, before reaching the fully-expanded form as the pH is lowered further still. Confirmation was provided by Ruddiford and Jennings[82]; in a study of the

Kerr effect on the protein, they observed sharp drops in the relaxation times at pH 4.1 and 3.6 as the pH-level was lowered.

Foster[23] developed this hypothesis further through consideration of electrophoretic and solubility behaviour of albumin at acid-pH levels, and titration and interaction experiments. He proposed that the molecule consists of four compact regions connected by relatively short segments of polypeptide chain. Above pH 4 close association of the sub-units through hydrogen bonds, salt links and hydrophobic bonds confers a globular structure on the molecule. This globule, the normal or N-form, consists of two F'-form globules bonded by electrostatic forces, and these two globules (corresponding to the two halves of the Harrington model) are each composed of two hydrophobically bonded F-form sub-units. At low pH-levels the molecule expands fully, and the four sub-units are joined only by short peptide links. Slayter[95] was able to demonstrate this expansion of bovine albumin at pH 1.9 by electron microscopy, but, as would be expected from the limitations of the technique, was not able to detect the presence of four globular regions. Nevertheless, since the expansion resulted in a thread-like form, rather than one resembling, for example, a ball and chain, it was concluded that the expansion was not confined to one section of the molecule, but involved the whole structure or, at least, several parts of it.

Weber and Young[101,102] pointed out that, if the albumin molecule were to consist of a mixture of globular and flexible parts,

"... an almost inescapable consequence ... is that a protease acting in acid solution should preferentially act on the peptide bonds of the linking segments, thus liberating a small number of globular units".

Acting on their hypothesis, Weber et al. treated bovine albumin briefly with pepsin at acid pH and were able to separate from the hydrolysate two fractions, one with an average molecular weight of 12 500 and the other of about 30 000. Although two components were found in the heavy fraction, they were so similar in composition as to suggest that they were both derived from a single entity. The light fraction was a mixture of molecular species with differing physical properties and amino acid compositions. Since this fraction formed about 55% of the recovered digest, Weber et al. concluded that the albumin molecule consists of a single large globule from the heavy fraction surrounded by two or three smaller globules from the light fraction, thus in principle supporting the Foster hypothesis.

Several workers have shown that the initial action of pepsin on bovine albumin is the production of fragments of intermediate size.[103–106] Weber et al. were suggesting that these fragments represented the Foster sub-units. It has always seemed to the author that this was an optimistic hope rather than "an almost inescapable consequence". Exposed linking segments should be attacked rapidly by a protease of wide specificity, but bonds elsewhere in the polypeptide chain could be attacked equally rapidly if they were accessible, for example, at the surface of the molecule; in acid conditions the expansion of the molecular structure could result in a large number of such bonds being accessible. The work of Braam *et al.* supports this criticism; these workers treated bovine albumin with pepsin at and below pH 3.6, and showed that, even after brief periods of reaction, at least ten well-defined fragments could be detected by polyacrylamide-gel electrophoresis.

Luzzati, Witz and Nicolaieff[107] proposed a different model from investigation of bovine albumin by low-angle X-ray scattering. At pH 5.3 the molecule was found to be compact; at pH 3.6 changes appeared which suggested that about 35% of the polypeptide chain had unwound to take up a loose configuration around a compact central globule. Bloomfield[108] in his own examination of the same data concluded that they were compatible with a variation of the Foster-Weber model if the molecule were to consist of two spherical globules of radius 19.0 Å separated by one of 26.6 Å radius. His model, however, accounted for only about 80% of the molecular weight, and he suggested that the remaining 20% formed the polypeptide segments linking the sub-units, *i.e.* as Luzzati *et al.* had concluded, the Foster-Weber model would have to be very loose in structure at low pH-levels if it were correct.

The albumin molecule has 16–18 intramolecular cystinyl residues which restrict the unfolding of the polypeptide chain. It appeared to the author that a consequence of the Foster-Weber model would be that, at low pH-levels, globular sub-units would be structured primarily by the presence of the disulphide cross-linkages of these residues. If the Weber hypothesis on the release of sub-units from the molecule by peptic proteolysis were correct, these fragments should contain sufficient cystinyl residues to allow them to retain a compact structure. On the other hand, if the Luzzati model were correct, brief peptic hydrolysis of albumin should produce lengths of unstructured polypeptide chain, together with a relatively undigested fragment containing a high proportion of the total cystinyl residues. Franglen and Swaniker,[109] therefore, treated human albumin briefly with pepsin after the manner of Weber and Young.[101] Five different fragments were isolated from the proteolysate, and three of these were found to have little or no cystinyl residues. A high proportion of the polypeptide chain was thus found to be significantly free of disulphide links, and it followed that at low pH-levels these parts would have a high degree of freedom of structure. It was also concluded that these cystinyl-free peptides probably represented the C- and N-terminal portions of the chain, *i.e.* Franglen *et al.* favoured the Luzzati model of albumin rather than that of Foster. Pedersen and Foster[110] also looked at the distribution of cystinyl residues amongst fragments of albumin after subtilisin proteolysis. They concluded that the grouping allowed the formation of four sub-units; nevertheless, long portions of the polypeptide chain appeared not to be structured by these residues.

There are, thus, two different models proposed for the general structure of albumin. Both are similar in assuming that at neutral pH-levels the molecule forms a compact globule. Foster and his colleagues suggest that this globule is made of four compact sub-units connected by short lengths of polypeptide chain; on lowering the pH-level the sub-units move away from one another, but still retain their local structure. According to Luzzati et al. and Franglen et al. the molecule consists of a highly organized core covered by a less organized coating of polypeptide chain; on lowering the pH-level the molecule expands through unwinding of this coating, but the core still retains a high degree of organization. Franglen et al. noted that albumin is remarkable in its ability to combine with a wide range of compounds,[111,112] and that this may be related to Karush's "configurational adaptibility", a process involving changes in the relative positions of the side-chains of the molecule.[113–115] Such changes would permit the protein to combine with many different substances by alterations of configuration leading to appropriate distributions of the surface patterns of reactive groups. Configurational adaptibility of this nature would occur most readily in those parts of a molecule where there are low levels of tertiary structure. Franglen et al. suggested, therefore, that a loosely organized surface structure in the albumin molecule may be vital to this protein's important physiological function in acting as a transport and detoxicating agent.

ACKNOWLEDGEMENT

The Author wishes to thank the Cancer Research Campaign of Great Britain for financial support of his and his colleagues' work mentioned in this Chapter.

REFERENCES

1. B. W. Low, J. Am. chem. Soc., 74, 4830 (1952)
2. B. W. Low and F. M. Richards, J. Am. chem. Soc., 74, 1660 (1952)
3. M. J. Hunter and F. C. McDuffie, J. Am. chem. Soc., 81, 1400 (1959)
4. A. Polson, Biochim. biophys. Acta, 140, 197 (1967)
5. G. Scatchard, A. C. Batchelder and A. Brown, J. clin. Invest., 23, 458 (1944)
6. J. L. Oncley, G. Scatchard and A. Brown, J. phys. colloid Chem., 51, 184 (1947)
7. P. A. Charlwood, Biochem. J., 51, 113 (1952)
8. J. M. Creeth, Biochem. J., 51, 10 (1952)
9. G. I. Loeb and H. A. Scheraga, J. phys. Chem., 60, 1633 (1956)
10. M. Champagne, J. Chim. phys., 54, 379 (1957)
11. S. E. Allerton, D. Elwyn, J. T. Edsall and P. F. Spahr, J. biol. Chem., 237, 85 (1962)

12. P. G. SQUIRE, P. MOSER and C. T. O'KONSKI, *Biochemistry, N.Y.*, **12**, 4261 (1968)
13. H. E. SCHULTZE and J. F. HEREMANS, *Molecular Biology of Human Proteins with Special Reference to Plasma Proteins, vol. 1*. (Elsevier Publishing Co., Amsterdam 1966)
14. R. A. PHELPS and F. W. PUTNAM, (ed. F. W. Putnam) *The Plasma Proteins, vol. 1*, p. 143. (Academic Press Inc., New York 1960)
15. H. HOCH and C. J. O. R. MORRIS, *Nature, Lond.*, **156**, 234 (1945)
16. K. O. PEDERSEN, *Archs Biochem. Biophys., Suppl. 1*, 157 (1962)
17. A. SAIFER, M. ROBIN and M. VENTRICE, *Archs Biochem. Biophys.*, **92**, 409 (1961)
18. L.-O. ANDERSSON, *Biochim. biophys. Acta*, **117**, 115 (1966)
19. R. A. KEKWICK and M. E. MACKAY, *The Separation of Protein Fractions from Human Plasma with Ether*. M.R.C. Spec. Rep. Ser. No. 286. H.M. Stationery Office, London (1954)
20. J. VALLANCE-OWEN, E. DENNES and P. N. CAMPBELL, *Lancet*, **2**, 336 (1958)
21. S. E. MICHAEL, *Biochem. J.*, **82**, 212 (1962)
22. M. DELAAGE, *Biochim. biophys. Acta*, **168**, 573 (1968)
23. J. F. FOSTER, (ed. F. Putnam) *The Plasma Proteins, vol. 1*, p. 179. (Academic Press Inc., New York 1960)
24. A. SAIFER and J. PALO, *Analyt. Biochem.*, **27**, 1 (1969)
25. P. F. SPAHR and J. T. EDSALL, *J. biol. Chem.*, **239**, 850 (1964)
26. J. F. FOSTER, *Chem. Soc., (Lond.), Spec. Publ.*, **23**, 25 (1968)
27. W. L. HUGHES, (eds., H. Neurath and K. Bailey) *The Proteins, vol. 2, part B*, chap. 21. (Academic Press Inc., New York 1954)
28. T. P. KING, *J. biol. Chem.*, **236**, PC5 (1961)
29. L. LIBENSON and M. JENA, *Archs Biochem. Biophys.*, **100**, 441 (1963)
30. L. LIBENSON and M. JENA, *Archs Biochem. Biophys.*, **104**, 292 (1964)
31. G. FRANGLEN, *Proc. R. Soc. Med.*, **60**, 1072 (1967)
32. W. L. HUGHES, *J. Am. chem. Soc.*, **69**, 1836 (1947)
33. W. L. HUGHES, *Cold Spring Harb. Symp. quant. Biol.*, **14**, 79 (1950)
34. W. L. HUGHES and H. M. DINTZIS, *J. biol. Chem.*, **239**, 845 (1964)
35. C.-B. LAURELL and J.-E. NILÉHN, *J. clin. Invest.*, **45**, 1935 (1966)
36. K. SCHMID, A. POLIS and S. TAKAHASHI, *Biochim. biophys. Acta*, **57**, 48 (1962)
37. R. HARTLEY, H. SOBER and E. PETERSON, *Biochemistry, N.Y.*, **1**, 60 (1962)
38. Š. ŠTOKROVÁ and J. ŠPONAR, *Colln Czech. chem. Commun. Engl. Edn*, **27**, 2516 (1962)
39. Š. ŠTOKROVÁ and J. ŠPONAR, *Colln Czech. chem. Commun. Engl. Edn*, **28**, 659 (1963)
40. J. ŠPONAR, I. FRIČ, Š. ŠTOKROVÁ and J. KOVÁRIKOVÁ, *Colln Czech. Commun. Engl. Edn*, **28**, 1831 (1963)
41. M. E. MACKAY and N. H. MARTIN, *Biochem. J.*, **65**, 284 (1957)
42. R. F. CHEN, *J. biol. Chem.*, **242**, 173 (1967)
43. L. R. WEITKAMP, D. C. SHREFFLER, J. L. ROBBINS, O. DRACHMANN, P. L. ADNER, R. J. WIEME, N. M. SIMON, K. B. COOKE, G. SANDOR, F. WUHRMANN, M. BRAEND and A. L. TARNOKY, *Acta genet., Basel*, **17**, 399 (1967)
44. L. R. WEITKAMP, G. FRANGLEN, D. A. ROKALA, H. F. POLESKY, N. E. SIMPSON, F. W. SUNDERMAN, H. E. BELL, J. SAAVE, R. LISKER and S. W. BOHLS, *Hum. Hered.*, **19**, 159 (1969)
45. L. R. WEITKAMP, E. B. ROBSON, D. C. SHREFFLER and G. CORNEY, *Amer. J. hum. Genet.*, **20**, 392 (1968)
46. L. R. WEITKAMP, J. H. RENWICK, J. BERGER, D. C. SHREFFLER, O. DRACHMANN, F. WUHRMANN, M. BRAEND and G. FRANGLEN, *Hum. Hered.*, **20**, 1 (1970)
47. D. P. EARLE, M. P. HUTT, K. SCHMID and D. GITLIN, *J. clin. Invest.*, **38**, 1412 (1959)
48. M. KNEDEL, *Blut*, **3**, 129 (1957)
49. A. L. TARNOKY and A. N. LESTAS, *Clinica chim. Acta*, **9**, 551 (1964)

50. M. S. ADAMS, *J. med. Genet.*, **3**, 198 (1966)
51. H. E. BELL, S. F. NICHOLSON and Z. R. THOMPSON, *Clinica chim. Acta*, **15**, 247 (1967)
52. A. TISELIUS, S. HJERTÉN and Ö. LEVIS, *Archs Biochem. Biophys.*, **65**, 132 (1956)
53. M. R. SALAMAN and A. R. WILLIAMSON, *Biochem. J.*, **122**, 93 (1971)
54. J. LEUTSCHER, *J. Am. chem. Soc.*, **61**, 2888 (1939)
55. D. G. SHARP, G. COOPER, J. ERICKSON and H. NEURATH, *J. biol. Chem.*, **144**, 139 (1942)
56. R. A. ALBERTY, *J. phys. colloid Chem.*, **53**, 114 (1949)
57. K. AOKI and J. F. FOSTER, *J. Am. chem. Soc.*, **78**, 3538 (1956)
58. K. AOKI and J. F. FOSTER, *J. Am. chem. Soc.*, **79**, 3385 (1957)
59. J. F. FOSTER, M. SOGAMI, H. A. PETERSEN and W. J. LEONARD, *J. biol. Chem.*, **240**, 2495 (1965)
60. M. SOGAMI and J. F. FOSTER, *J. biol. Chem.*, **283**, PC 2245 (1963)
61. H. A. PETERSEN and J. F. FOSTER, *J. biol. Chem.*, **240**, 2503 (1965)
62. R. H. MCMENAMY and Y. LEE, *Archs Biochem. Biophys.*, **122**, 635 (1967)
63. D. S. GOODMAN, *J. Am. chem. Soc.*, **80**, 3892 (1958)
64. M. SOGAMI and J. F. FOSTER, *Fedn Proc. Fedn Am. Socs. exp. Biol.*, **26**, 827 (1967)
65. W. E. MOORE and J. F. FOSTER, *Biochemistry, N.Y.*, **7**, 3409 (1968)
66. K. P. WONG and J. F. FOSTER, *Biochemistry, N.Y.*, **8**, 4096 (1969a)
67. K. P. WONG and J. F. FOSTER, *Biochemistry, N.Y.*, **8**, 4104 (1969b)
68. E. J. WILLIAMS and J. F. FOSTER, *J. Am. chem. Soc.*, **81**, 865 (1959)
69. L.-O. ANDERSSON, *Int. J. Protein Research*, **1**, 151 (1969)
69a. F. ANTONI, S. BOZSÓKY, T. DÉVÉNYI, A. LENDVAI and B. SZÖRÉNYI, *Acta physiol. hung.*, **9**, 309 (1956), *through Chem. Abstr.*, **50**, 17102 (1956)
69b. T. PETERS, A. C. LOGAN and C. A. STANFORD, *Biochim. biophys. Acta*, **30**, 88 (1958)
69c. K. KUSAMI, *J. Biochem., Tokyo*, **44**, 375 (1957)
70. B. JIRGENSONS and T. IKENAKA, *Makromol. Chem.*, **31**, 112 (1959)
71. E. O. P. THOMPSON, *J. biol. Chem.*, **208**, 565 (1954)
72. T. IKENAKA, *J. Am. chem. Soc.*, **82**, 3180 (1960)
73. T. PETERS and C. HAWN, *J. biol. Chem.*, **242**, 1566 (1967)
74. W. T. SHEARER, R. A. BRADSHAW, F. R. N. GURD and T. PETERS, *J. biol. Chem.*, **242**, 5451 (1967)
75. W. F. WHITE, J. SHIELDS and K. C. ROBBINS, *J. Am. chem. Soc.*, **77**, 1267 (1955)
76. T. L. K. LOW, *The Amino Acid Sequences of Porcine and Bovine Serum Albumins.* Ph.D. Dissertation presented to the University of Texas at Austin. (1970)
77. A. WITTER and H. TUPPY, *Biochem. biophys. Acta*, **45**, 429 (1960)
78. P. URNESS and P. DOTY, *Adv. Protein Chem.*, **16**, 401 (1961)
79. J. A. SCHELLMAN and C. SCHELLMAN, (ed. H. Neurath) *The Proteins, vol. 2*, p. 1. (Academic Press Inc., New York 1964)
80. P. CALLAGHAN and N. H. MARTIN, *Biochem. J.*, **83**, 144 (1962)
81. B. JIRGENSONS, *Makromol. Chem.*, **91**, 74 (1966)
82. C. L. RUDDIFORD and B. R. JENNINGS, *Biochim. biophys. Acta*, **126**, 171 (1966)
83. J. L. ONCLEY, H. M. DINTZIS and N. R. S. HOLLIES, Abstract of papers, 122nd meeting, American Chemical Society, p. 12P. (1952)
84. H. N. RITLAND, P. KAESBERG and W. W. BEEMAN, *J. chem. Phys.*, **18**, 1237 (1950)
85. W. W. BEEMAN, J. W. ANDEREGG and S. SHULMAN, *Phys. Rev.*, **87**, 186 (1952)
86. D. P. RILEY and G. OSTER, *Disc. Faraday Soc.*, **11**, 107 (1951)
87. G. WEBER, *Biochem. J.*, **51**, 155 (1952)
88. G. WEBER, *Disc. Faraday Soc.*, **13**, 33 (1953)
89. K. O. PETERSEN, *Disc. Faraday Soc.*, **13**, 49 (1963)
90. P. DOTY and R. F. STEINER, *J. chem. Phys.*, **20**, 85 (1952)
91. J. T. EDSALL, H. EDELHOCH, R. LONTIE and P. R. MORRISON, *J. Am. chem. Soc.*, **72**, 4641 (1950)
92. H. GUTFREUND, *Trans. Faraday Soc.*, **50**, 628 (1954)

93. W. F. HARRINGTON, P. JOHNSON and R. H. OTTEWILL, *Biochem. J.*, **62**, 569 (1956)
94. A. CHATTERJEE and S. N. CHATTERJEE, *J. mol. Biol.*, **11**, 432 (1965)
95. E. M. SLAYTER, *J. mol. Biol.*, **14**, 443 (1965)
96. B. J. ADKINS and J. F. FOSTER, *Biochemistry, N.Y.*, **4**, 634 (1965)
97. B. J. ADKINS and J. F. FOSTER, *Biochemistry, N.Y.*, **5**, 2579 (1966)
98. C. TANFORD, *Proc. Iowa Acad. Sci.*, **59**, 206 (1952)
99. J. T. YANG and J. F. FOSTER, *J. Am. chem. Soc.*, **76**, 1588 (1954)
100. C. TANFORD, J. G. BUZZELL, D. G. RANDS and S. A. SWANSON, *J. Am. chem. Soc.*, **77**, 6421 (1955)
101. G. WEBER and L. B. YOUNG, *J. biol. Chem.*, **139**, 1415 (1964a)
102. G. WEBER and L. B. YOUNG, *J. biol. Chem.*, **239**, 1424 (1964b)
103. E. ANNAU, *Nature, Lond.*, **183**, 190 (1959)
104. M. SCHLAMOWITZ, L. U. PETERSON and F. C. WISSLER, *Archs Biochem. Biophys.*, **92**, 58 (1961)
105. F. SMET, R. LONTIE and G. PREAUX, (ed. H. Peeters) *Protides of the Biological Fluids*, p. 119. (Elsevier Publishing Co., Amsterdam 1963)
106. W. G. M. BRAAM, B. J. M. HARMSEN and G. A. J. VAN OS, *Biochim. Biophys. Acta*, **236**, 99 (1971)
107. V. LUZZATI, J. WITZ and A. NICHOLAIEFF, *J. mol. Biol.*, **3**, 379 (1961)
108. V. BLOOMFIELD, *Biochemistry, N.Y.*, **5**, 684 (1966)
109. G. FRANGLEN and G. R. E. SWANIKER, *Biochem. J.*, **109**, 107 (1968)
110. D. M. PEDERSON and J. F. FOSTER, *Biochemistry, N.Y.*, **8**, 2357 (1969)
111. H. BENNHOLD, (ed. H. Peeters) *Protides of the Biological Fluids*, p. 58. (Elsevier Publishing Co., Amsterdam 1961a)
112. H. BENNHOLD, *Bull. schweiz. Akad. med. Wiss.*, **17**, 62 (1961b)
113. F. KARUSH, *J. phys. Chem.*, **56**, 70 (1952)
114. F. KARUSH, *J. Am. chem. Soc.*, **76**, 5536 (1954)
115. G. MARKUS and F. KARUSH, *J. Am. chem. Soc.*, **80**, 89 (1958)

Chapter 10

Nutrition and Plasma Proteins

J. S. Garrow

Clinical Research Centre
Watford Road, Harrow,
Middlesex HA1 3UJ

Because the plasma proteins are so much easier to sample than other body proteins much of the early work on the effects of protein nutrition centred on changes in plasma protein. This accessability was technically an advantage, but it also tended to distort the true picture of the nutritional role of the circulating and fixed tissue proteins respectively. Thus Whipple and his colleagues[1] concluded that during protein depletion there was "raiding" of tissue proteins to maintain plasma protein, and that "the blood proteins in these experiments take priority over the organ and tissue proteins". This conclusion is hardly justified in view of the techniques used to produce protein depletion: the dogs were fed a protein-free diet and plasma proteins were removed by plasmapheresis. If thereafter plasma protein was made at the expense of tissue protein this is no more than a demonstration of homeostasis. Had it been feasible to remove progressively portions of liver no doubt this organ too would have been regenerated at the expense of other proteins which were not specifically depleted, and it would have been equally logical to conclude that liver protein had priority over other tissues in time of protein shortage. The true pricture is that protein depletion in an intact animal, however, it is produced, affects all the body proteins since they are in dynamic equilibrium[2] although the extent and timing of protein loss varies greatly from one organ to another.[3] It is convenient, therefore, to consider the effects of nutrition on plasma proteins from three aspects: first, the interrelationship of plasma and tissue protein stores in times of nutritional stress of various kinds; second, the diagnostic value of plasma protein concentrations in assessing nutritional status; and third, the dynamic changes in plasma protein metabolism which occur in response to changes in diet. Only studies in intact mammals especially man, will be considered, since elegant and ingenious studies on isolated perfused organs are described elsewhere.

10.1. THE NUTRITIONAL INTERRELATIONSHIP OF PLASMA AND TISSUE PROTEIN STORES

According to the rather small number of whole body analyses available in the literature,[4-7] a normal adult contains about 160 g of protein per kg body weight, and of this about 2% is intravascular plasma protein. It is clear that, since dietary protein requirements in the adult are about 0.5 g/kg per day, the total intravascular protein mass represents only about one week's dietary requirements. The tissue proteins must therefore provide the majority of the aminoacids required during long periods of protein deprivation. The data in Table 10.1 show the effects

TABLE 10.1

The loss of total body protein, and of intravascular plasma protein, in dogs on a protein-free diet with and without plasmapheresis

Dog	Days on protein-free diet	Protein removed by plasmapheresis (g)	Intravascular plasma protein (g)	Plasma protein loss (%)	Tissue protein loss (%)
A	0	0	19.8	0	0
	22	0	18.0	9.2	7.9
	57	44	9.6	51.5	31.0
B	0	0	24.4	0	0
	22	0	17.8	27.0	8.9
	61	42	12.2	50.0	28.7
C	0	0	20.7	0	0
	29	0	18.5	10.7	15.5
	64	35	9.9	52.1	29.6
D	0	0	18.4	0	0
	29	0	16.8	8.8	17.7
	68	54	9.1	50.6	37.7

of prolonged feeding of protein-free, but calorically adequate, diets to dogs in the laboratory of the late J. B. Allison.[8] After 3 to 4 weeks of protein-free diet some dogs had lost relatively more intravascular plasma protein, whilst others had lost more tissue protein: this variability in response from one animal to another is also remarked by Whipple and is colleagues.[1] After about 9 weeks on the protein-free diet, with bleeding to accelerate depletion during the 5th and 6th week, the dogs had lost more than half of their intravascular protein and about a third of their tissue protein. This degree of depletion is about the maximum compatible with survival in the adult dog.

For obvious reasons there are no comparable results in human subjects. A most interesting study was that of Hoffenberg[9] who showed that 14 normal adult volunteers on a low protein diet for 3 to 6 weeks decreased both intravascular and extravascular albumin pool masses, and that there was a slight shift of albumin from the extravascular to the intravascular compartment. The experimental diet provided 10 g of protein per day, and although no measurements of nitrogen balance were made one can infer from the data of Calloway and Spector[10] that the daily loss of body protein would be about 30 g. Over a period of 3 to 6 weeks, therefore, the loss would be of the order of one kilogram from a total body protein mass of about 10 kg, in other words a 10% loss. This calculation is, of course, very rough, but the answer obtained is plausible when compared to the protein loss recorded in other experiments in human undernutrition.[11] While the loss to body protein was of the order of 10%, intravascular albumin mass was reduced by 13%, and intravascular gamma globulin by 18%, while the extravascular albumin and globulin pools were reduced by 19% and 23% respectively. These results, therefore, support the general view that in moderate protein depletion, the loss to plasma protein is, in general, rather greater than that to tissue proteins. In severe human malnutrition, estimates of the degree of depletion of tissue and plasma protein are necessarily crude. Malnourished children have been intensively studied in well equipped metabolic units in many countries, and the changes in plasma protein mass, and in whole body nitrogen balance during recovery from malnutrition are well documented. However an almost insuperable difficulty arises when one tries to express degrees of protein depletion in a quantitative way: there is no suitable standard to which one can compare a severely malnourished child. This problem has been discussed at length elsewhere[12] and a case can be made for comparing the malnourished child with a normal child of the same height. On this basis children who die of malnutrition have lost some 20–38% of total body protein.[13] On the other hand, in children who recover from malnutrition, it can be calculated retrospectively that the initial deficit in circulating protein was 34% and in intravascular albumen 37%.[14] The two methods calculations are not comparable, nor should we assume that the children who died were similar in depletion to those who recovered. The best we can do from the available data is to conclude that, in severely malnourished children, the percentage deficit in plasma and tissue protein is roughly of the same magnitude. It is unlikely that better data will become available, since there is at present no method of measuring the degree of tissue protein loss in a living malnourished child.[15] However, it is clear from animal experiments that starvation initially causes a profound loss of liver protein[3]—the so-called labile or reserve protein—and that subsequently the loss is born by tissues of lower metabolic activity. We therefore cannot assume that the deficit in plasma protein is proportional to that in tissue protein during the development of severe human malnutrition, and again it is unlikely that evidence on this point will ever be obtained, since it would involve making serial measurements of plasma protein mass and of cumulative nitrogen deficit in a child who was becoming progressively more malnourished. Anyone with the facilities to make these quite difficult measurements would presumably also have the facilities and inclination to treat the child, so one hopes that the experiment will not be done.

10.2. PLASMA PROTEIN CONCENTRATION AS AN INDICATOR OF NUTRITIONAL STATUS

There is no evidence that, if protein intake is adequate, there is any correlation between plasma protein concentration and dietary protein, nor, on physiological grounds, would such a correlation be expected. A study of 40 pre-school children who ate 30–103 g protein per day[16] showed no significant relationship between protein intake and any of the plasma protein fractions, although there was a positive correlation coefficient, significant at the 5% level, between protein intake and haemoglobin concentration.

In protein deficiency states, plasma protein concentration, and especially plasma albumin concentration, is one of the simplest and most sensitive tests. South African paediatricians have used plasma albumin as an index of malnutrition in children who also had gastro-enteritis,[17] since in such children the other useful criterion, deficit in body weight, is invalidated by dehydration. Obviously, if a child is both dehydrated and hypoproteinaemic it is reasonable to conclude that if it were not dehydrated, it would be even more hypoproteinaemic. How-ever, a recent committee report on methods of assessing protein nu-tritional status[18] concludes that "all the biochemical tests which have been described leave much to be desired . . ." (including plasma protein concentrations) and they note in particular that, although protein and albumin concentrations are consistently low in kwashiorkor, they may be normal in marasmus. Furthermore, factors other than protein nutrition may affect serum protein concentrations: thus malnourished children with hookworm infestation are more hypoproteinaemic than similarly malnourished but uninfested children.

Despite these limitations, workers in Guatemala, Jamaica, West Africa, Mexico, Uganda[14] and other countries are agreed that in protein malnutrition there is a large reduction in albumin concentration, and a lesser reduction in beta-globulin. Reports on other fractions separated by paper electrophoresis are not unanimous. While it is encouraging to find such international agreement on anything, the argument is to some extent a circular one: having decided that hypoproteinaemia is a car-dinal feature of protein malnutrition, everyone finds that all malnour-ished children are hypoproteinaemic. With the kwashiorkor type of malnutrition, in which growth retardation and oedema are always found[19] (usually in association with skin and hair changes) and hepa-tomegaly due to fatty infiltration, hypoproteinaemia is also constantly found. The difficulty arises, and is interminably debated, about the child who is very small and wasted, but who has normal skin and hair, no hepatomegaly, and no oedema. These marasmic children may have normal plasma protein concentrations, but if such a child weighs 5 kg at the age of one year, has a history of semistarvation since birth, and on being fed a normal diet gains weight at the astonishing rate of 60 g or more per day[20] it is undeniable that the child was malnourished.

International attempts to agree on the diagnostic features of marasmus have failed, because, although kwashiorkor has common features in most countries[19] marasmus has not. Thus the features which distinguish the two syndromes in Uganda,[21] one of which is the hypoproteinaemia of kwashiorkor, do not apply in Jamaica[22] or Cape Town.[23] Generally in Africa, the Middle East[24] and in South America[25] marasmus occurs acutely in very young children who have never been adequately breast fed, whereas kwashiorkor occurs in the weanling child at about two years of age, when for some months a starchy pap or sugar has been substituted for the mother's milk. On the other hand in Central America, and the West Indies,[14] and in India,[26] there is no clear demarcation in age distribution between the two types of malnutrition, and indeed, relatively few cases are either classical marasmus or kwashiorkor: the majority lie in an intermediate range of the clinical spectrum.[22]

The whole concept of a "test of nutritional status" in man is an invitation to circular and futile argument. In farm animals nutrition has a well defined and measurable objective: the production of meat, eggs, milk, or some such human foodstuff, with maximum efficiency and minimum cost. Obviously the objectives in human nutrition are different, and it is by no means certain that the diet which promotes the most rapid growth in children leads to maximum health and longevity in the adult. In comparison with their colleagues in animal husbandry, human nutritionists are painfully unsure what they are trying to achieve, and virtually incapable of finding out whether or not they have succeeded. In the face of such uncertainty it is rather ridiculous to describe plasma protein concentrations (or any other biochemical measurement) as a "test of nutritional status". If a child with kwashiorkor has a low albumin concentration, this tells us nothing that clinical examination would not have revealed. It is tempting, but not necessarily correct, to assume that a child with no clinical evidence of malnutrition, but with a marginal reduction in albumin concentration, has "subclinical malnutrition". At this point the discussion passes from the realm of biochemistry and physiology into that of semantics.

10.2.1. CHANGES IN SPECIFIC PROTEINS IN MALNUTRITION

In recent years, investigation of the effects of malnutrition on plasma proteins has concentrated increasingly on protein fractions with known physiological functions: antibody globulins, lipoproteins and other plasma proteins with transport functions, and plasma enzymes. Although, as discussed above, estimates of the degree of protein depletion from plasma concentrations are not very helpful, indications of disordered function are more relevant. In one series[27] it was shown that

11

the mortality among children with severe primary malnutrition was not correlated with plasma protein concentration, but was very significantly correlated with the concentration of serum glutamic oxalacetic transaminase which indicated the degree of liver damage.[28]

Circulating Enzymes

A large volume of work has been done on enzymes circulating in the plasma of malnourished children in the hope that specific points of metabolic breakdown would thus be identified, but on the whole the results have been disappointing from this point of view. The literature up to 1960 is excellently reviewed by Waterlow[14] and since that time attention has shifted more towards the measurement of intracellular enzymes in leucocytes and biopsy samples of liver.[29] Even with this more direct approach on the sites of enzyme production the results are very difficult to interpret: empirically there is no pattern of enzyme change which is a reliable indicator of the nature of the malnutrition in an individual child, or of the probable clinical outcome and theoretically there is ambiguity about an abnormal enzyme concentration. If the enzyme is deficient, does this mean that under the stress of protein deficiency there is not enough of this valuable enzyme available, or merely that there is so little substrate available for the reaction with which the enzyme is concerned that there is no need for it in normal concentrations? Conversely, if the concentration is abnormally high, does this mean that there is an adequate supply of the enzyme, or that this point in the metabolic path is overloaded, or even that the enzyme has built up in the absence of a co-factor—as for example phosphorylase does in fructosaemia? In the absence of firm answers to questions such as these isolated enzyme measurements are unhelpful. Theoretically sounder, but technically more exacting, is the measurement of a series of enzymes along a metabolic pathway, combined with measurements of the intermediary metabolites.[30]

Difficult as they are to interpret, the changes in serum enzymes in children with kwashiorkor are shown in Table 10.2. These data are

TABLE 10.2

Abnormalities in serum enzyme concentration in children with kwashiorkor. (after Whitehead 1968)

Enzyme	Abnormality
Amylase	Much reduced
Lipase	Much reduced
Alkaline phosphatase	Increased or reduced
Cholinesterase	Reduced
Creatine kinase	Reduced
Ornithinecarbamyl transferase	Reduced
Lactate dehydrogenase	Increased or normal
Isocitrate dehydrogenase	Increased or normal
Malate dehydrogenase	Increased or normal

taken from a review, by Whitehead[31] of the findings in the Medical Research Council unit in Uganda.

Changes in Immunoglobulins

Clinically, the problem of malnutrition is greatly complicated by associated infection.[32] Many factors affect this association: infection, especially gastroenteritis, predisposes to worse nutrition; the social conditions which lead to malnutrition are favourable to the transmission of infection; and the malnourished patient is less able to resist infection than a well nourished one. The susceptibility of malnourished children to infection has caused several workers to investigate the ability of malnourished children to produce antibodies, but the findings are inconclusive. The older work, in which the globulin fractions were separated by electrophoresis, has been mentioned above, so only those studies in which IgG, IgA and IgM were measured by specific methods will be further discussed.

Brown and Katz[33] found lower concentrations of IgG in 20 children with kwashiorkor, than in 5 controls, but in 7 marasmic infants Najjar et al.[34] found increased levels of IgG, IgM and IgA. Keet and Thom[35] found normal levels of IgG and IgM, but increased IgA in children with kwashiorkor. Two recent papers report serial measurements of immunoglobulins in patients recovering from malnutrition. Watson and Freesemann[36] found in 24 malnourished children that the concentration of all three immunoglubulins was initially higher than in normal healthy children, and during 6 weeks of convalescence IgG showed a further significant increase, while IgA significantly decreased. However, in Ibadan, McFarlane et al.[37] followed over 80 children with kwashiorkor for 7 weeks of treatment, and concluded that there was no significant abnormality in any of the immunoglobulins, nor was there a consistent trend during recovery. From these data, taken as a whole, it seems unlikely that malnourished children have any impairment of antibody production, and the differences in findings between one series and another is probably explained by the variations in type and severity of infection.

Specific Metal-binding Proteins in Malnutrition

It has been shown that malnourished children have reduced blood or tissue levels of iron, copper, selenium and zinc, as well as a profound potassium deficiency. The general problem of electrolyte metabolism in malnutrition has been discussed elsewhere,[12] but it is relevant to this chapter to discuss the extent to which deficiency in transport protein can be contributing to trace metal deficiencies.

Edozien and Udeozo[38] found in 35 children with kwashiorkor a reduction in serum iron (40.6 $\pm$ 21.5 micrograms % compared with 62.7 $\pm$ 17.0 in normal controls) and also in iron binding capacity

(119.3 ± 54.4 compared with 273.0 ± 59.0). It is interesting, therefore, to note that Srikantia found ferritin circulating in the plasma of 6 children with kwashiorkor, and 2 adults with nutritional oedema, and postulated that the oedema might be due to the antidiuretic effect of ferritin.[39] It is not clear if the presence of ferritin in the plasma simply reflects damage to the cells of the gut macosa, to which ferritin is normally confined, or if this is some sort of compensation for a gross deficiency of the normal plasma transport protein, transferrin (syn. siderophylin). There is good evidence that transferrin is reduced in concentration in children with kwashiorkor. Antia et al.,[40] say "This may well be the most profound defect known of acquired disturbance of production of a plasma protein". Certainly in their series the transferrin was reduced proportionately much more than total plasma protein, but a later publication, from the same laboratory,[37] reported that 37 children with "severe" (on clinical criteria) kwashiorkor had transferrin levels on admission of 0.72 ± 0.65 mg/ml, and after 7 weeks of treatment this had increased to 1.81 ± 0.41: this increase on recovery is similar in magnitude to that of albumin, which increased from 1.96 ± 0.54 g% to 3.00 ± 0.51: in both cases the mean value increased by about twice the standard deviation.

Similar observations have been made about serum and tissue copper concentrations in malnourished children. Serum copper is reduced in malnourished children in Central Africa,[38] Guatamala,[42] India[43] and Egypt,[44] and there is also evidence of reduction of copper in skin, hair and liver[45] but there is no simple relationship between the degree of depletion in different organs. In Chile, anaemia due to copper deciency has been reported in children with both malnutrition and gastroenteritis.[46] The carrier protein, caeruloplasmin, is at low concentration in the serum of children with kwashiorkor, and in McFarlane's series[37] increased with treatment from 0.22 ± 0.11 mg/ml to 0.37 ± 0.06. Caeruloplasmin is therefore a candidate for inclusion in a test battery for biochemical assessment of malnutrition.[47] However the philosophical problem remains, however sophisticated such a test battery may become, either it is demonstrating something which we already know clinically, namely that the child is malnourished; or else in a clinically normal child we are to believe there is subclinical malnutrition. The logical dangers of a diagnosis of subclinical malnutrition have been discussed above.

Lipoproteins in Malnutrition

There is a refreshing objectivity about the study of plasma lipoproteins in malnourished children, because some children do, and some do not, have fatty infiltration of the liver. When the liver is heavily infiltrated, as much as one third of the total body fat may be in the liver,[13] and 30–50% of the wet weight of the liver is fat.[48] Minimal clinical skill is required to assess the size, or change in size, of a liver so grossly enlarged, so in the case of fat transport we are able to follow the progress of a recognisable abnormality day by day with no trauma to the child, and no diagnostic facilities other than a moderate skill in palpation of the abdomen. Some workers insist[49] "palpation of the liver is unreliable in diagnosing fatty liver in kwashiorkor", and say that needle biopsy is required for accurate estimation of liver fat. This is certainly true of minor degrees of fatty infiltration. The important point, however, is that we cannot quantitate malnutrition in children, but we can quantitate fat in livers.

Until recently, explanations for the fatty liver of kwashiorkor were very unsatisfactory. However, the situation has become considerably clearer with two recent publications from South Africa[49] and Uganda.[50] Truswell and his colleagues in South Africa showed that during recovery from kwashiorkor the concentration of beta lipoprotein cholesterol increased twofold in the first week or so, while alpha lipoprotein cholesterol remained virtually unchanged. Fasting serum triglycerides also increased even more rapidly, reached a peak at about 5 days, and then declined. They also showed that the children with the more severe fatty liver had lower initial concentrations of beta lipoproteins and triglyceride. They interpreted these findings as consistent with the hypothesis that reduced synthesis of low density lipoproteins is a major cause of fatty liver in kwashiorkor. The results of Flores and his colleagues in South America, who studied both experimentally protein depleted rats[51] and malnourished children[52] are also in favour of this explanation. The publication from Uganda is a valuable development of this hypothesis, since Coward and Whitehead[50] classify their series of malnourished children according to the albumin concentration (rather than by the liver fat, as used by Truswell[49]) and they also conclude that the fatty liver is a result of insufficient beta lipoprotein being available for transport of fat away from the liver. Coward and Whitehead extended their investigations to include screening of children at an outpatient clinic to try to trace the development of biochemical changes in those at tisk to severe protein calorie malnutrition, and they report two cases in whom they have serial readings over a period of more than a year. By combining these longitudinal data with a cross-sectional survey, and data from children in hospital who are recovering from malnutrition, they build up a picture which is more comprehensive than any other in the literature, showing the natural history of the development and resolution of kwashiorkor. If one makes the assumption that the development of protein calorie malnutrition is characterised by a progressive fall in serum albumin concentration then the sequence of changes in beta lipoprotein and cholesterol is as follows: cholesterol is the first to decrease, so that when the albumin concentration has fallen to 2.5 g/100 ml the cholesterol is only 70% of the initial value, while beta lipoprotein is virtually unchanged. By implication this early decrease in cholesterol indicates loss of alpha lipoprotein. As the albumin falls further to 1.5 g/100 ml, which is the level found in frank kwashiorkor, beta lipoprotein falls to about 70% of the initial value, cholesterol and triglyceride are only half the initial value, and alpha lipoprotein is virtually undetectable. During recovery, as in Truswell's series,[49] triglycerides increase very rapidly, followed by beta lipoproteins a few days later, and albumin, cholesterol, and presumably alpha lipoprotein later still. In marasmic children, who do not have fatty livers, none of these changes are found.

Here, then, is an unusually well documented effect of malnutrition on a plasma protein fraction. We can see not only that the beta lipoprotein is reduced in quantity, but also that its function—that of transporting fat from the liver—is not being performed, and conversely, as the child recovers, beta lipoprotein is rapidly regenerated and plasma triglyceride increases and the fat is once again mobilised. This is not in itself a completely satisfactory account of why the fatty liver develops, as Coward and Whitehead themselves observe: it is still not clear, for example, why a child who is having less than his calorie requirements (although relatively more calories than proteins) should be so misguided as to deposit a significant fraction[22] of these dietary calories as fat in the liver. A possible explanation is that the severe protein deficiency greatly inhibits protein turnover and metabolic rate and hence requirements for dietary calories, and that the virtual absence of alpha lipoprotein means that the excess calories cannot be deposited as subcutaneous fat. The only possible fate for the excess calories is then to be deposited as liver fat, but this is speculation for which there is no direct evidence.

10.3. THE EFFECTS OF NUTRITION ON PLASMA PROTEIN TURNOVER RATES

Up to this point the discussion has concerned static measurements of plasma proteins: the amount or concentration present at the time of measurement. However changes in amount or concentration reflect changes in flux. If the amount decreases it must be because synthesis has decreased, or catabolism increased, or some combination of these; and a change in flux may be quite undetectable by static measurements, if the pool size remains constant but the rate of turnover changes. It is generally true that in the intact animal, changes in flux can be measured only if the protein under study is isotopically labelled. These isotope tracer techniques are of two kinds: non-reutilisable, as for example when protein is labelled in vitro with radioactive iodine and then injected intravenously, and reutilisable, as when a labelled aminoacid is incorporated into the protein *in vivo*. The interpretation of the former class of experiment is much simpler, and will be considered first.

10.3.1. CATABOLIC RATES OF IODINATED PROTEINS

The simplest study of all is to give an iodinated protein intravenously and to observe the rate of disappearance of the iodine label, from the plasma, over a period of a week or so. However, this technique gives quite misleading results if the plasma protein pool is not in a steady state, and in children recovering from malnutrition it is expanding rapidly. Some early workers, falling into this trap, observed that the rate of decrease of iodinated albumin (cpm. per mg.) was similar in malnourished and recovered children[53,54] and concluded wrongly that albumin turnover rate was not affected in malnutrition. McFarlane[55] drew attention to the erroneous conclusions which may be drawn when pulse labelling techniques are applied to non-steady systems, and in the same year (1962) Picou and Waterlow in Jamaica[56] and Cohen and Hansen in South Africa[57] showed that in fact the rate of synthesis of albumin is greatly reduced in the malnourished child. In order to draw correct conclusions from tracer experiments using iodinated proteins it is necessary to measure the rate of excretion of iodine released from protein during catabolism. The similarity between the apparent half life of plasma albumin in normal and malnourished children arose because, in the malnourished child, there were two factors operating which combined to produce an apparently normal disappearance curve: albumin catabolic rate was reduced, but during the early stages of recovery the albumin pool was rapidly expanding. The former factor tended to reduce the rate of disappearance of iodinated protein, but the latter caused the labelled protein to be diluted by an influx of newly synthesised unlabelled protein. When observations are made only on the plasma these two factors confuse interpretation, but if the iodine excretion rate is measured the true catabolic rate is measured unaffected by fluctuations in pool size.

In the studies of experimental protein depletion of normal adults, Hoffenberg et al.[9] showed that the "synthesis-plus-transfer" rate for albumin decreased from 8.9 ± 0.4 g per day on the control diet to 5.8 ± 0.3 g per day on the low protein diet. "Synthesis-plus-transfer" means the amount of new albumin appearing in the intravascular pool either from de novo synthesis or by transfer from the extravascular pool. In the same subjects the decrease in gamma globulin synthesis plus transfer rate from 6.1 ± 0.9 to 5.0 ± 0.6 g per day was not statistically significant. The mathematical analysis of these data was by the "equilibrium time" method[58] which is designed for reasonably steady state conditions. Recently, however, a technique has been developed which does not require steady state conditions.[59]

A more complex analysis was attempted by James and Hay,[60] who took advantage of the fact that it is possible with a whole body counter to measure the total body radioactivity in a child with an accuracy of about $\pm 1\%$ over a period of 1 month, and hence to obtain separate estimates of loss of isotope from the plasma and from the whole body pool. With these data they were able to follow changes in synthesis and catabolic rate in malnourished and recovered children as they were changed from a high to a low protein intake. In both groups of children on a high protein diet the synthetic rate was high, but whereas in the recovered children the catabolic rate was almost equally high, and hence the pool size steady, in the malnourished children the catabolic rate was reduced (166 mg/kg/day instead of 219 mg/kg/day) and hence the pool size was increasing. When the protein intake was reduced the synthetic rate promptly decreased, (the effect is detectable after 1 day, and maximal by the second week), while the catabolic rate decreased more gradually: thus a new equilibrium was reached with a smaller pool size.

From the investigations quoted above of experimentally protein depleted adults,[9] and malnourished children studied during the process of recovery,[60] it is possible to draw conclusions about the mechanisms which control the adaptive responses in albumin metabolism to protein depletion and repletion. It is convenient to consider first the effects on catabolic rate, since this is the process more directly measured by tracer studies using iodinated albumin. It is clear from the work of James and Hay[60] that dietary intake has a greater effect than nutritional status on catabolic rate, since well-nourished children showed a fall in catabolic rate after one week on a diet supplying 0.7–1.0 g protein per kg per day. This diet is insufficient to support normal growth in a child about one year old, but it cannot be maintained that one week of such a diet would produce protein depletion. The reduction in catabolic rate occurs before there is any fall in serum albumin concentration, so it cannot be an effect of hypoalbuminaemia, nor does it occur immediately the protein intake is reduced, but only after a delay of 3–5 days. It is also clear that neither during depletion nor repletion is albumin catabolic rate directly linked to synthetic rate, since both studies show that the change in synthetic rate occurs much more rapidly than that in catabolic rate when dietary intake is altered. Also in proteinuria, the synthetic rate is normal although the catabolic rate is low. It seems probable, therefore, that of the two processes which operate to defend the intravascular albumin pool in times of protein restriction the first, namely transfer of albumin from the extravascular pool, operates rapidly and independently of dietary intake: this process operates equally well if plasma albumin is removed by plasmapheresis. The second adaptive process, namely change in turnover rate, is more complex and related to dietary intake rather than protein stores. The two components of this process appear to be independently controlled: synthesis alters rapidly with dietary change and catabolism only after a time lag which possibly indicates a hormone-mediated control system.

The other point which comes out from these studies is the remarkable capacity for synthesis which is maintained, even in states of severe protein deficiency. This aspect is linked with the question of re-utilisation of amino-acids, which is discussed further in the section below.

10.3.2. STUDIES WITH REUTILISABLE ISOTOPE TRACERS

The work described above has clearly shown that the metabolism of plasma proteins is markedly affected by nutritional deprivation, but that discussed in the first section of this chapter, also showed that there is continuous interaction between the protein metabolic pools of plasma and tissues. It is an inherent limitation of studies using non-reutilisable labels, such as *in vitro* iodination, that they cannot be expected to yield information about any protein other than that which was so labelled. If, on the other hand, an aminoacid with a suitable isotopic tracer is administered to an intact animal its subsequent distribution might, in principle, shed light on the distribution of synthesis between different organs and tissues. Given a sufficiently naive approach,[61] one can interpret data concerning urine and plasma radioactivity as showing a shift in synthesis from the more active proteins (such as plasma proteins) during periods of protein depletion. Fig. 10.1 shows the incorporation of ^{35}S methionine into mixed plasma proteins in a dog during progressively more severe protein depletion, and Fig. 10.2 shows similar curves in a malnourished child before and after treatment. There is no doubt that, during protein depletion, the plasma protein specific activity, or the percentage of a given dose of labelled aminoacid which appears in intravascular plasma protein, is dramatically increased. However, as McFarlane[55] has shown, this need not be in any way a reflection of altered metabolic rate: exactly the same results could be obtained if the turnover rates of all the proteins remained the same, but the size of the metabolic pools was altered. The point is that one cannot calculate the rate of synthesis of any protein from the rate of incorporation of isotope into that protein unless the specific activity of the precursor amino acid is known. In the case of pulse labelling of whole animals the precursor specific activity cannot be accurately known: it is changing rapidly in the plasma, and, in different tissues, the intracellular specific activity at any point in time will depend both on the plasma free aminoacid specific activity at that time and also on the permeability of the tissue to free aminoacids. This varies from tissue to tissue, and perhaps from time to time in the same tissue: liver, for example, is much more permeable to free aminoacids than is muscle.[62] Therefore if studies with labelled aminoacids are to be used to give quantitative results a quite sophistiated mathematical analysis is needed.

Readers who are familiar with the recent literature on aminoacid tracer techniques may feel that it is unnecessary to explain yet again that it is impossible in the intact animal to calculate protein synthetic rates from rates of aminoacid incorporation without knowledge of precursor specific activity. However, at the time of writing, there was a paper in a current journal which tries to do just that. Walker *et al.*[63] report albumin synthesis rates in patients with hypoproteinaemia: these were measured from the rate of incorporation of seleno-methionine after a

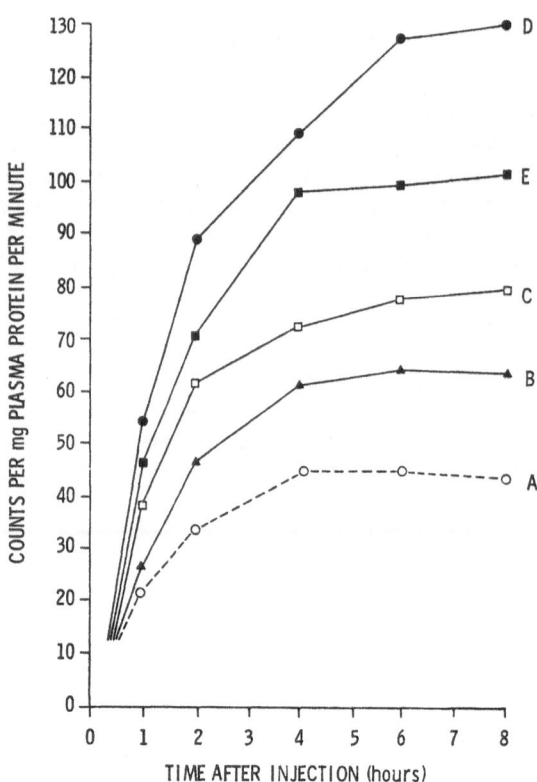

FIG. 10.1. Incorporation of radioactivity into plasma protein in dogs given ^{35}S-methionine intravenously. Curve A, normal dog in nitrogen balance; B, normal dog on protein free diet; C, dog depleted of 17.7% of body protein; D, depleted of 21.1% of body protein; E, depleted of 37.7% of body protein.

single intravenous injection of the aminoacid labelled with ^{75}Se. The authors state "The rate of incorporation of an aminoacid into a specific protein directly reflects the rate of synthesis of that protein", and quote as authority McFarlane[64] and Jarnum.[65] In the relevant passage McFarlane[64] makes it clear that the quoted statement is true only if the precursor specific activity is constant, and adds "In practice, the precursor specific activity following a single injection or feeding of an aminoacid is far from constant . . .". Jarnum[65] says on the page referred to: "The rate at which an aminoacid is incorporated into plasma protein is a *relative* measure of the synthetic rate of the particular protein concerned. However, the value of this method is limited, because a reduction of total plasma protein mass due to continuous abnormal protein loss (as in nephrosis) causes *per se* an increased rate of incorporation, even if no increase in synthesis occurs". Despite these

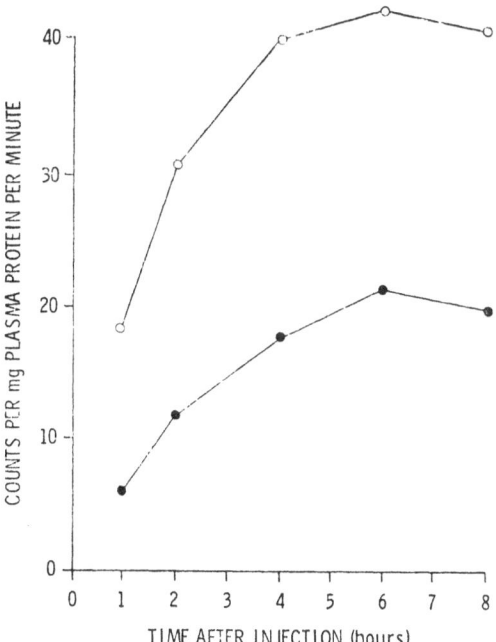

FIG. 10.2. Incorporation of radioactivity into plasma protein of a child after intravenous injection of ^{35}S-methionine. Upper curve, malnourished child; lower curve, the same child after 31 days of protein repletion.

warnings, Walker et al.[63] interpret increased incorporation rates in patients with abnormal protein loss (including nephotics) as evidence of increased synthesis rates. Their choice of selenomethionine as the marker aminoacid adds further confusion to the interpretation of their data, because there is marked reutilisation of the selenium label in children,[66] and measurements of protein turnover rate measured in the same child, at the same time, by simultaneous doses of selenomethionine and ^{15}N-glycine give quite different results.[66]

It is obvious, on reflection, that aminoacid incorporation rates into plasma proteins must be affected also by the turnover rate of other tissue proteins. Suppose that in a normal animal, 6 hours after the injection of a labelled amino acid, 5% of the radioactivity had been excreted, 10% of the radioactivity was in the plasma proteins, and the remaining 85% was incorporated in other tissues. Suppose that in a similar animal a highly malignant tumour is implanted which does not affect the quantity or turnover rate of any other protein in the body, but which has itself a very high protein turnover rate. If the same dose of labelled aminoacid is injected into this tumour-bearing animal, and the same 5% of activity is excreted in the first 6 hours, it is obvious that

the distribution of the remaining activity in the plasma and tissues cannot be the same as in the normal animal, although pool sizes and turnover rates are the same, because the radioactivity taken up by the tumour
will not be available to label the normal tissues. Despite the apparent
obviousness of this conclusion, and the frequency with which it is
restated in authoritative reviews, the point is evidently still not clear
even to the editorial board of a reputable international scientific
journal.

No attempt will be made here to review the valid methods for mathematical analysis of data derived after administration of labelled amino
acids. The chapter by Waterlow[69] in the series of volumes entitled
"Mammalian Protein Metabolism" is an excellent account of the effects
of nutrition on human protein metabolism, and is easily understood by
non-mathematicians. For the more mathematical the report of a
symposium on "Radioisotope techniques in the study of protein metabolism"[68] makes fascinating reading: it is particularly interesting to
note in the discussion sections that, even among the foremost experts,
there is some divergence of opinion about the validity of different
methods of analysis. These differences, however, are relatively minor,
so an attempt will be made below to state the conclusions reached with,
where necessary, a comment on the possible magnitude of analytical
artefact.

For the reasons given above the single-shot-plasma-activity, type of
experiment[61,68] is quantitatively meaningless. An alternative type of
study is that of Sprinson and Rittenberg who used the rate of excretion
of ^{15}N in urine, after giving ^{15}N glycine, to calculate the protein turnover rate in man. Their mathematical analysis was based on two nitrogen pools in the body in dynamic equilibrium: a metabolic pool of
free aminoacid which the ^{15}N glycine labelled and from which the urea
was derived, and a protein pool, which was a weighted mean of all
body proteins. This model was refined by San Pietro and Rittenberg,[69]
who inserted a urea pool between the metabolic pool and the urine.
This type of experiment is by no means meaningless[70] and has been
used, with modifications, in the study of protein metabolism during
protein restriction and malnutrition.[71,72,73] However this type of
experiment cannot, by its nature, yield information about the nutritional interrelationships between plasma and tissue protein: the metabolic pool in a stochastic model is a mathematical abstraction, whereas the plasma albumin pool is an anatomical reality. It is not possible
to show that calculations based on one type of pool reflect metabolic
behaviour in the other.

There are two techniques by which reutilizable markers can be made
to yield information about the turnover rates of specific proteins. The
easier to understand is the continuous infusion method which has been
developed by Waterlow and Stephen, the theory of which is explained
in his review.[67] If, instead of giving an isotopically labelled aminoacid

as a single shot, it is given as a continuous infusion over a period of about 24 hours, the specific activity of protein precursor reaches a more or less constant plateau level: the rate of entry of labelled aminoacid into the metabolic pool then equals the rate of loss, if the pool size remains constant. The mathematical analysis now becomes relatively simple, and during the period of plateau, the requirement for quantitative estimation of protein turnover rate—that the precursor specific activity should be both known and constant—is met. Under these conditions it is to be expected that the specific activity of plasma protein would increase linearly with time at least for the first few hours, and this was found to be so.[74] It was also found that, by this technique the lysine turnover in normal adult men gave results which agreed well with other methods.[75] Picou and Taylor-Roberts[76] used this method to measure rates of synthesis and catabolism in malnourished and recovered children, and found, rather surprisingly, that the malnourished children had a normal net rate of protein synthesis, but a considerably increased rate of nitrogen turnover. This paradoxical result can be explained by the increased efficiency of reutilisation of aminoacids in malnourished children: a greater proportion than normal of aminoacids released by catabolism of protein find their way back into protein again, instead of being deaminated to form urea.

It is perhaps not generally realised what a large reutilisation of aminoacids occurs even in well nourished subjects. If an adult takes in 50 g per day of protein in his diet, and excretes an equivalent amount of urea nitrogen the "exogenous" protein metabolism (in the old Folin nomenclature) is only 50 g per day. However the studies of San Pietro and Rittenberg[69] and others indicate that the total protein turnover is about 300 g per day in a normal adult: therefore about 250 g of this protein is made with aminoacid which has been released into the metabolic pool during turnover of tissue proteins.

Increased reutilisation in malnourished children can be inferred from the results of tracer experiments, but there is also direct evidence of this form of adaptation. Stephen and Waterlow[77] have shown that aminoacid activating enzymes decrease, and argininosuccinase increases, in the liver of children as they recover from malnutrition. Thus, in the malnourished child, the probability of an aminoacid being reutilised is enhanced, and that of it being degraded to urea is reduced: the combined effect of these changes must be to favour the salvage rate of aminoacids for further protein synthesis, despite a poor dietary intake.

The second method by which labelled aminoacid may be used to give quantitative estimates of synthesis rate, is that of McFarlane.[78] This method meets the requirement that both the rate of incorporation of radioactivity into protein and the specific activity of the precursor should be known, by taking advantage of the unusual properties of the guanidine carbon of the aminoacid arginine. Arginine itself is used for protein synthesis, and so is as valid a tracer as any other aminoacid.

The respect in which arginine differs is that it is a part of the urea cycle, and that in the process of urea formation the guanidine carbon of arginine is transferred to the urea molecule. The effect of this is that the probability is very low of arginine being released into the metabolic pool by catabolism of protein, and being reincorporated into protein again, still with the same guanidine carbon atom. In other words, if this carbon atom of arginine is isotopically labelled we have effectively a non-reutilisable, aminoacid—at least the radio-active label is not reutilisable, and this is what is measured. Further-more, since the labelled atom is stripped off to form urea, one can infer the specific activity of the arginine at any point in time from the specific activity of the urea being formed at that time. Still more con-veniently, it happens that it is not even necessary to administer labelled arginine: the same urea cycle reactions can be used to label the ap-propriate atom of arginine *in vivo* (admittedly at rather low yield) if carbonate containing radioactive carbon is given. The above descrip-tion is perhaps unduly simplified, but more extensive critiques of the methodology are available in the reviews already cited. It is sufficient for the present argument to note that the method gives results which agree well with other methods. Kirsch et al.[79] used a combination of ^{14}C-carbonate and ^{131}I-albumin to measure simultaneously the rates of synthesis and catabolism of albumin in rats. In animals fed the normal diet (20 % protein) rates of synthesis and catabolism were equal, as one would expect. When some of the rats were transferred to a protein free diet, synthetic rate fell immediately and catabolic rate fell after about 5 days. On protein repletion synthesis rate rose immediately and cata-bolic rate more slowly. The excellent concordance between this well controlled experiment, and the results derived by less direct methods on malnourished children establishes, in so far as that is possible in clinical medicine, that the general adaptive changes described above are common both to malnourished children and experimentally depleted rats.

10.3.3. ADAPTATION IN THE DISTRIBUTION OF PROTEIN SYNTHESIS BETWEEN TISSUES

An animal which is deprived of adequate dietary protein is in a similar position to a government faced with the prospect of national insolvency. It is possible to make cuts "across the board" in the current treasury jargon: to decide that the flux of money (or aminoacid) will be reduced by a certain percentage in each department (or tissue). From the viewpoint of survival, in both cases of the analogy, it is usually possible to devise a better strategy by cutting less essential services more heavily in order to spare as much as possible the more vital ones.

In 1966 Waterlow and Stephen[80] published a paper which was, in effect, a collection of experiments designed to investigate the way in which a rat on a low protein diet reorganised its protein metabolism by altering the distribution of protein synthesis. In particular, attention was paid to the incorporation of L-^{14}C-lysine into liver and plasma proteins (representing proteins of high metabolic activity), and muscle and skin (representing the relatively inert structural proteins). To obtain valid measures of synthesis rate they measured the specific activity of free aminoacid in the tissues. This type of experiment presents formidable difficulties both of technique and interpretation, but since their conclusions are supported by independent evidence obtained by other methods, both before and since their publication, we can assume that these conclusions are substantially correct.

It appears that in protein depletion the plasma proteins suffer cuts in synthesis rate almost immediately the dietary intake is reduced: this is largely made good to the intravascular pool by transfer from the extravascular pool. Probably similar changes in synthesis take place in other tissues of high turnover rate such as liver, gut and exocrine glands, but it is technically difficult to study protein turnover in an organ which is a heterogeneous collection of cells. This may be termed the short-term adaptation, and it is the dynamic equivalent of the changes in organ protein content found after a day or a week of fasting.[3] It is obvious that sacrifice of this so called "labile" protein is biologically convenient as a temporary measure, but biologically disastrous as a long-term policy.

It is not possible to state exactly at what stage of protein depletion the short-term adaptation gives way to long-term effects. In different species the pattern of response varies widely[81] and the quantity of "labile" protein is only about 1 % of the total body protein in children[82] or adults.[83] We may therefore assume that in man, the change from short-term to long-term adaptation takes place earlier than in most other species.

The long-term adaptation is characterised by an increase in the catabolic rate of the lower priority proteins such as skin and muscle. This has been shown by direct measurements on skeletal muscle by Millward[84] using a ^{14}C-carbonate technique, as well as by the long-term changes in body composition of rats,[80] and the clinical features of malnourished children.[85] The long-term changes will therefore make more endogenous aminoacid available for the synthesis of higher priority proteins and this, together with enhanced reutilization, will tend to support the continued synthesis of plasma protein. There is fragmentary evidence that terminally the whole process of adaptation may break down,[61] but this is a phenomenon which it is obviously difficult to study in a systematic and well-controlled way.

10.4. SUMMARY AND CONCLUSIONS

In the intact animal it is difficult to define exactly the effects of nutrition on plasma proteins, since the protein in the circulation is in metabolic equilibrium with a fifty times larger mass of protein in the tissues. Variations within the range of adequate diets have no detectable effect on the plasma proteins, but under conditions of protein deprivation a series of adaptive changes take place. The intravascular albumin pool is defended by transfer from the extravascular pool, but there is no good evidence that plasma proteins take priority over visceral tissue proteins in times of dietary protein restriction. The reduction in total albumin pool size occurs because, when protein intake is reduced below maintenance requirements, there is a very rapid decrease in the rate of albumin synthesis, whereas the rate of catabolism declines more slowly.

The concentration of albumin, transferrin and caeruloplasmin all decrease in malnutrition and may be used as an indicator of the degree of depletion, but they are neither so accurate nor so sensitive for this purpose as the changes in turnover rate which can be shown by tracer studies. However, in clinical practice, tracer studies are impractical, so plasma concentrations serve as a measure of malnutrition, which is itself a very ill-defined state. Gamma globulins are remarkably little affected by nutritional state.

Changes in plasma lipoproteins are of great interest in kwashiorkor, because recent work has shown that the fatty liver which is characteristic of this condition is probably a consequence of the diminished capacity of the plasma to transport fat, owing to the low concentration of lipoprotein carrier. On the other hand, marasmic children, who are also severely malnourished but do not have a fatty liver, have normal lipoprotein concentrations.

To understand the effects of nutrition on plasma proteins it is necessary to study the dynamic equilibrium between plasma and tissue proteins: with improved experimental and mathematical techniques this is becoming increasingly possible.

REFERENCES

1. G. H. WHIPPLE, L. L. MILLER and F. S. ROBSCHEIT-ROBBINS, *J. exp. med.*, **85**, 277 (1947)
2. R. SCHOENHEIMER, "The Dynamic State of Body Constituents", Harvard University Press, Cambridge, Mass. (1942) p. 25
3. T. ADDIS, L. J. POO and W. LEW, *J. biol. chem.* **115**, 111 (1936)
4. H. H. MITCHELL, T. S. HAMILTON, F. R. STEGGARDA and H. W. BEAN, *J. biol. chem.* **158**, 625 (1945)
5. E. M. WIDDOWSON, R. A. McCANCE and C. M. SPRAY, *Clin. Sci.*, **10**, 113 (1951)
6. R. M. FORBES, A. R. COOPER and H. H. MITCHELL, *J. biol. chem.*, **203**, 359 (1953)
7. R. M. FORBES, H. H. MITCHELL and A. R. COOPER, *J. biol. chem.*, **223**, 969 (1956)
8. J. S. GARROW, "Distribution of protein synthesis in malnourished children", M.D. Thesis, University of St. Andrews, (1957)
9. R. HOFFENBERG, E. BLACK and J. F. BROCK, *J. clin. Invest.*, **45**, 143, (1966)
10. D. H. CALLOWAY and H. SPECTOR, *Amer. J. clin. Nutr.*, **2**, 405 (1954)

11. E. G. BENEDICT, W. R. MILES, P. ROTH and H. M. SMITH, Publs. Carnegie Instn. No. 280 (1919)
12. J. S. GARROW, R. SMITH, E. E. WARD, "Electrolyte Metabolism in Severe Infantile Malnutrition", (Pergamon Press, Oxford 1968)
13. J. S. GARROW, K. FLETCHER and D. HALLIDAY, J. clin. Invest., 44, 417 (1965)
14. J. C. WATERLOW, J. CRAVIOTO and J. M. L. STEPHEN, Advances in Protein Chemistry, 15, 131 (1960)
15. D. HALLIDAY, Brit. J. Nutr., 26, 147 (1971)
16. J. CRUMRINE and B. A. FRYER, J. Amer. diet. Ass. 57, 509 (1970)
17. A. S. TRUSWELL, J. D. L. HANSEN, C. FREESEMAN and T. F. SMIDT, S. Afr. med. J., 37, 527 (1963)
18. Committee Report, Amer. J. clin. Nutr., 23, 807 (1970)
19. J. C. WATERLOW and N. S. SCRIMSHAW, Bull. W. H. O., 16, 458 (1957)
20. A. ASHWORTH, Brit. J. Nutr., 23, 835 (1969)
21. R. F. A. DEAN, (ed. D. Gairdner) "Recent Advances in Pediatrics", (J. A. Churchill, London 3rd ed., 1965) p. 234
22. J. S. GARROW, Arch. Latinamer, Nutr., 14, 145 (1966)
23. J. D. L. HANSEN, (ed. R. A. McCance and E. M. Widdowson) "Calorie deficiencies and protein deficiencies", (Churchill, London 1968) p. 39
24. D. S. MCLAREN, Lancet, ii, 485 (1966)
25. F. MONCKEBERG, (ed. R. A. McCance and E. M. Widdowson) "Calorie deficiencies and protein deficiencies" (Churchill, London, 1968) p. 91
26. C. GOPALAN, (ed. R. A. McCance and E. M. Widdowson) "Calorie deficiencies and protein deficiencies" (Churchill, London, 1968) p. 50
27. J. S. GARROW and M. C. PIKE, Brit. J. Nutr., 21, 155 (1967)
28. A. E. M. MCLEAN, Lancet ii, 1292 (1962)
29. D. PINEDA, (ed. R. A. McCance and E. M. Widdowson) "Calorie deficiencies and protein deficiencies" (Churchill, London, 1968) p. 75
30. J. METCOFF, Ann. Rev. med., 18, 377 (1967)
31. R. G. WHITEHEAD, (ed. R. A. McCance and E. M. Widdowson) "Calorie deficiencies and protein deficiencies" (Churchill, London, 1968) p. 115
32. L. J. MATA, J. J. URRUTIA and B. GARCIA "Nutrition and Infection", Ciba Foundation Study Group, No. 31 (Churchill, London 1967)
33. R. E. BROWN and M. KATZ, E. Afr. med. J., 42, 221 (1965)
34. S. S. NAJJAR, M. STEPHAN and R. Y. ASFOUR, Arch. dis. childh., 44, 120 (1969)
35. M. P. KEET and H. THOM, Arch. dis. childh., 44, 600 (1969)
36. C. E. WATSON and C. FREESEMANN, Arch. dis. childh., 45, 282 (1970)
37. H. MCFARLANE, S. REDDY, A. COOKE, O. LONGE, M. D. ONABAMIRO and J. E. HOUBA, Trop. Geogr. Med., 21, 61 (1970)
38. J. C. EDOZIEN and I. O. UDEOZO, J. Trop. Pediatr., 6, 60 (1960)
39. S. BAEZ, A. MAZUR and E. SHORR, Amer. J. Physiol., 162, 198 (1950)
40. A. V. ANTIA, H. MCFARLANE and J. F. SOOTHILL, Arch. dis. childh., 43, 459 (1968)
41. R. G. WHITEHEAD, Arch. dis. childh., 42, 479 (1967)
42. M. E. LAHEY, M. BEHAR, F. VITERI and N. S. SCRIMSHAW, Pediatrics, 22, 72 (1958)
43. C. GOPALAN, V. REDDY and V. S. MOHAN, J. Pediatr., 63, 646 (1963)
44. H. H. STANDSTEAD, A. S. SHUKRY, A. S. PRASAD, M. K. GABR, A. E. HIFNEY, N. MOKHTER and W. J. DARBY, Amer. J. Clin. Nutr., 17, 15 (1965)
45. I. MACDONALD and P. J. WARREN, Brit. J. Nutr., 15, 593 (1961)
46. G. G. GRAHAM, A. CORDANO and J. M. BAERTL, 6th Int. Cong. Nutr., Edinburgh. p. 523
47. H. MCFARLANE, M. I. OGBEIDE, S. REDDY, K. J. ADCOCK, H. ADESHINA, J. M. GURNEY, A. COOKE, G. D. TAYLOR and J. A. MORDIE, Lancet, i, 392, (1969)
48. J. C. WATERLOW and G. BRAS, Brit. Med. Bull., 13, 107 (1957)
49. A. S. TRUSWELL, J. D. L. HANSEN, C. E. WATSON and P. WANNENBURG, Am. J. clin. Nutr., 22, 568 (1969)

50. W. A. COWARD and R. G. WHITEHEAD, *Br. J. Nutr.*, **27**, 383 (1972)
51. H. FLORES, W. SIERRALTA and F. MONCKEBERG, *J. Nutr.*, **100**, 375 (1970)
52. H. FLORES, N. PAK, A. MACCIONI and F. MONCKEBERG, *Br. J. Nutr.*, **24**, 1005 (1970)
53. D. GITLIN, J. CRAVIOTO, S. FRENK and E. L. MONTANO, R. RAMOS-GALVAN, *J. clin. Invest.*, **37**, 682 (1958)
54. J. S. GARROW and J. C. WATERLOW *Clin. Sic.*, **18**, 35 (1959)
55. A. S. McFARLANE, J. S. GARROW and J. C. WATERLOW, "Radioisotopes in Tropical Medicine", I.A.E.A., Vienna (1962)
56. D. PICOU and J. C. WATERLOW, *Clin. Sci.*, **22**, 459 (1962)
57. S. COHEN and J. D. L. HANSEN, *Clin. Sci.*, **23**, 351 (1962)
58. C. M. E. MATTHEWS, *Phys. Med. Biol.*, **2**, 36 (1957)
59. A. S. McFARLANE, A. KOJ, *J. clin. Invest.*, **49**, 1903 (1970)
60. W. P. T. JAMES, A. M. HAY, *J. clin. Invest.*, **47**, 1958 (1968)
61. J. S. GARROW, *J. clin. Invest.*, **38**, 1241 (1959)
62. A. NEUBERGER and F. F. RICHARDS, (ed. H. N. Munro, J. B. Allison), "Mammalian Protein Metabolism" (1964) p. 243
63. W. A. WALTER, R. A. ULSTROM and J. T. LOWMAN, *J. Pediatr.*, **78**, 812 (1971)
64. A. S. McFARLANE, (ed. H. N. Munro and J. B. Allison) "Mammalian Protein Metabolism", Vol. 1, (Academic Press, New York, 1964) p. 298
65. S. JARNUM, "Radioisotope techniques in the study of Protein Metabolism", I.A.E.A. Vienna (1965)
66. J. C. WATERLOW, J. S. GARROW and D. J. MILLWARD, *Clin. Sci.*, **36**, 489 (1969)
67. J. C. WATERLOW, (ed. H. N. Munro) "Mammalian Protein Metabolism", (Academic Press, New York 1969) Vol. 3, p. 325
68. Symposium Report no. 45, "Radioisotopes techniques in the study of Protein Metabolism, I.A.E.A., Vienna (1965)
69. A. SAN PIETRO and D. RITTENBERG, *J. Biol. Chem.*, **201**, 457 (1953)
70. J. C. WATERLOW, *Nutrition Reviews*, **28**, 115 (1970)
71. P. RICHARDS, A. METCALFE-GIBSON, E. E. WARD, O. WRONG, B. J. HOUGHTON, *Lancet*, **ii**, 845 (1967)
72. W. W. C. READ, D. S. McLAREN, M. TCHALIAN and S. NASSAR, *J. clin. Invest.*, **48**, 1143 (1969)
73. W. W. C. READ, D. S. McLAREN and M. TCHALIAN, *Clin. Sci.*, **40**, 375 (1971)
74. J. C. WATERLOW and J. M. L. STEPHEN, *Clin. Sci.*, **33**, 489 (1967)
75. J. C. WATERLOW, *Clin. Sci.*, **33**, 507 (1967)
76. D. PICOU and T. TAYLOR-ROBERTS, *Clin. Sci.*, **36**, 283 (1969)
77. J. M. L. STEPHEN and J. C. WATERLOW, *Lancet*, **i**, 118 (1968)
78. A. S. McFARLANE, *Biochem. J.*, **87**, 228 (1963)
79. R. KIRSH, L. FRITH, E. BLACK and R. HOFFENBERG, *Nature*, **217**, 578 (1968)
80. J. C. WATERLOW and J. M. L. STEPHEN, *Brit. J. Nutr.*, **20**, 461 (1966)
81. E. M. WIDDOWSON and R. A. McCANCE, *Brit. J. Nutr.*, **10**, 363 (1956)
82. H. CHAN, *Brit. J. Nutr.*, **22**, 315 (1968)
83. V. R. YOUNG, M. A. HUSSEIN and N. S. SCRIMSHAW, *Nature*, **218**, 568 (1968)
84. D. J. MILLWARD, *Clin. Sci.*, **39**, 577 (1970)
85. J. S. GARROW, *Proc. Nutr. Soc.*, **28**, 242 (1969)

Index